Victor Stearns

10-06

The Bonded Electrical Resistance
Strain Gage

The Bonded Electrical Resistance Strain Gage

An Introduction

WILLIAM M. MURRAY
Professor Emeritus
Massachusetts Institute of Technology

WILLIAM R. MILLER
Professor Emeritus
The University of Toledo

New York   Oxford
OXFORD UNIVERSITY PRESS
1992

Oxford University Press

Oxford New York Toronto
Delhi Bombay Calcutta Madras Karachi
Kuala Lumpur Singapore Hong Kong Tokyo
Nairobi Dar es Salaam Cape Town
Melbourne Auckland

and associated companies in
Berlin Ibadan

Copyright © 1992 by Oxford University Press, Inc.

Published by Oxford University Press, Inc.,
200 Madison Avenue, New York 10016

Oxford is a registered trademark of Oxford University Press

All rights reserved. No part of this publication may be reproduced,
stored in a retrieval system, or transmitted, in any form or by any means,
electronic, mechanical, photocopying, recording, or otherwise,
without the prior permission of Oxford University Press.

Library of Congress Cataloging-in-Publication Data
Murray, William M.
The bonded electrical resistance strain gage :
an introduction / by William M. Murray and William R. Miller.
p. cm. Includes bibliographical references and index.
ISBN 0-19-507209-X
1. Strain gages. 2. Electric resistance—Measurement.
I. Miller, William R. (William Ralph), 1917– . II. Title.
TA413.5.M87 1992 624.1′76′0287—dc20 91-41369

2 4 6 8 9 7 5 3 1
Printed in the United States of America
on acid-free paper

PREFACE

Experimental stress analysis is an important tool in the overall design and development of machinery and structures. While analytical techniques and computer solutions are available during the design stage, the results are still dependent on many assumptions that must be made in order to adapt them to the problems at hand. Only when the design is fixed, the prototypes are constructed, and testing is underway, can the problem areas be realistically determined, and this must be done through experimental means.

One method of finding the weaknesses, and a method which is used extensively, is through the use of the electrical resistance strain gage. Strain gages are relatively low in cost, easily applied by a reasonably skilled technician, do not require extensive investment in instrumentation (for the general user), and yet they yield a wealth of information in a relatively short time. The information and its validity is, of course, dependent on the training and knowledge of the engineer who plans the tests and reduces the data. The latter statement becomes painfully apparent when one finds a user trying to interpret data from a single strain gage applied in an unknown biaxial stress field.

In 1988, the authors decided to edit Dr. Murray's notes, which were developed over his extensive career, and to write an introductory text on electrical resistance strain gages. The text is directed at senior and first-year graduate students in the engineering disciplines, although students from other fields (geology, engineering physics, etc.) will also benefit.

The prerequisites for a strain gage course are the following: (1) The basic courses in resistance of materials. (2) An elementary course in electrical circuits. (3) At least one course in mechanical or structural design is desirable. It follows that the more experience students have in analysis and design, the more they will benefit from an experimental course. It is in the laboratory and in experimental courses that students really develop a sense of security in, and a better understanding of, the theory they have been exposed to in their analytical studies.

The development of stress and strain transformation equations and the corresponding Mohr's circles, as well as the stress–strain relationships, are covered in Chapter 2. Depending on the student's preparation, the instructor may use this chapter for a rapid review or eliminate it entirely. The authors, however, have found it beneficial to spend at least several periods on the material.

Basic electrical circuits are examined in Chapters 3 through 5. An elementary circuit consisting of a single strain gage and its response to strain is first considered, followed by the potentiometric circuit and the Wheatstone bridge. In the development of the expressions for output voltage, as the strain gage's resistance changes with increasing loading, is the effect of circuit nonlinearity. The equations are developed so that the student can easily handle the intervening algebra between steps and thereby see the nonlinearity terms unfold. It is important that students recognize this and understand, when recording large strains, how to correct the indicated strains to obtain the actual strains. The effect of resistance in both the power supply and indicating meter is also accounted for.

Lead-line resistance is considered in the Wheatstone bridge circuits. The circuits are the full bridge, the half bridge with four wires, the half bridge with three wires, the quarter bridge with three wires, and the quarter bridge with two wires. The equations are developed so that the nonlinearity effects are apparent.

Sensitivity variation in order to obtain a desired output is next discussed in Chapter 6. Equations are developed, including nonlinearity effects, for the desensitization of single gages, half-bridge circuits, and full-bridge circuits.

Chapter 7 is devoted to the lateral, or transverse, effect on strain gages, along with a discussion of the methods used to determine the gage factor and the transverse sensitivity factor of strain gages. This is followed by Chapters 8 and 9 on strain gage rosettes and data reduction. It is shown how to reduce rosette data by both analytical methods and graphical methods. This is followed by considering transverse effects, using information from Chapter 7, in rosette data reduction.

Chapter 10 discusses how strain gages may be used to measure both normal stresses and shearing stresses directly, while Chapter 11 considers the effect of temperature on strain gage readings. Temperature-induced strains are discussed, followed by an examination of self-temperature-compensated gages and their thermal output curves when the gages are bonded to several different materials. One can see how to correct the indicated strain not only for the temperature-induced strain, but also for the gage factor variation resulting from temperature change.

Several types of strain-gage transducers are covered in Chapter 12. Among them are the axial-force load cell, the torque meter, the shear meter, and the pressure transducer. The purpose is to introduce the student to several types of transducers that could be made and calibrated for his use in the laboratory.

At the time of Dr. Murray's death on August 14, 1990, the major portion of the manuscript had been completed. If there are errors or discrepancies, the fault is not his but mine. In completing the text, I gathered together all of the source material in order to give proper credit; I sincerely hope none has been overlooked.

A textbook is not the work of one or several people alone. All of us are influenced not only by our contemporaries but by those who have preceded us (one has only to think of Professor Otto Mohr to realize this). Therefore, I want to acknowledge our debt to all of these people, not the least of whom were our students. I want especially to thank Martha Watson Spalding of Measurements Group, Inc. for her cooperation in furnishing a considerable amount of material. I also want to acknowledge the assistance of the following companies: BLH Electronics, Inc.; Eaton Corporation, Transducer Products; Electrix Industries, Inc.; Hartrun Corporation; Measurements Group, Inc.; Stein Engineering Services, Inc.; and Texas Measurements, Inc.

W. R. Miller

CONTENTS

1. Fundamental Concepts for Strain Gages, 3
    1.1 Introduction, 3
    1.2 Characteristics Desired in a Strain Gage, 4
    1.3 General Considerations, 5
    1.4 Analysis of Strain Sensitivity in Metals, 14
    1.5 Wire Strain Gages, 24
    1.6 Foil Strain Gages, 29
    1.7 Semiconductor Gages, 32
    1.8 Some Other Types of Gages, 33
    1.9 Brittle Lacquer Coatings, 36

2. Stress–Strain Analysis and Stress–Strain Relations, 42
    2.1 Introduction, 42
    2.2 Basic Concepts of Stress, 43
    2.3 Biaxial Stresses, 45
    2.4 Mohr's Circle for Stress, 54
    2.5 Basic Concepts of Strain, 61
    2.6 Plane Strain, 62
    2.7 Mohr's Circle for Strain, 68
    2.8 Stress–Strain Relationships, 72
    2.9 Application of Equations, 77
    2.10 Stress and Strain Invariants, 81

3. Elementary Circuits, 90
    3.1 Introduction, 90
    3.2 Constant-Voltage Circuit, 91
    3.3 Constant-Current Circuit, 94
    3.4 Advantages of the Constant-Current Circuit, 96
    3.5 Fundamental Laws of Measurement, 97

4. The Potentiometric Circuit, 100
    - 4.1 Introduction, 100
    - 4.2 Circuit Equations, 101
    - 4.3 Analysis of the Circuit, 106
    - 4.4 Linearity Considerations, 119
    - 4.5 Temperature Effects, 129
    - 4.6 Calibration, 141

5. Wheatstone Bridge, 146
    - 5.1 Introduction, 146
    - 5.2 Elementary Bridge Equations, 149
    - 5.3 Derivation of Elementary Bridge Equations, 157
    - 5.4 General Bridge Equations, 172
    - 5.5 Effect of Lead-Line Resistance, 180
    - 5.6 Circuit Calibration, 193
    - 5.7 Comments, 195

6. Sensitivity Variation, 205
    - 6.1 Introduction, 205
    - 6.2 Analysis of Single Gage Desensitization, 207
    - 6.3 Analysis of Half-Bridge Desensitization, 218
    - 6.4 Analysis of Full-Bridge Sensitivity Variation, 227

7. Lateral Effects in Strain Gages, 234
    - 7.1 Significance of Strain Sensitivity and Gage Factor, 234
    - 7.2 Basic Equations for Unit Change in Resistance, 236
    - 7.3 Determination of Gage Factor and Transverse Sensitivity Factor, 242
    - 7.4 Use of Strain Gages Under Conditions Differing from those Corresponding to Calibration, 246
    - 7.5 Indication from a Pair of Like Strain Gages Crossed at Right Angles, 248

8. Strain Gage Rosettes and Data Analysis, 253
    - 8.1 Reason for Rosette Analysis, 253
    - 8.2 Stress Fields, 253
    - 8.3 Rosette Geometry, 256
    - 8.4 Analytical Solution for the Rectangular Rosette, 258

- 8.5 *Analytical Solution for the Equiangular or Delta Rosette, 267*
- 8.6 *Rosettes with Four Strain Observations, 275*
- 8.7 *Graphical Solutions, 281*

9. Strain Gage Rosettes and Transverse Sensitivity Effect, 291
    - 9.1 *Introduction, 291*
    - 9.2 *Two Identical Orthogonal Gages, 291*
    - 9.3 *Two Different Orthogonal Gages, 294*
    - 9.4 *Three-Element Rectangular Rosette, 296*
    - 9.5 *The Equiangular or Delta Rosette, 301*

10. Stress Gages, 310
    - 10.1 *Introduction, 310*
    - 10.2 *The Normal Stress Gage, 310*
    - 10.3 *The SR-4 Stress–Strain Gage, 316*
    - 10.4 *Electrical Circuit for Two Ordinary Gages to Indicate Normal Stress, 320*
    - 10.5 *The V-Type Stress Gage, 321*
    - 10.6 *Application of a Single Strain Gage to Indicate Principal Stress, 326*
    - 10.7 *Determination of Plane Shearing Stress, 327*

11. Temperature Effects on Strain Gages, 337
    - 11.1 *Introduction, 337*
    - 11.2 *Basic Considerations of Temperature-Induced Strain, 337*
    - 11.3 *Self-Temperature-Compensated Strain Gages, 343*
    - 11.4 *Strain Gage–Test Material Mismatch, 349*
    - 11.5 *Compensating Gage, 353*

12. Transducers, 360
    - 12.1 *Introduction, 360*
    - 12.2 *Axial-Force Transducers, 363*
    - 12.3 *Simple Cantilever Beam, 368*
    - 12.4 *Bending Beam Load Cells, 372*
    - 12.5 *Shear Beam Load Cell, 375*
    - 12.6 *The Torque Meter, 378*
    - 12.7 *The Strain Gage Torque Wrench, 380*
    - 12.8 *Pressure Measurement, 382*

13. Strain Gage Selection and Application, 390
    - *13.1 General Considerations, 390*
    - *13.2 Strain Gage Alloys, 391*
    - *13.3 Grid Backing Materials, 393*
    - *13.4 Gage Length, Geometry, and Resistance, 394*
    - *13.5 Adhesives, 396*
    - *13.6 Bonding a Strain Gage to a Specimen, 398*

Answers to Selected Problems, 402

Index, 405

The Bonded Electrical Resistance
Strain Gage

1
FUNDAMENTAL CONCEPTS FOR STRAIN GAGES

1.1. Introduction

The constant demand for improvement in the design of machine and structural parts has led to the development of various experimental techniques for determining stress distributions. These experimental methods are employed for both the checking of theoretical predictions, and the evaluation of stresses in situations where mathematical approaches are unavailable or unsuited.

However, since stress cannot be measured directly, the experimental procedures, of necessity, make their approach through some type of strain measurement. The measured strains are then converted into their equivalent values in terms of stress. In order to achieve this ultimate objective, some type of strain-indicating device or measuring device is required.

In addition to their uses for stress analysis, strain gages also find wide application in sensing devices and control devices. In these applications, the strain in some mechanical part is used as an indication of force, bending, torque, pressure, acceleration, or some other quantity related to strain.

Even the most casual survey of the literature relating to the measurement of mechanical strain will yield information on a wide variety of devices which have been developed for this purpose. In addition to photoelasticity, brittle lacquer (1, 2, 3),[1] and X-rays, one finds all sorts of mechanical, optical, and electrical strain gages and extensometers, and various combinations thereof, which have been developed for one purpose or another, frequently with regard to some very specific application. It is very obvious that the development of a single instrument possessing all the optimum characteristics, for all applications, is unlikely. However, a good approach to the ultimate is still possible.

The brittle lacquer marketed as Tens-Lac (1, 2) is no longer available, although Stresscoat (3) can be obtained. These references, however, give a good description of the use of brittle lacquers in experimental stress analysis.

---

[1] Numbers in parentheses refer to References at the end of a chapter.

1.2. Characteristics desired in a strain gage

If we set out to devise a general-purpose strain gage, we would probably make a list of all possible desired characteristics. Some of these include, not necessarily in their order of importance, the following:

1. Ability to measure strains precisely under static and dynamic conditions.
2. Small size and weight. The small size permits mounting the instrument in confined locations, or to obtain reasonably precise indications in regions of high stress gradient. Small weight is required so that the inertia effects in the gage will be negligible under dynamic conditions.
3. The possibility of remote observation and recording. This is very much a relative requirement, since remote might mean anything from a few feet in the laboratory to thousands of miles, as in the case of a rocket or missile with radio transmission (telemetering) of the signal to the location of the observer.
4. Independence of the influence of temperature. This is probably the most difficult requirement of all. Very satisfactory results can be achieved over small temperature excursions, but when the temperature may fluctuate up or down in the range from about $-400°F$ to $+1500°F$ (-240 to $815°C$), the problem becomes exceedingly difficult.
5. Easy installation. In order to be commercially attractive, a strain gage should be sufficiently easy to install so that relatively unskilled people can be trained, in a short space of time, to perform this operation satisfactorily and reliably.
6. Stability of calibration. It is extremely desirable that the calibration should be stable over the entire range of operating conditions.
7. Linear response to strain. Although not absolutely essential, this is very desirable. Small deviations from linearity can frequently be brought within tolerable limits by combination (opposition) with the inherent nonlinearity of the electrical circuit of which the gage forms a part. For larger departures from linearity, the electrical circuit can be specially designed to provide automatic compensation (4, 5). When large-scale computers are employed to condition and process the strain gage indications, provided that the relation between strain and gage indication is known, this function can be directly programmed into the machine.
8. Low cost. This is another relative consideration that depends upon the work at hand. Generally speaking, the cost of modern strain gages is relatively insignificant in comparison with the other costs associated with an important project.
9. Dependability. Unless the strain gage indications can be depended upon, its use becomes very limited. Fortunately, the strain gages

available today are very dependable when used under the conditions for which they were intended.
10. The possibility of operation as an individual strain gage, or in multiple arrangements, to determine quantities that are indicated by the simultaneous observation of strains at more than one location. This means that, for certain applications, we should be able to use strain gages in multiple arrangements to perform automatic computation of some quantity that is related to strains at several locations.

No one has yet developed a strain gage possessing all of these desired characteristics. However, one can generally say that bonded electrical resistance strain gages (wire, foil, or semiconductor) come much nearer than any other device to satisfying all these requirements.

1.3. General considerations

Basic principle

In common with photoelasticity and stresscoat, the basic principle underlying the operation of electrical resistance strain gages has been known for a long time. However, the application of the principle to strain measurement (on a commercial scale) is much more recent. In 1856 Lord Kelvin (6) reported his observations that certain electrical conductors he had been studying exhibited a change in electrical resistance with change in strain.

The change of electrical resistance resulting from mechanical strain represents the basic principle upon which electrical resistance strain gages operate. For semiconductor gages, the detail of the means by which strain changes the resistance seems to be well understood, but for metallic conductors (wire or foil), we are still a long way from a complete understanding of what takes place within the material.

Definition of strain sensitivity

When a conductor is trained in the axial direction, its length will change, and, if unrestrained laterally, its cross-sectional area will also change (the Poisson effect). The increase in length, shown in Fig. 1.1, is accompanied by a decrease in the cross-sectional area, and vice versa. In addition, the specific resistivity of the material may change. These three influences, the change in length, the change in cross-sectional area, and the change in specific resistivity, combine to produce a change in the overall electrical resistance of the conductor. The amount of the resistance change, in relation to the change in length of the conductor, is an index of what is called the strain sensitivity of the material of the conductor. This relationship is expressed as a dimensionless ratio called the *strain sensitivity factor*. For a straight

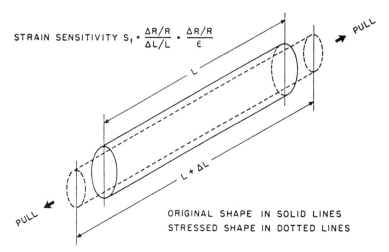

FIG. 1.1. Schematic diagram of strained conductor (tensile effect shown).

conductor of uniform cross section, this is expressed as

$$\text{Strain sensitivity factor} = \frac{\text{unit change in resistance}}{\text{unit change in length}}$$

$$= \frac{\text{unit change in resistance}}{\text{strain}}$$

In symbols, this can be written as

$$S_t = \frac{\Delta R/R}{\Delta L/L} = \frac{\Delta R/R}{\varepsilon} \tag{1.1}$$

where S_t = strain sensitivity (factor) of the conductor and is dimensionless; this is a physical property of the material

R = resistance in ohms

L = length in inches

$\Delta R, \Delta L$ = corresponding changes in resistance and length, respectively, in ohms and inches

$\varepsilon = \Delta L/L$ = strain along the conductor (dimensionless)

Examination of Eq. (1.1) and the definitions of the symbols will raise a question regarding the values that should be used for R and L in calculating the strain sensitivity. Do these symbols represent the following?

FUNDAMENTAL CONCEPTS FOR STRAIN GAGES 7

1. The initial resistance, R_0, and the initial length, L_0, when the conductor is stress free? In which case the denominator, ε, corresponds to nominal strain based on L_0.
2. Any corresponding values of resistance and length which may prevail after a certain amount of initial load has been applied?
3. The instantaneous values of resistance and length which prevail during infinitely small changes of length and resistance. In which case, as $\Delta L \to 0$, in the limit,

$$S_t = \frac{dR/R}{dL/L} \qquad (1.2)$$

In Eq. (1.2) the denominator, $\varepsilon = dL/L$, is what is sometimes called the true strain (as contrasted with the nominal strain), and the value of S_t obtained in this manner is sometimes called the instantaneous sensitivity factor, since it refers to the resistance and length in the stretched condition for which both R and L are variable (7).

Except for the special case in which R happens to be directly proportional to L, theoretically, these three modes of interpretation will yield different results for the value of S_t, the strain sensitivity factor. This means that we are confronted with the problem of having to decide upon which particular procedure we should follow. For the special case in which the resistance is directly proportional to the length, $R = \bar{K}L$, where $\bar{K}$ is a constant. Thus, $\Delta R = \bar{K}(\Delta L)$, and hence

$$S_t = \frac{\Delta R/R}{\Delta L/L} = \frac{\bar{K}(\Delta L)/\bar{K}L}{\Delta L/L} = 1 \qquad (1.3)$$

Since $R = \rho L/A$, therefore $\bar{K} = \rho/A$, which means that to fulfill this condition, the specific resistivity, ρ, will have to be proportional to the area of the cross section.

Elastic strains in metals

For small strains with correspondingly small changes in resistance, such as might be expected in metals when strained within the elastic limit, there is no problem. Here L_0 and L will be nearly equal and, likewise, R_0 and R will be so nearly alike it will make no noticeable difference in the value of S_t, whether it is computed on the basis of L_0 and R_0, or from the values of L and R which correspond to the elastic limit. This is a great convenience for the following reasons:

1. The initial resistance, R_0, and the initial length, L_0, provide good references from which the changes ΔR and ΔL can be readily determined.

2. The strain sensitivity, S_t, can be determined from the slope of the curve which is established by plotting $\Delta R/R_0$ against $\Delta L/L_0$.
3. The analyses of the basic electrical circuits which are used with strain gages, developed in following chapters, show that the output, or indication, is given in terms of $\Delta R/R_0$.

Plastic strains in metals

When a metal conductor is strained beyond the elastic limit into the plastic range, the changes in resistance and length (from the initial values) will ultimately become so large that there will be a considerable difference between R and R_0, and also between L and L_0.

When this happens, the previous approximate method of determining S_t from the values of R_0 and L_0 will no longer be satisfactory. It will be necessary to compute the instantaneous value of S_t from the instantaneous values of R and L, according to Eq. (1.2).

At first glance, this might appear to be a formidable task, but fortunately this is not so. We determine a series of corresponding values of R and L as the conductor is being stretched (or compressed), and then plot the logarithm of the dimensionless ratio, R/R_0, against the logarithm of the dimensionless ratio, L/L_0. The slope of the line thus drawn represents the instantaneous value of the strain sensitivity factor, S_t. Further discussion will be found later in the chapter.

Semiconductor materials

The relatively high strain sensitivity of silicon and germanium has made these semiconductor materials attractive for strain gage sensing elements. For silicon, which is the preferred material, the theoretical value of S_t lies in the range between -150 and about $+175$. Furthermore, by suitable processing (doping), silicon can be produced with any arbitrarily specified value of S_t within this range. For commercial strain gages, in order to achieve a suitable compromise between response to strain and response to temperature, it is usual to process the material for strain sensitivities in the range of about -100 to about $+120$.

The resistance–strain relation for silicon is somewhat more elaborate than that for metallic conductors. It is nonlinear, and very noticeably influenced by temperature. Dorsey (8, 9) gives the following expression for unit change in terms of strain:

$$\frac{\Delta R}{R_{0(T_0)}} = \left(\frac{T_0}{T}\right)(GF')\varepsilon + \left(\frac{T_0}{T}\right)^2 (C'_2)\varepsilon^2 \qquad (1.4)$$

where ΔR = change in resistance from $R_{0(T_0)}$ (ohms)

$R_{0(T_0)}$ = resistance (ohms) of the unstressed material (prior to being mounted as a strain gage) at temperature T_0, in Kelvin

T_0 = temperature at which $R_{0(T_0)}$ was determined (Kelvin)

T = temperature (Kelvin)

ε = strain (dimensionless)

GF', C_2' = constants for the particular piece of material (dimensionless)

Equation (1.4) indicates the following characteristics regarding the relation between unit change in resistance and strain for silicon:

1. The strain sensitivity factor, which corresponds to the slope of the curve of $\Delta R/R_{0(T_0)}$ vs. ε, will be a variable whose value will depend upon both the strain level and the temperature.
2. Since the relationship expressed in Eq. (1.4) represents a parabola, one can expect the degree of nonlinearity to vary with strain and temperature.
3. At constant temperature, T_0, Eq. (1.4) reduces to

$$\frac{\Delta R}{R_{0(T_0)}} = (GF')\varepsilon + (C_2')\varepsilon^2 \qquad (1.5)$$

Hence, for this special condition shown in Fig. 1.2, GF' corresponds to the slope of the curve, or the sensitivity factor, for $\varepsilon = 0$, and C_2' represents the nonlinearity constant which determines the degree of departure of the curve from the slope at the point $\Delta R = 0$, $\varepsilon = 0$, for which the resistance equals $R_{0(T_0)}$. Baker (10) also expresses Eq. (1.5) in essentially the same form.

Over a limited range of strain, for example about 600 microstrain (1 microstrain = 1 μin/in), and particularly at strain levels where the slope of the curve changes more gradually, the variable strain sensitivity can be approximated by a constant that corresponds to the average value, and good results may be expected from this. For larger ranges of strain, or for more precise indications, more elaborate methods must be employed.

When the temperature varies, the whole problem of relating resistance changes to strain becomes more complicated. This is due to the fact that changes in temperature, as indicated in Eq. (1.4), produce changes in the sensitivity. In addition, the value of $R_{0(T_0)}$ will also change with variations in the reference temperature, T_0.

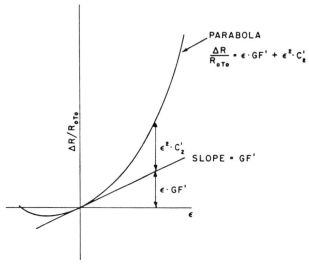

FIG. 1.2. Schematic diagram for $\Delta R/R_{0(T_0)}$ vs. ε at constant temperature, T_0. (Resistance = $\Delta R/R_{0(T_0)}$ when $\Delta R = \varepsilon = 0$.)

Desired properties of strain-sensitive materials

1. Linear relation between unit change in resistance and change in strain (i.e., constant sensitivity).
2. Negligible effect from temperature.
3. High strain sensitivity factor.
4. Moderately high resistance.
5. Ability to be connected to lead wires easily.
6. Low cost.
7. Availability.
8. Absence of creep and hysteresis.

One cannot expect to find all the desirable characteristics in any particular material without some adverse properties, too. In general, the selection of a material for the sensing element of a strain gage will result in a compromise depending upon the intended use of the gage.

Properties of some metals

In view of the previous discussion of strain sensitivity, and the properties desired in strain sensing materials, let us look at some typical characteristics as represented by a few metals. These are indicated in Figs. 1.3 and 1.4, taken from the work of Jones and Maslen (11). In each case, the percent change in resistance, based on R_0, has been plotted against percent strain, on the basis of $\Delta L/L_0$. The slopes of the lines represent S_t, and the different general relationships are indicated as follows:

1. The same linear relation between $\Delta R/R_0$ and $\Delta L/L_0$ in both the elastic and plastic ranges. This condition is represented by annealed copper, as well as annealed copper–nickel alloys like Ferry. This means that the strain sensitivity factor will be the same in the plastic range as it is in the elastic range. This characteristic is highly desirable because it eliminates all concern about the possibility of a change in gage factor in the event the sensing element of a strain gage might be strained beyond its elastic limit. In consequence, this type of material is well suited for gages which will be required to measure high elastic strains, or both elastic and plastic strains.
2. Nonlinear relationship such as exhibited by nickel.
3. Relationship approximated by two straight lines indicating a change of strain sensitivity with the transition from elastic to plastic conditions. Some materials, such as minalpha, manganin, and hard silver–palladium, show a lower strain sensitivity at low strains than at high strains.
4. The same general relationship as indicated in Item (3), but with the difference that the higher strain sensitivity corresponds to the lower strains, as shown by rhodium–platinum.

For the relations indicated in Items (3) and (4), the change in slope as yielding sets in is not abrupt, as suggested by the graphs, but follows a smooth transition from the elastic to the pastic range.

Numerical values of the strain sensitivity factor

Table 1.1 presents typical strain sensitivity values for a number of metals at low strain, together with corresponding information with respect to the effects of temperature changes (12).

A more elaborate tabulation, which includes some of the pure metals and a number of alloys (with approximate compositions), is given in the Appendix of this chapter. Where possible, information for sensitivities in both the elastic and plastic strain ranges, and for material in the cold worked and annealed conditions, has been included.

Approximate compositions of some of the alloys in Table 1.1 are given in Table 1.2.

A study of the literature and of the tabulated data in the Appendix at the end of the chapter yields the following observations regarding material properties:

1. Different values of strain sensitivity for hard and annealed conditions of the same material suggests that the degree of cold working, and the heat treatment, have an influence. This is of particular importance in relation to the effects of temperature and temperature compensation.

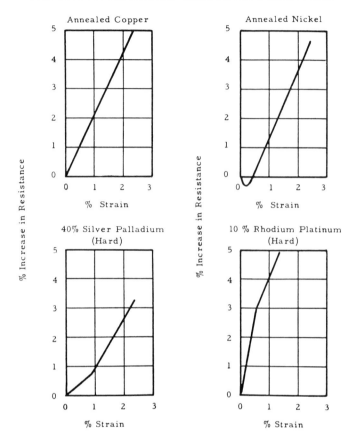

FIG. 1.3. Typical examples of resistance change vs. strain (From ref. 11 with permission of HMSO.)

2. Differences in sensitivity for different lots of nominally the same material suggest that differences in impurities, and in trace elements, exert an influence on the physical properties. This is also of importance with respect to temperature effects.
3. For nearly all the metals investigated, the strain sensitivity factor appears to approach a value of 2.0 in the plastic range.

For large strains (up to 30 percent), Weibull (13) has reported some very interesting detailed experimental results on the relation between changes in length and resistance for 0.45-mm diameter Copel wire. This is a 55 percent copper, 45 percent nickel alloy.

From the data in Table 1.3, the values of R/R_0, L/L_0, $\Delta R/R_0$, and $\Delta L/L_0$, have been computed. Plots of $\ln(R/R_0)$ vs. $\ln(L/L_0)$ and $\Delta R/R_0$ vs. $\Delta L/L_0$ are shown in Fig. 1.5 for comparative purposes. From the slope of the logarithmic plot, which is represented by a straight line, the value of the

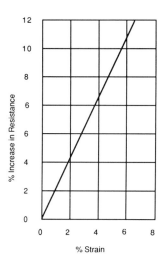

Fig. 1.4. Resistance change vs. strain for annealed Ferry wire (60/40 cupronickel). (From ref. 12.)

Table 1.1. Typical strain sensitivity factors

Material	Strain sensitivity factor (for small strains)	Stress in lb/in^2 equivalent to influence of temperature change of 1°C for installation on steel material[a]
Manganin	0.47	−400
Nickel	−12.1 (nonlinear)	−13 500
Nichrome	2.1	2 100
Phosphor bronze	1.9	7 800
5% Iridium–Platinum	5.1	11 600
Advance	2.1 (selected material)	±30
Copel	2.4	−200
Monel	1.9	8 000
Isoelastic	3.6	5 000

Source: reference 12.
[a] One should note that these figures can only be considered as semiquantitative indications because they will vary with heat treatment and cold working of the material and also with temperature level.

Table 1.2. Composition of alloys

Material	Composition
Advance and Copel	45% Ni; 55% Cu
5% Iridium–platinum	5% Ir; 95% Pt
Isoelastic	36% Ni; 8% Cr; 52% Fe; 0.5% Mo; +(Mn, Si, Cu, V) = 3.5%
Manganin	4% Ni; 12% Mn; 84% Cu
Nichrome V	80% Ni; 20% Cr

Table 1.3. Weibull's observations from static test on Copel wire

Initial diameter = 0.45 mm;	initial length = 125 mm
ΔL (mm)	R (ohms)
0.00	0.376
6.25	0.414
12.50	0.455
18.75	0.497
25.00	0.542
31.25	0.588
37.50	0.635

Source: reference 13. Reprinted by permission. © 1948 Macmillan Magazines Ltd.

strain sensitivity factor is found to be

$$S_t = \frac{dR/R}{dL/L} = 2 \tag{1.6}$$

Weibull does not state the metallurgical condition of the wire, but from the magnitude (60 percent) of the elongation reported for one of his specimens, it is assumed that the material was in the annealed condition. He also reports essentially comparable results for a dynamic test on 0.45-mm diameter wire with a length of 101 mm. The maximum strain reached 34 percent with a velocity of 6.2 m/sec for the moving head of the testing device.

The 0.45 mm (0.0177 in) wire diameter which Weibull investigated is somewhat larger than the 1-mil (0.001-in) size normally employed for bonded strain gages. With the smaller diameter, smaller ultimate elongation is expected because minor variations in diameter will have, relatively, much greater influence. Shoub (14) reports elongations up to 22 percent for specially annealed constantan wire of 0.001 in diameter. His results indicate a straight-line relationship, with a slope of 2.02, for the plot of log (R/R_0) vs. log (L/L_0). This confirms Weibull's observations.

1.4. Analysis of strain sensitivity in metals

The general case

Figure 1.6 shows a metal conductor of uniform cross section (not necessarily rectangular, although this is shown) referred to the axes X, Y, and Z. We want to establish an expression for the ratio of unit change in resistance in the X direction to the unit change in length, in terms of strains ε_x, ε_y, and ε_z (in the directions of the X, Y, and Z axes, respectively) and the material property of the conductor.

FUNDAMENTAL CONCEPTS FOR STRAIN GAGES

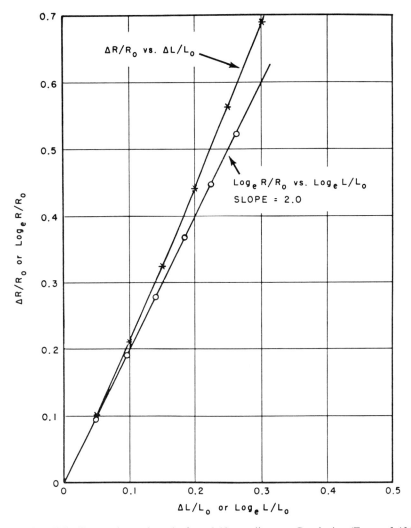

FIG. 1.5. Weibull's experimental results from 0.45-mm diameter Copel wire. (From ref. 13.)

The expression for the resistance in the X direction can be written as

$$R = \rho \frac{L}{A} \tag{1.7}$$

where R = resistance in length L (ohms)
ρ = specific resistivity of the material (ohms-in)
L = length (in)
A = area of the cross section (in^2)

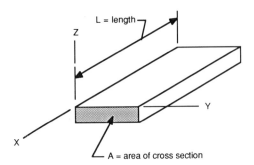

FIG. 1.6. Metal conductor referred to X, Y, and Z axes.

By multiplying the numerator and denominator of the right-hand term by the length L, Eq. (1.7) can be rewritten as

$$R = \rho \frac{L^2}{V} \tag{1.8a}$$

where $V = LA =$ volume (in³). By taking the logarithm of both sides, Eq. (1.8a) becomes

$$\ln R = \ln \rho + 2 \ln L - \ln V \tag{1.8b}$$

Differentiation of Eq. (1.8b) results in

$$\frac{dR}{R} = \frac{d\rho}{\rho} + 2\frac{dL}{L} - \frac{dV}{V} \tag{1.9}$$

Equation (1.9) expresses the unit change in resistance in terms of the unit changes in resistivity, length, and volume.

We now postulate that the unit change in resistivity can be related to the unit change in volume as follows:

$$\frac{d\rho}{\rho} = m\frac{dV}{V} \tag{1.10}$$

where $m =$ a function of the material properties and the two ratios of the transverse to the longitudinal strain. For the elastic strains, and fixed values of the two strain ratios, some materials exhibit a constant value of the function m. This relation is stated by Biermasz et al. (15), who gives credit for it to Bridgeman. Meier (16) uses the same relation in a slightly different form.

By substituting the value of $d\rho/\rho$ given by Eq. (1.10) into Eq. (1.9), we may write

$$\frac{dR}{R} = m\frac{dV}{V} + 2\frac{dL}{L} - \frac{dV}{V}$$

or

$$\frac{dR}{R} = 2\frac{dL}{L} + (m-1)\frac{dV}{V} \tag{1.11}$$

Dividing all terms of Eq. (1.11) by dL/L, we obtain

$$\frac{dR/R}{dL/L} = 2 + (m-1)\frac{dV/V}{dL/L} \tag{1.12}$$

Equation (1.12) indicates that, for plastic deformation (which takes place at constant volume, so that $dV = 0$), the value of the instantaneous strain sensitivity can be expected to be 2 for any strain condition.

Since $dL/L = \varepsilon_x$, and because $dV/V = (\varepsilon_x + \varepsilon_y + \varepsilon_z)$, Eq. (1.12) can be expressed in terms of the strains as follows:

$$S_t = \frac{dR/R}{dL/L} = \frac{dR/R}{\varepsilon_x} = 2 + (m-1)\left[1 + \frac{\varepsilon_y}{\varepsilon_x} + \frac{\varepsilon_z}{\varepsilon_x}\right] \tag{1.13}$$

Special case of a uniform straight wire

For the special case of a straight wire of any uniform cross section, which is free to contract or expand laterally due to the Poisson effect, the ratios of lateral to axial strain are given by the expression

$$\frac{\varepsilon_y}{\varepsilon_x} = \frac{\varepsilon_z}{\varepsilon_x} = -v \tag{1.14}$$

where $v =$ Poisson's ratio of the material.

When the values of the strain ratios, given for this special case by Eq. (1.14), are substituted into Eq. (1.13) for strain sensitivity, we arrive at

$$S_t = \frac{dR/R}{dL/L} = \frac{dR/R}{\varepsilon_x} = 2 + (m-1)(1-2v) \tag{1.15}$$

For small changes, such as encountered within the elastic ranges of metals, Eq. (1.15) can be modified to read

$$S_t = \frac{\Delta R/R_0}{\Delta L/L_0} = 2 + (m-1)(1-2v) \tag{1.16}$$

18   THE BONDED ELECTRICAL RESISTANCE STRAIN GAGE

Equations (1.15) and (1.16) indicate two interesting characteristics in regard to the strain sensitivity of a wire.

1. If the material property is such that $m = 1$, then, regardless of the value of Poisson's ratio, the strain sensitivity factor of the metal will be 2. This means the strain sensitivity will be the same in the elastic and plastic ranges, even though there will be a change in v as one proceeds from elastic to plastic conditions. Conversely, this also tells us that only those materials whose strain sensitivity is 2 can have the same sensitivity in both the elastic and plastic ranges.
2. For perfectly plastic deformation, which takes place at constant volume, $dV = 0$ and $v = 0.5$. Therefore, no matter what the value of m is, the strain sensitivity factor for plastic deformation will be 2, as previously indicated by Eq. (1.12). This means that, for plastic deformation, all metals should exhibit a strain sensitivity factor of 2. This is substantiated by the results of tests, as indicated in the tabulation presented in the Appendix of this chapter, for which, in almost all cases, the strain sensitivities in the high strain ranges approximate a value of 2.

The slight deviation of some of the values from 2 is probably due to the effect of a certain amount of elastic strain which will be present during the plastic deformation. The few cases involving larger deviations from 2 likely correspond to rather incomplete or gradual plastic deformation, and possibly the influence of some type of work hardening.

Equations (1.15) and (1.16) can now be converted into a more familiar form customarily found in the literature. Expansion of the second term on the right-hand side of these equations results in the expression

$$S_t = \frac{dR/R}{dL/L} = 2 - 1 + 2v + m(1 - 2v) \tag{1.17}$$

In order to write Eq. (1.17) in a different form, the change in the volume of the wire as it is strained axially can be considered. The unstrained wire volume is

$$V = AL \tag{a}$$

Taking the logarithm of both sides and then differentiating yields

$$\frac{dV}{V} = \frac{dA}{A} + \frac{dL}{L} \tag{b}$$

As the wire is strained, its length increases by dL, but due to the Poisson effect its diameter decreases by $(-v\, dL/L)D$, where D is the wire diameter.

FUNDAMENTAL CONCEPTS FOR STRAIN GAGES

The final wire diameter is

$$D_f = D\left(1 - v\frac{dL}{L}\right) \tag{c}$$

The change in area can now be written as

$$dA = \frac{\pi}{4}(D_f^2 - D^2) = \frac{\pi D^2}{4}\left[\left(1 - v\frac{dL}{L}\right)^2 - 1\right]$$

$$= A\left[-2v\frac{dL}{L} + \left(v\frac{dL}{L}\right)^2\right] \tag{d}$$

If the higher-order term in Eq. (d) is neglected, then we can write

$$\frac{dA}{A} = -2v\frac{dL}{L} \tag{e}$$

Substituting the value of dA/A given by Eq. (e) into Eq. (b) gives

$$\frac{dV}{V} = -2v\frac{dL}{L} + \frac{dL}{L} = (1 - 2v)\frac{dL}{L} \tag{f}$$

Thus, Eq. (f) can be expressed as

$$(1 - 2v) = \frac{dV/V}{dL/L} \tag{g}$$

From Eq. (1.10) we can write

$$m = \frac{d\rho/\rho}{dV/V} \tag{h}$$

If the values of $(1 - 2v)$ and m from Eqs. (g) and (h), respectively, are substituted in Eq. (1.17), then

$$S_t = \frac{dR/R}{dL/L} = 2 - 1 + 2v + \left(\frac{d\rho/\rho}{dV/V}\right)\left(\frac{dV/V}{dL/L}\right)$$

or

$$S_t = \frac{dR/R}{dL/L} = (1 + 2v) + \frac{d\rho/\rho}{dL/L} \tag{1.18}$$

For small changes, as encountered with elastic strains, we can write

$$S_t = \frac{\Delta R/R_0}{\Delta L/L_0} = (1 + 2v) + \frac{\Delta \rho/\rho_0}{\Delta L/L_0} \qquad (1.19)$$

Equation (1.18) is of particular interest, not just because it represents a more familiar form of the expression for the strain sensitivity factor, but for two other reasons as well.

1. The relationship given in Eq. (1.18) can be derived independently of the relation given by Eq. (1.10).
2. For any particular metal, Eq. (1.18) indicates the portions of the strain sensitivity factor which are the result of geometrical change and resistivity change, respectively. The value $(1 + 2v)$ corresponds to the geometrical change, while $(d\rho/\rho)/(dL/L)$ corresponds to the resistivity change.

We see that when plastic deformation takes place, since $v = 0.5$ and $d\rho = 0$, Eq. (1.18) also indicates a value of 2 for S_t.

Small strain vs. large strain

Let us now look into the detail of the difference between the expressions for the instantaneous and approximate values of the strain sensitivity factors. The instantaneous value of S_t is

$$S_t = \frac{dR/R}{dL/L} \qquad (1.20)$$

while the approximate value of S_t is

$$S_t = \frac{\Delta R/R_0}{\Delta L/L_0} \qquad (1.21)$$

For small strains (less than 1 percent), as developed in the elastic range of metals, both expressions will yield, for all practical purposes, the same result. However, since it will be more convenient to evaluate the strain sensitivity, and subsequently to compute strains, on the basis of changes from the initial condition, we wish to know the magnitude of the largest strain that can be handled in this manner without running into intolerably large errors.

Returning to Fig. 1.5, we see a comparison, based on Weibull's experimental observations, between the plot of $\Delta R/R_0$ vs. $\Delta L/L_0$ and the logarithmic plot of $\ln(R/R_0)$ vs. $\ln(L/L_0)$. The logarithmic plot shows a straight line with a slope, S_t, of 2.0, whereas the plot of $\Delta R/R_0$ vs. $\Delta L/L_0$ gives a long radius curve whose initial slope (for $\Delta R = \Delta L = 0$) is 2.0, but for which the slope increases slightly as the changes in length and resistance build up.

Examination of Fig. 1.5 reveals that, for a graph of this size and within the limits of error in plotting the points, the curve of $\Delta R/R_0$ vs. $\Delta L/L_0$ can be represented by a straight line up to values of about 10 to 15 percent of $\Delta L/L_0$. For larger strains the departure from linearity, although not serious, can be noticed. However, we observe that the slope of the line (the indicated value of S_t) is slightly greater than that of the logarithmic plot. This explains why one can use post-yield gages up to strain levels in the range of 10 percent or more, on the basis of $\Delta R/R_0$ and $\Delta L/L_0$, without introducing noticeable errors as a result of making a linear approximation.

As these comments have been developed from experimental observations, we can now examine the situation from a theoretical point of view. We start by developing the relation between resistance and length from Eq. (1.20) on the assumption that S_t is a constant. We can rewrite Eq. (1.20) in the following form:

$$\frac{dR}{R} = S_t \frac{dL}{L} \quad (1.22)$$

Equation (1.22) can also be expressed as

$$d(\ln R) = S_t \, d(\ln L) \quad (1.23)$$

Integrating Eq. (1.23) results in

$$\ln R = S_t \ln L + C \quad (1.24)$$

where C = constant of integration.

Since the initial values of resistance and length, R_0 and L_0, will be known, the constant of integration can be written as

$$C = \ln R_0 - S_t \ln L_0 \quad (1.25)$$

Substituting the value of C from Eq. (1.25) into Eq. (1.24) gives us

$$\ln R - \ln R_0 = S_t(\ln L - \ln L_0)$$

This expression can be modified to read

$$\ln\left(\frac{R}{R_0}\right) = S_t \ln\left(\frac{L}{L_0}\right) \quad (1.26)$$

Equation (1.26) tells us that the plot of $\ln(R/R_0)$ vs. $\ln(L/L_0)$ will give a straight line whose slope is equal to S_t. This has been verified experimentally by both Weibull (13) and Shoub (14).

From Eq. (1.26) we can express the relation between resistance and length of a metal conductor that has been strained in the plastic range as

$$\frac{R}{R_0} = \left(\frac{L}{L_0}\right)^{S_t} \tag{1.27}$$

Since the value of S_t for plastic strain has been predicted theoretically as 2.0, as shown by Eq. (1.12), and because this value has been corroborated by the experiments of Weibull (13) and Shoub (14), this is the number that will be used for the exponent in Eq. (1.27). Thus, Eq. (1.27) can now be written as

$$\frac{R}{R_0} = \left(\frac{L}{L_0}\right)^2 \tag{1.28}$$

Because $R = R_0 + \Delta R$ and $L = L_0 + \Delta L$, Eq. (1.28) can be converted into terms of ΔR, ΔL, R_0, and L_0. Thus,

$$\frac{R_0 + \Delta R}{R_0} = \left[\frac{L_0 + \Delta L}{L_0}\right]^2$$

or

$$1 + \frac{\Delta R}{R_0} = \left[1 + \frac{\Delta L}{L_0}\right]^2 \tag{1.29}$$

Expanding the right-hand side of Eq. (1.29) results in

$$\frac{\Delta R}{R_0} = 2\frac{\Delta L}{L_0} + \left(\frac{\Delta L}{L_0}\right)^2 \tag{1.30}$$

Equation (1.30) presents the theoretical relationship between $\Delta R/R_0$ and $\Delta L/L_0$ for a metal conductor subjected to plastic strain. It provides the following information:

1. $\Delta R/R_0$ is a nonlinear function at $\Delta L/L_0$.
2. For positive values of ΔL (tension), $\Delta R/R_0$ will always be larger than $2(\Delta L/L_0)$.
3. The slope of the curve at the origin is 2.
4. The deviation from the tangent (slope = 2) through the origin is given by $(\Delta L/L_0)^2$.

Item 4 indicates both the deviation from linearity and the deviation from the relation involving the instantaneous values of R and L.

It is noteworthy that when $\Delta L/L_0$ is 10 percent, the deviation from linearity is only 5 percent. This is illustrated in Fig. 1.7, which shows a plot of theoretical values of $\Delta R/R_0$ vs. $\Delta L/L_0$, as computed from Eq. (1.30).

If an approximate linear relation is set up by using the secant from the origin to some point on the curve, then the error will be zero at the point of intersection with the curve, and at all other points the error will be less than that represented by the deviation of the secant from the initial tangent. This is due to the fact that the curve lies between the secant and the tangent

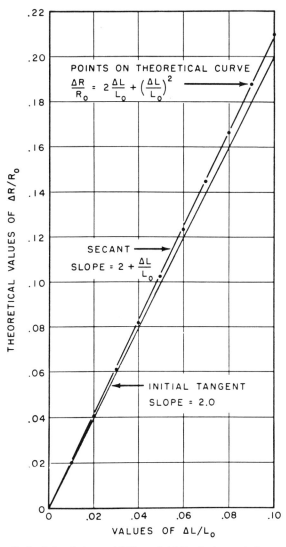

FIG. 1.7. Theoretical relation between $\Delta R/R_0$ and $\Delta L/L_0$ for large strains.

through the origin. For example, when $\Delta L/L_0$ equals 10 percent, the expected error, at any point, will never be more than 5 percent, as a maximum. In general it will probably not exceed 2.5 percent, except for relatively low strain values where the numerical magnitude of the error will be of less importance. Examination of Fig. 1.7 will help to clarify these points.

From Eq. (1.30) an expression can be written for the value of the strain sensitivity factor:

$$S_t = \frac{\Delta R/R_0}{\Delta L/L_0} = 2 + \frac{\Delta L}{L_0} \tag{1.31}$$

The value of S_t varies in accordance with the value of $\Delta L/L_0$ and corresponds to the slope of the secant from the origin to the point whose coordinates are $(\Delta R/R_0, \Delta L/L_0)$ on the curve.

1.5. Wire strain gages

The unbonded wire strain gage

One of the early wire gages was the unbonded type. In this type of instrument, the strain-sensitive wire is mounted, under tension, on mechanical supports (pins) in such a manner that a slight relative motion of the supports will cause a change in strain. This, in turn, produces a change in electrical resistance. This resistance change is then a measure of the relative displacement of the supports and, in turn, may represent a strain or some other quantity.

With the unbonded type of gage, the fact that the strain-sensitive wires must be carried on some sort of mechanical mount gives rise to certain difficulties in connection with attachment. Discrepancies, due to inertia, may be introduced when dynamic observations are made. The procedure of making observations at an appreciable distance from the surface on which strain is to be determined may sometimes be open to question.

The bonded wire strain gage

The first major improvement in the wire resistance strain gage came with the realization that many of the difficulties with the unbonded wire gage could be eliminated by bonding a very fine strain-sensitive wire directly to the surface on which strain is to be measured. The filament has to be electrically insulated and the bonding perfect for the strain-sensitive element to follow the strain on the surface to which it is attached. Only conductors of small diameter are suitable, since the force necessary to strain the sensing element must be transmitted through its surface by shear in the cement, or bonding agent. Unless the surface area per unit length is large relative to the cross-sectional area, the shearing stress in the cement will be too high to permit faithful following of the strains in the surface to which the conductor is attached.

Since the surface area (per unit length) of small-diameter wires is enormously greater than the cross-sectional area (for 0.001-in diameter wire, the ratio is 4000 to 1), the bonding agent is able to force the filament to take up the necessary strain without excessive stress in itself. Suitable cements can actually force the small conductor into the plastic range (and back again) when necessary.

Chronologically, the second major development, and that which has actually been responsible for making the bonded strain gage commercially attractive, is represented by the concept of premounting the strain-sensing element on some suitable carrier that can be attached to a surface relatively easily. Originally, the strain gage wire was cemented directly to the surface on which strain was to be measured, and the glue or cement acted as insulation. As far as operation was concerned, this procedure was satisfactory, but from the point of view of gage installation, it was inconvenient. The attachment of the gage required an inordinate amount of skill and time on the part of the installer if consistent results were to be obtained. The introduction of a paper, plastic, metal, or other type of carrier upon which the strain-sensing wire could be premounted, under controlled factory conditions, represented a tremendous improvement. With this form of premounted filament strain gage, much less skill and time are required to achieve satisfactory installations giving good and consistent results.

Most bonded wire strain gages are made from wire of approximately 0.001 in diameter, or less, and in resistances varying from about 50 ohms to several thousand ohms. The filaments are mounted on carriers made of materials selected for the particular applications for which the gages are to be employed.

Since a length of several inches of wire is usually needed to produce the necessary total resistance, and because the desired gage length is almost always less than the required length of wire, it is necessary to arrange the wire in some form of grid in order to economize on space, and thereby to permit reduction of the gage length to a suitable size. Figure 1.8 shows diagrams of typical grid configurations for wire gages. There are, of course, variations of these typical designs, as manufacturers' literature shows (17,18).

The flat grid is probably the most useful form. When the gage is on a flat surface, the centre line of the entire sensing element lies in one plane that is parallel to the surface of attachment. Due to the end loops, there is some response to strain at right angles to the direction of the grid axis. Usually the filament consists of one continuous length of wire; however, for some self-temperature-compensated gages, two elements, which possess opposing, or compensating, temperature characteristics are joined together.

An alternate type of construction originated as an expedient for manufacturing gages of short gage length (0.250 in or less) prior to the development of the techniques now used to make short flat grid gages. In

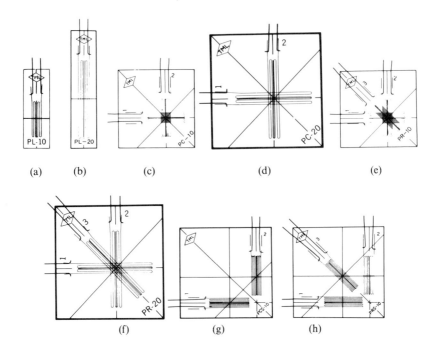

FIG. 1.8. Typical wire strain gages. (a, b) Single element gages. (c, d) Two-element stacked rectangular rosettes. (e, f) Three-element stacked rectangular rosettes. (g) Two-element rectangular rosette. (h) Three-element rectangular rosette. (From ref. 18.).

the wrap-around construction, the sensing element is wound tightly around a small flat carrier which is then encased between two cover sheets providing insulation and protection. An alternative procedure is to wind the sensing element on a small tubular mandrel (like a soda straw) that is then flattened and encased between the cover sheets.

For the various types of bonded wire strain gages, the strain is determined from the relation

$$\varepsilon = \frac{\Delta R/R}{G_F} \quad (1.32)$$

where ε = strain in the direction of the gage axis

$\Delta R/R$ = unit change in resistance

G_F = manufacturer's gage factor

Due to the geometrical differences between a straight wire and a strain gage grid, the value of the manufacturer's gage factor, G_F, is generally slightly lower than the strain sensitivity factor, S_t, of the wire from which the grid is constructed. Furthermore, the magnitude of G_F will vary slightly with variations in grid design.

Gages containing a single continuous filament which is wound back and forth will respond slightly to the effect of lateral strain which is sensed by the end loops. This means that Eq. (1.32), although generally applicable, is subject to some error when the strain field in which the gage is actually used differs from that of calibration. Usually the error caused by the response to lateral strain can be neglected, but there are a few situations in which it becomes appreciable. The magnitude of the error caused by lateral effects and, where necessary, the means of correcting for this error, are discussed in detail in a later chapter.

Some specific examples of the relation between strain and unit change in resistance for complete wire gages are shown in Fig. 1.9. In each case the slope of line relating the percent change in resistance to the percent strain represents the gage factor. One will note that the advance wire (constantan type) gage has the same gage factor for both elastic and plastic strains, whereas the isoelastic and nichrome gages both show a change in gage factor as one proceeds from elastic to plastic conditions. One should not be alarmed about this change in gage factor because we are usually interested in measuring elastic strains in metals, and these occur well below the change points shown in the diagrams. This is especially so in the case of isoelastic wire (whose change point occurs at approximately 0.75 percent strain), because this material is usually chosen to take advantage of its high gage factor for measuring very small strains.

Wire gages were used until the early 1950s, when foil gages were introduced. Some wire gages are still used today and can be purchased from several manufacturers.

Weldable wire gages

The first weldable wire gage was developed in the mid-1950s (19). Subsequent development for a quarter-bridge circuit used a single filament of nickel–chromium wire that was chemically etched so that its center length was approximately 1 mil in diameter. The wire was then folded in half and inserted into a stainless steel tube. The tube was filled with a metallic oxide powder which was compacted so that it not only electrically isolated the filament but mechanically coupled it to the tube in order to transmit strain. The construction is shown in Fig. 1.10.

In order to minimize the apparent strain due to temperature changes, the nickel–chromium filament is heat treated. Since different levels of heat treatment result in different values of the thermal coefficient of resistivity, it is possible to make this change equal in magnitude but of opposite polarity to the thermal coefficient of expansion.

To achieve temperature compensation, a separate compensating, or dummy, gage can be mounted on a stress-free piece of material identical to

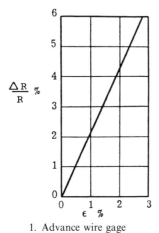

Gage factors			
Diagram No.	1	2	3
Low-strain	2.14	3.73	2.45
High-strain	2.14	2.46	2.00
Change point (percent strain)	–	0.76	0.45

1. Advance wire gage

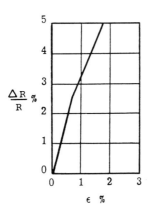

2. Isoelastic wire gage

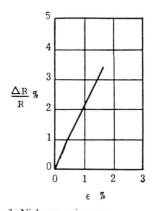

3. Nichrome wire gage

FIG. 1.9. Typical gage characteristics in tension. (From ref. 11, with permission of HMSO.)

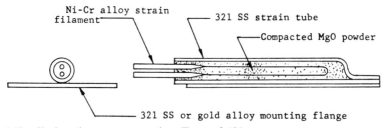

FIG. 1.10. Single active gage construction. (From ref. 19.)

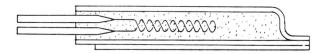

FIG. 1.11. 'True' dummy gage construction. (From ref. 19.)

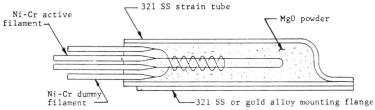

FIG. 1.12. Ni–Cr half-bridge gage construction. (From ref. 19.)

the material on which the active gage is mounted. The two gages are then arranged into a half-bridge circuit. This is a satisfactory method providing the material on which the dummy gage is mounted is completely stress free and that the dummy gage's temperature is identical to the active gage. Because these conditions do not always prevail, a 'true' dummy gage was developed. The dummy gage filament, identical to the active gage filament, is wound in a tight helix of the proper pitch angle. Since it is embedded in a strain tube with compacted magnesium oxide powder, the same as the active gage, it has the same heat-transfer characteristics. Therefore, the dummy gage can be used with a compensated active gage to minimize the apparent strain. The dummy gage is shown in Fig. 1.11.

The next step was to incorporate the single active gage and the 'true' dummy gage into one strain tube and mounting flange assembly. This results in a half-bridge gage rather than a quarter-bridge gage. The half-bridge gage is shown in Fig. 1.12.

The early weldable wire strain gage has resulted in a line of both quarter- and half-bridge gages (20). Two wire types are used for the filament. The first is a nickel–chromium that is temperature compensated and used for static measurements up to 600°F (315°C). Because of excessive drift above 600°F, the gages are used only for dynamic tests between 600°F and 1500°F (815°C). The second wire type is platinum–tungsten that can be used for static measurements up to 1200°F (650°C). Since this wire cannot be heat treated for temperature compensation, the half-bridge gage is recommended.

1.6. Foil strain gages

General characteristics

The foil gage operates in essentially the same manner as a wire gage. However, the sensing element consists of very thin metal foil (about 0.0002 in thick) instead of wire. In contrast to the wire gage, in which the sensing element possesses a uniform cross section throughout its entire length, the cross section of the sensing element of the foil gage may be somewhat variable from one end to the other. One of the most important advantages of the foil gage is that the ratio of contact surface area to the volume of the resistance element is relatively high, whereas in the wire gage, due to the circular cross section, this ratio is a minimum.

The early foil gages, introduced in England in 1952, were made from foil cemented to a lacquer sheet. The desired grid design for the strain gage was printed on the foil with an acid-resisting ink and the sheet was then subjected to an acid bath which removed all metal except where the printed design protected it. During the intervening years, a tremendous amount of very fruitful research has been carried on with respect to foil gages. The well-established alloys have been improved and new ones developed. In addition, there has been a vast improvement in the photographic techniques currently used in the photoetching process employed to manufacture foil gages. The degree of precision with which gages can now be produced, and the sharpness of definition of the boundaries of line elements, have made it possible to make gages possessing a uniform gage factor for a large range of gage lengths (previously, gage factor varied slightly with gage length). The result of these improvements has been to extend the advantages of the foil gage to a much wider variety of applications, including those at very low and very high temperatures, and especially for very precise transducers.

Foil gages are available in various gage lengths from 1/64 in to 6 in, and in a wide variety of grid configurations, including single gages, two-, three-, and four-element rosettes, half bridges, and full bridges. Figure 1.13 shows a few of the available designs. Standard alloys such as constantan, isoelastic, nichrome, karma, and platinum–tungsten, as well as a number of special proprietary alloys, are used in the sensing elements.

In general, foil gages exhibit a slightly higher gage factor and lower transverse response than their equivalent in wire. Since they are thinner, they conform more easily to surfaces with small radius of curvature, which means they are easier to install in fillets. As a result of their greater contact area, they can dissipate heat more readily and, in consequence, it is possible to use higher operating currents (applied voltage) with foil gages. The relatively large contact area, especially at the ends of the grid, reduces shearing stress in the bonding agent, and consequently, foil gages show comparatively little creep and hysteresis. Depending upon the carrier, the alloy, and its metallurgical condition, foil gages (generally the larger sizes) will measure strains precisely into the range of 10 to 15 percent. In terms of fatigue, suitable gages have exhibited life in excess of ten million cycles at strains of ± 1500 μin/in. Foil gages can be obtained on carriers of paper, epoxy, phenolic, glass reinforced resins, and other plastics.

By judicious choice of alloy and by careful control of the metallurgical condition (cold working and heat treatment), it is possible to produce foil

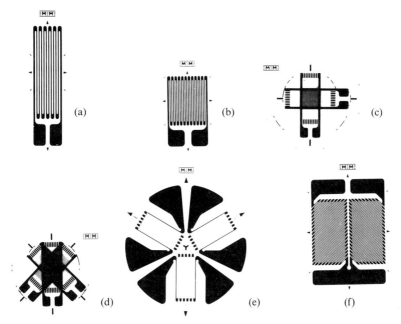

FIG. 1.13. Foil strain gages. (a, b) Single-element gages. (c) Stacked two-element rectangular rosette. (d) Stacked three-element rectangular rosette. (e) Three-element delta rosette. (f) Two-element rectangular rosette torque gage. (Courtesy of Measurements Group, Inc.

with its coefficient of linear expansion and resistance–temperature characteristic very closely matched to the coefficient of linear expansion of some arbitrarily selected material. By this means, it has been possible to produce temperature-compensated foil gages whose response (within certain limits) is, for practical purposes, independent of temperature, within a given temperature range.

Weldable foil strain gages

For situations in which the conventional installation techniques may not be applicable, weldable foil gages are available (18, 20, 21). Single-element gages and T-rosettes (two-element) are made by premounting gages on a carrier of stainless steel shim stock approximately 0.005 in thick. Surface preparation of the specimen requires solvent cleaning and abrasion with silicon-carbide paper or a small hand grinder. The unit is then attached to the specimen with a small spot welder designed specifically for this purpose.

Sensing elements of constantan, nichrome, and high-temperature alloys are available. The normal operating temperature ranges from $-320°F$ to $570°F$ (-195 to $300°C$) for static observations, although under some conditions a single-loop wire (typically nichrome V) encased in a stainless steel tube may be used to $925°F$ ($495°C$) or higher.

1.7. Semiconductor gages (4, 8, 9, 22–25)

Within certain limitations, semiconductor gages can be used in the same manner as metallic gages. However, the semiconductor gage is really a much more elaborate device whose optimum use requires a knowledge of all the variables involved, and the degree to which they influence the performance of the instrument. The comparison between the uses of metal and semiconductor gages is somewhat parallel to the difference between playing checkers and playing chess. Both are good games, but chess has a much broader range of opportunities for making moves and, correspondingly, many more possibilities of getting into trouble unless one considers all the variables carefully.

The main attraction of the semiconductor is, of course, the high strain sensitivity of silicon, which is the favored material for the sensing element. This means a relatively large resistance change per unit of strain, which characteristic is helpful for both high and low values of strain.

1. For high strains, the large response enables one to drive indicating devices directly without intermediate amplification. This provides a simplification which is accompanied by reduced weight and expense.
2. For low strains, which produce exceedingly small changes in resistance of metal gages, the semiconductor gages will develop unit changes about 50 times greater, with the result that the indications of $\Delta R/R$ can be measured conveniently and precisely.

As contrasted with the above advantages, one must also recognize, and be able to cope with, certain disadvantages.

1. The unit change in resistance (which is based on the initial resistance, R_0, of the unstressed sensor at temperature T_0) is a nonlinear function of the strain, although for some special conditions it can be taken as linear for small strain excursions.
2. The large resistance change per unit of strain, which is the very thing that makes the semiconductor gage attractive, may also present a minor problem due to the fact that, in the process of installation, the resistance of the gage may be altered considerably from the value which prevailed in the unstressed condition of the sensing element. On this account, it is necessary to determine the gage resistance following installation so that, if necessary, an appropriate correction can be made for the gage factor.
3. The resistance of the gage will change with change in temperature.
4. The strain sensitivity, or gage factor, will change with change in temperature.

Investigation of silicon reveals that both the strain sensitivity and the temperature sensitivity (change of resistance with temperature) vary considerably with the quantity of impurity which is present. It is also observed that high sensitivity to strain is accompanied by high sensitivity to changes in temperature. This suggests that some compromise between strain sensitivity and temperature response may be desirable, and perhaps essential, depending upon the particular application.

Fortunately, by suitable doping (introduction of controlled amounts of impurities) during the manufacturing process, the strain and temperature sensitivities can be varied and adjusted (although not independently) to meet specified requirements. Therefore, by suitable procedures in the manufacturing process, it is possible to achieve a desired compromise which will result in much improved temperature characteristics at the expense of a modest reduction in strain sensitivity. Practical considerations indicate that a good balance is achieved when the gage factor is about 120.

Since semiconductor gages are available with both positive and negative gage factors, another approach, although perhaps a more difficult one, is to take advantage of the characteristics of the electrical circuit of which the gage forms a part, and to employ two similar gages with gage factors of opposite sign.

Due to the relatively large number of variables involved, and consequently the somewhat more complex procedure required for converting resistance change into terms of strain, it seems unlikely, at least for the present, that semiconductor gages will replace metallic gages for purposes of stress analysis, except perhaps, under special circumstances involving the determination of very small strains.

However, for transducers, in which gages can be installed under carefully controlled factory conditions, and subsequently calibrated in complete bridges, the high output of the semiconductors makes them exceedingly attractive. It seems that semiconductor strain gages will achieve greatest success and optimum utility in this type of application.

1.8. Some other types of gages

Temperature gages

Examination of the characteristics of metal and semiconductor strain gages reveals that changes in resistance occur not only as a result of changes in strain, but also from changes in temperature. Although the response to temperature may complicate the determination of strain, it nevertheless provides the possibility of making, and using, temperature sensors with essentially the same techniques as those which are employed in the making and using of strain gages.

The choice of material for the sensing element, of course, will be different for these two applications. When it is desired to measure strain, with a minimum influence from temperature changes, a copper–nickel alloy of the constantan type is frequently employed for temperatures in the range from about $-250°F$ to about $500°F$ ($155-260°C$). For lower or higher temperatures, it is necessary to select another type of alloy (26). However, for a temperature sensor, it is preferable to choose a material, such as nickel, platinum, or an iridium–platinum alloy, which possesses a much greater response to changes in temperature. For semiconductor materials, the processing is varied to produce the preferred characteristics for either strain or temperature sensing.

For a number of years, bonded wire temperature sensors have been commercially available, followed more recently by foil temperature gages (27, 28). Foil temperature gages have several advantages over wire-wound sensors in that they are less expensive, not as fragile, and their time–temperature response is similar to that of a strain gage. Standard strain gage instrumentation may also be used with them.

For convenience in making observations, sensors and their signal-conditioning networks have been designed to produce signals corresponding to indications of 10 or 100 microstrain per degree Fahrenheit. Therefore, when the strain indicator is referenced to some temperature, one is able to obtain a direct reading of all other temperatures within the working range of the system. For example, if a temperature sensor and network is used that provides an indication of 10 microstrain per degree Fahrenheit, the initial balance of the indicator may be adjusted so that the reading will be 750 microstrain when the sensor is actually 75°F (24°C). Then, for any subsequent observation, the temperature in Fahrenheit will be represented by the indicator reading divided by 10. If a subsequent reading turns out to be 830, then the temperature at the sensor is 83°F (28°C).

The obvious advantage of this method of determining temperature lies in the fact that a standard strain indicating (and recording) system can be employed, without any modification at all, for the measurement of temperature at strain gage locations, or elsewhere, by the simple procedure of switching the temperature sensor (with its conditioning network) in and out of the indicating circuit just as if it were another strain gage.

Crack measuring gages

Another instrument incorporating certain features of the strain gage is known commercially as the Krak Gage. Its main purpose is to monitor the progression of cracks which usually develop as a result of fatigue caused by repeated stressing. If the progress of a crack is watched, a part can be taken out of service before a disaster occurs, which is a very valuable consideration in the aircraft and many other industries (29).

A schematic diagram of the gage, shown in Fig. 1.14, is produced by Hartrun Corporation in a variety of different sizes (30). It possesses certain characteristics which are like those of the strain gage, but its use is very different. Basically, the Krak Gage consists of a constantan foil sensor 5 μm thick mounted on an epoxy-phenolic or cast epoxy carrier, depending on the operating temperature. The carrier and the gage are cemented to the test

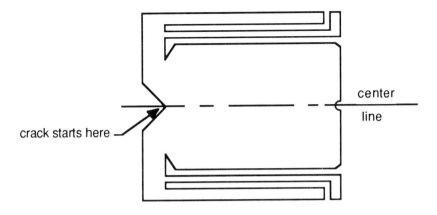

FIG. 1.14. Schematic diagram of a crack measuring gage. (From ref. 30.)

piece, or machine part, by the usual strain gage bonding procedure at a location where a crack is expected to start, or may already have started. The positioning of the gage is such that it will be cracked under its centerline in step with the material underneath it. The gage is energized with a constant current, usually in the range between 0 and 100 milliamperes, and the change in potential drop between its two inner leads is a measure of the distance by which the crack has advanced. Since these gages have a resistance of about 1 ohm before the crack commences, they cannot be used with ordinary strain gage equipment.

Another crack detection gage is the CD-Series produced by Micro-Measurements (31). This gage is used to indicate the presence of a crack, and crack growth rate may be monitored by using several CD gages at a location. The CD gage overcomes two of the limitations suffered by the use of thin copper wires. These are the possibility of a crack progressing beyond the wire without breaking it, and also the failure of the wire by fatigue when located in a region subjected to cyclic strain of large magnitude.

The gages consist of a single strand of high-endurance beryllium–copper wire on a tough polyimide backing. A rigid high-modulus adhesive is used to bond the sensor to the polyimide backing. A crack that is growing underneath the gage indices local fracture of the sensing wire and opens the electrical circuit. Bonding of the gages to a structure or a machine can be accomplished with conventional strain gage adhesives that are compatible with polyimide backing.

Friction gages

For stress probing, especially for vibrating stresses, when a number of observations are to be made quickly without taking time to install a larger number of strain gages, a very useful type of gage has evolved (18).

This is a conventional 120-ohm foil gage to which strain is transmitted by friction. The gage is bonded to one face of a rubber sheet, then emery powder is cemented over the gage face to provide a frictional surface. To the other side of the rubber is cemented a metal backing plate. The gage is pressed against the test member so that the emergy powder contacts the test surface, where the friction is great enough to transmit the surface strains to the sensing element of the strain gage. This device can be moved quickly and easily from place to place, thus enabling one to make a rapid survey with a minimum amount of equipment.

Embedment gages

Embedment gages and transducers are designed and used primarily to measure curing and loading strains in concrete. They may also be used, however, with resins, ice, asphalt, and other materials. There are several variations of these gages and transducers.

One is a polyester mold gage that can be supplied as a single gage, a two-element rectangular rosette, or a three-element rectangular rosette. Standard wire gages and lead wires are hermetically sealed between thin resin plates, thus waterproofing the unit. The unit is then coated with a coarse grit to enhance bonding between it and concrete. Excellent electrical insulation is exhibited even after several months of embedment (18).

A transducer is available in either half- or full-bridge arrangements, thus giving temperature compensation. The gages in this transducer are made of a special alloy foil encased in a low-elastic-modulus material in order to prevent swelling and to minimize loading effects. A quarter bridge is also available for temperature measurement (18).

Another embedment gage uses nickel–chrome wire in a quarter bridge and comes in gage lengths of 2, 4, and 6 in (20). The gage wire is enclosed in a 0.040-in diameter stainless steel tube and is insulated by compacted magnesium oxide powder. End disks with three equally spaced holes are attached at each end of the stainless steel tube for anchoring the gage. Anchoring is accomplished by tying wires through the holes in the disks, then pulling them radially outward and tying them to the structure or to reinforcing bars. The wires are pulled taut but should not load the gage along its axis or apply a torque. The gage length should be four times the size of the largest aggregate in order to provide strain averaging, and it is important that concrete should contact the gage along its entire length for optimum bonding and strain transfer. Figure 1.15 shows this gage.

1.9. Brittle lacquer coatings (3)

Brittle lacquer coatings have been mentioned earlier as a tool in experimental stress analysis, and so a few comments are in order, since these coatings have been used quite extensively. Their development has evolved over the years, having its beginning in the observation that brittle oxide coats on hot-rolled

FIG. 1.15. Typical embedment gage. (Courtesy of Eaton Corporation.)

steel cracked when a member was loaded. This led, in the early part of the 1900s, to the use of varnish, lacquer, or molten resins on machine or structural members. When loaded in the elastic region, the coating cracked in a direction normal to the maximum principal strain direction. In the 1930s, Greer Ellis developed a brittle lacquer while at the Massachusetts Institute of Technology. It was subsequently marketed in 1938 under the name of Stresscoat by Magnaflux Corporation of Chicago, Illinois.

Brittle lacquers are sensitive to both temperature and relative humidity. For this reason, they are made in a number of formulations for specific temperature and relative humidity conditions. When planning a test, one must anticipate the temperature and relative humidity at the time the test is to be conducted, and then choose the coat accordingly. When properly chosen, the threshold strain of the coating will be approximately 500 μin/in. If, however, the temperature or the relative humidity increases, the threshold strain will increase and perhaps produce no cracks within the loading range. Conversely, a decrease in temperature or relative humidity will decrease the threshold strain, resulting, in the worst case, in the coating becoming crazed (cracking into a random pattern).

The application of a brittle lacquer consists of several steps. The test member is first thoroughly cleaned to insure that it is free of scale, dirt, and oils. The member is next sprayed with a coat of aluminum powder in a carrier solvent and allowed to dry for at least 30 minutes. This undercoat forms a reflective coating that enables one to see cracks in the lacquer more easily. Next, the brittle lacquer is applied in a number of thin coats until its thickness is approximately 0.003 to 0.006 in thick. At the same time that the test member is coated, at least four calibration bars should be prepared in the same manner and kept with the test member. The entire group, test member and calibration bars, is then allowed to cure for at least 18 hours before testing.

The test member is loaded in increments, and at each incremental load the brittle coat is examined for cracks. The tips of the cracks in each area where they appear may be outlined with a felt-tip pen. As the load is increased, the crack growth at each area is marked, as well as noting other areas where new cracks appear. This process is continued until the maximum load is reached. When the yield point of the material is attained in any area of the test member, the brittle coat will flake off.

Although the brittle coat cracks only under tensile strains, it can also be used to determine compressive strains. To accomplish this, the full load is applied to the test member and held for at least 3 hours after the tensile strains are determined. During the hold time the brittle coat creeps and relaxes. The load is then removed as quickly as possible, with the coat then reacting to the compressive strains as though they were tensile strains.

At the start of the test, a calibraton bar is loaded into a cantilever fixture and one end deflected a known amount. The bar is subjected to strains along its length, and the minimum strain at which a crack is observed is recorded; this is the threshold strain. As the test progresses, particularly over a period of time in which the temperature or relative humidity may change and thus change the threshold strain, other calibration bars can be tested at intervals in order to determine whether or not the threshold strain has changed.

The brittle coat can be treated to enhance the cracks. The cracks can be recorded by photographing, marking a drawing, or some other means. If further testing is to be done with strain gages (as is usually the case), the coat can be stripped off if the same member is to be used and strain gages applied. Since the principal strain directions are known, two strain gages (a two-element rectangular rosette) may be applied in these directions and the principal stresses computed. The advantages of the brittle coat are:

1. The brittle coat and its crack pattern allow one to see the strain (and stress) distribution over most of the entire test member.
2. When strain gages are applied in the directions of the principal strains in the various areas on the test member, only two gages are required rather than the three that would be necessary if the principal strain directions were unknown. This results in a saving of both time and money.
3. The method is relatively inexpensive and is extremely useful for a preliminary investigation prior to a detailed strain gage study.

Although brittle lacquers have been used extensively in order to observe the strain distribution on the surface of a member, their main use has been as an aid in the placement of strain gages. One should be aware, however, of the use of photoelastic coatings that can be applied to a structure. They give full-field data that accurately identify areas of high strain, and constitute a nondestructive test. The member, unlike brittle lacquer tests, can be tested a number of times, with the results being recorded on film or video tape. For more information, one should consult either manufacturers of photoelastic equipment or any of several books on the subject.

Appendix 1

Approximate strain sensitivities of some metals (11)

	Hard drawn			Annealed		
Metal	Sensitivity in low range	Sensitivity in high range	Change point (strain percent)	Sensitivity in low range	Sensitivity in high range	Change point (strain percent)
Silver	2.9	2.4	0.8	3.0	2.3	0.2
Platinum	6.1	2.4	0.4	5.9	2.3	0.3
Copper	2.6	2.2	0.5	2.2	2.2	–
Iron	3.9	2.4	0.8	3.7	2.1	0.5
Nickel	Negative	2.7	–	Negative	2.3	–
Ferry (60/40 Cu–Ni)	2.2	2.1	0.5	2.2	2.2	–
Minalpha (Manganin)	0.8	2.0	0.6	0.6	1.9	–
10 percent iridium–platinum	4.8	2.1	0.4	3.9	1.9	0.3
10 percent rhodium–platinum	5.5	2.4	0.5	5.1	2.0	0.4
40 percent silver–palladium	0.9	1.9	0.8	0.7	2.0	0.5

REFERENCES

1. "Brittle Coating for Stress Analysis Testing," Bulletin S-109, Measurements Group, Inc., P.O. Box 27777, Raleigh, NC 27611, 1978. (Now out of print.)
2. "General Instructions for the Selection and Use of Tens-Lac Brittle Lacquer and Undercoating," Instruction Bulletin 215-C, Measurements Group, Inc., P.O. Box 27777, Raleigh, NC 27611, 1982. (Now out of print.)
3. "Using Stresscoat," Electrix Industries, Inc., P.O. Box J, Roundlake, IL 60073.
4. Sanchez, J. C. and W. V. Wright, "Recent Developments in Flexible Silicon Strain Gages," in *Semiconductor and Conventional Strain Gages*, edited by Mills Dean III and Richard D. Douglas, New York, Academic Press, 1962, pp. 307–345.
5. Mack, Donald R., "Linearizing the Output of Resistance Temperature Gages," *SESA Proceedings*, Vol. XVIII, No. 1, April 1961, pp. 122–127.
6. Thomson, W. (Lord Kelvin), "On the Electrodynamic Qualities of Metals," *Philosophical Transactions of the Royal Society of London*, Vol. 146, 1856, pp. 649–751.
7. Sette, W. J., L. D. Anderson, and J. G. McGinley, "Resistance–Strain Characteristics of Stretched Fine Wires," David Taylor Model Basin, Report No. R-212, Sept. 1945.
8. Dorsey, James, "Semiconductor Strain Gages," *The Journal of Environmental Sciences*, Vol. 7, No. 1, Feb. 1964, pp. 18–19.
9. Dorsey, James, *Semiconductor Strain Gage Handbook*, Part 1. BLH Electronics, 75 Shawmut Road, Canton, MA 02021. (Now out of print.)
10. Baker, M. A., "Semiconductor Strain Gauges," in *Strain Gauge Technology*, edited by A. L. Window and G. S. Holister, London and New Jersey, Applied Science Publishers Inc., 1982, p. 274. Copyright Elsevier Science Publishers Ltd. Reprinted with permission.

11. Jones, E. and K. R. Maslen, "The Physical Characteristics of Wire Resistance Strain Gauges," R. and M. No. 2661 (12,357), A.R.C. Technical Report, Her Majesty's Stationery Office, London, 1952, Reproduced with the permission of the Controller of Her Britannic Majesty's Stationery Office.
12. de Forest, A. V., "Characteristics and Aircraft Applications of Wire Resistance Strain Gages," *Instruments*, Vol. 15, No. 4, April 1942, pp. 112–114, 136–137.
13. Weibull, W., "Electrical Resistance of Wires with Large Strains." *Nature*, Vol. 162, pp. 966–967. Copyright © 1948 Macmillan Magazines Limited.
14. Shoub, H., "Wire-Resistance Gages for the Measurement of Large Strains," David Taylor Model Basin, Report No. 570, March 1950.
15. Biermasz, A. J., R. G. Boiten, J. J. Koch, and G. P. Roszbach, "Strain Gauges—Theory and Application," Philips Technical Library, Philips Industries, Eindhoven, Netherlands, 1952.
16. Meier, J. H., "On the Transverse-strain Sensitivity of Foil Gages," *Experimental Mechanics*, Vol. 1, No. 7, July 1961, pp. 39–40.
17. "Strain Gages, SR-4," BLH Electronics, Inc., 75 Shawmut Road, Canton, MA 02021, 1985 Edition.
18. "TML Strain Gauges," E-101 V and E-101 Y, Tokyo Sokki Kenkyujo Co., Ltd., Tokyo, Japan, 1988. Distributed by Texas Measurements, Inc., P.O. Box 2618, College Station, TX 77841.
19. Gibbs, Joseph P., "Two Types of High-temperature Weldable Strain Gages: Ni-Cr Half-bridge Filaments and Pt-W Half-bridge Filaments," *Proceedings of the Second SESA International Congress on Experimenal Mechanics*, Washington, DC, Sept. 28 to Oct. 1, 1965, pp. 1–8.
20. "Weldable and Embeddable Integral Lead Strain Gages," Applications and Installation Manual, Eaton Corp., Ailtech Strain Gage Products, 1728 Maplelawn Rd., Troy, MI 48084, 1985.
21. "Catalog 500: Part A—Strain Gage Listings; Part B—Strain Gage Technical Data," Measurements Group, Inc., P.O. Box 27777, Raleigh, NC 27611, 1988.
22. Sanchez, J. C., "The Semiconductor Strain Gage—A New Tool for Experimental Stress Analysis," in *Experimental Mechanics*, edited by B. E. Rossi, New York, The Macmillan Company, 1963, pp. 255–274.
23. Vaughn, John, *Application of B & K Equipment to Strain Measurements*, Bruel & Kjaer, Naerum, Denmark, 1975, Ch. 10.
24. "Semiconductor Strain Gages," SR-4 Application Instructions, BLH Electronics, Inc., 75 Shawmut Road, Canton, MA 02021, 1986.
25. Dorsey, James, "Data-reduction Methods for Semiconductor Strain Gages," *Experimental Mechanics*, Vol. 4, No. 6, June 1964, pp. 19A–26A.
26. Weymouth, L. J., "Strain Measurement in Hostile Environment," *Applied Mechanics Reviews*, Vol. 18, No. 1, Jan. 1965, pp. 1–4.
27. "Cryogenic Linear Temperature Sensor," Product Bulletin PB-104-3, Mesasurements Group, Inc., P.O. Box 27777, Raleigh, NC 27611, 1983.
28. "Temperature Sensors and LST Matching Networks," Product Bulletin PB-105-7, Measurements Group, Inc., P.O. Box 27777, Raleigh, NC 27611, 1984.
29. Liaw, Peter K., W. A. Logsdon, L. D. Roth, and H. R. Hartmann, "Krak-Gages for Automated Fatigue Crack Growth Rate Testing: A Review," *ASTM Special Technical Publication No. 877*, 1989, pp. 177–196. Copyright ASTM. Reprinted with permission.

30. Hartmann, H. R. and R. W. Churchill, "Krak-Gage, a New Transducer for Crack Growth Measurement," presented at SESA Fall Meeting, Keystone, CO, Oct. 1981.
31. "CD-Series Crack Detection Gages," Product Bulletin PB-118, Measurements Group, Inc., P.O. Box 27777, Raleigh, NC 27611, 1984.

2

STRESS–STRAIN ANALYSIS AND STRESS–STRAIN RELATIONS

2.1. Introduction

The material in Chapter 2 should be familiar from courses in mechanics of materials and design, and so serves as a review. The notation and sign convention for both stress and strain follow that generally given in the theory of elasticity.

Strain gages are applied to a surface that is usually stress free in a direction normal to the strain gage surface. For this reason, the transformation equations for plane stress are developed instead of the more complicated triaxial stress state. The necessary equations are derived that enable us to transform from one coordinate system to another. Furthermore, we can compute the principal stresses and determine their orientation relative to a chosen coordinate system.

Since we cannot determine stress experimentally by direct measurement, we resort to measuring strain on a surface through the use of a strain-measuring device. In order to make use of the experimentally determined strains, transformation equations for plane strain are generated that are similar in form to the transformation equations for plane stress. Here we see that the orientations of the principal strains are identical to the orientations of the principal stresses for the chosen coordinate system.

Although all of the necessary values wanted may be handled through calculation, it is often desirable to determine the values graphically. To accomplish this, Mohr's circle for stress and for strain are generated. These diagrams allow us to visualize the transformation from one coordinate system to another, and, if they are accurately drawn, will give satisfactory answers. With the availability of hand-held calculators, though, it is much easier to draw the diagrams freehand, observe the required orientations, and then calculate the answers. In drawing the circles, note the definition for positive shearing stress and shearing strain.

You will observe that material properties do not enter into the development of the transformation equations. The transformation equations for stress are based on the static equilibrium of an element, while the transformation equations for strain are based on the geometry of small deformations of the element. In order to relate the two, material properties

2.2. Basic concepts of stress

When a solid body is acted upon by a system of forces, which may be either external or internal, or both external and internal, it is said to be subjected to stress. In general, this means that forces are transmitted from one elemental particle to another within all or part of the body. How these forces are distributed on the external surfaces, or throughout the interior body, is of vital importance, since the ability of the body material to withstand the action of the forces depends upon the force intensity prevailing at each point within the material.

Usually we think of stress as the effect of forces on part, or all, of the surface of a body, or internally as the influence which the forces acting on one side of a section (usually a plane section) through the body exert upon the material on the other side of the section.

Since, from practical considerations, the forces which act on solid bodies must, of necessity, be distributed over areas (or throughout the volume), we must be rather specific regarding our meaning of the term stress. It is sometimes used to indicate total force, and under other conditions implies force per unit area. Both usages are correct, but every now and then the exact meaning is somewhat loosely implied.

To be technically correct, one should say "total stress" when referring to force, and "intensity of stress" or "unit stress" when force per unit area is meant. However, when only one of the two meanings is required in a particular discussion, it is quite common to use the word "stress" by itself with the words total, intensity of, or unit, being understood. For our purposes, the term stress will be used to indicate force per unit area.

Figure 2.1a shows such a body acted upon by forces P_1, P_2, P_3, and P_4. An imaginary cutting plane AB is passed through the body and the upper portion of the body is removed. In order for the lower portion to remain in equilibrium, a system of forces, representing the effect of the upper part of the body, acts upon the cut surface as shown in Fig. 2.1b. One of the elemental forces is represented by the force ΔP acting on the incremental area ΔA. If all such forces are summed over the entire area, the resultant will be a force (not normal, in general, to plane AB) having the proper magnitude and direction to maintain equilibrium.

We now turn our attention to the force ΔP and define stress at a point as

$$\text{Stress} = \lim_{\Delta A \to 0} \left[\frac{\Delta P}{\Delta A} \right] \tag{2.1}$$

Since the loading on the body in Fig. 2.1 is complex, we expect the stress to vary in intensity from point to point on the cut surface. Thus, when we speak

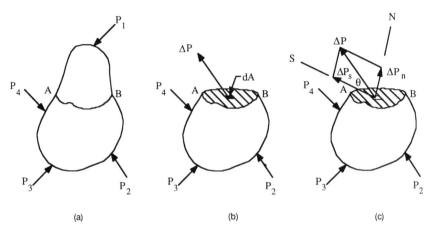

FIG. 2.1. Body in equilibrium acted upon by external forces.

of stress, we must define the point on the cut surface on which it is acting. Furthermore, ΔP will generally not be normal to the surface but will be inclined at some angle θ as shown in Fig. 2.1c. The line N is normal to the surface and the line S lies on the surface, and so ΔP can be resolved into two components along lines N and S, namely ΔP_n and ΔP_s. Using the definition of stress given by Eq. (2.1), we now have the total stress resolved into two components; we call the stress directed along N the normal stress, σ, and the stress directed along S the shearing stress, τ. The normal stress σ will be tensile ($+$) if it tends to separate the material on opposite sides of the section, or it will be compressive ($-$) if it tends to push together the material on opposite sides of the section. The shearing stress τ has a tendency for the material on one side of the section to slide by the material on the other side of the section.

When the force acting on the area is distributed uniformly over the area, each element of the area will be subjected to the same intensity of loading, and the magnitude of the stress at every point will be the same as the average value, which is computed by dividing the total force by the whole area. Thus, for uniformly distributed stress,

$$\text{Stress (at each point)} = \frac{\text{total force}}{\text{whole area}} = \frac{P}{A} \qquad (2.2)$$

We will generally be working with plane stress. Suppose, in Fig. 2.1c, that all the elemental forces ΔP were contained in planes parallel to the plane defined by lines N and S. The normal stresses and the shearing stresses would also lie in these planes, and so no stresses would appear in planes normal to the plane containing lines N and S. This condition gives us the plane stress state.

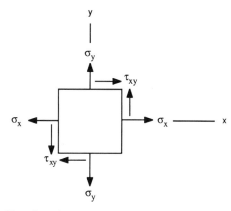

FIG. 2.2. Element subjected to plane stress.

2.3. Biaxial stresses

Since most of our problems are plane stress problems, we want to develop the transformation equations for this stress state. Figure 2.2 shows the plane stress state, where $\sigma_z = \tau_{xz} = \tau_{yz} = 0$. The sign convention for the stresses is the following: (1) the normal stress is positive (tensile) it it is directed outward from the plane, negative (compressive) if directed inward toward the plane; and (2) the shear stresses are positive when directed in a positive coordinate direction on a plane whose outward normal is directed in a positive coordinate direction, or when directed in a negative coordinate direction on a plane whose outward normal is directed in a negative coordinate direction. In Fig. 2.2, all stresses are positive according to the sign convention just stated.

Suppose the element in Fig. 2.2 has a cutting plane, AB, passed through it as shown in Fig. 2.3. We now want to determine the stresses in the new $x'y'$ system, where x' is normal to plane AB and y' lies in plane AB. The normal stress acting on plane AB is designated as $\sigma_{x'}$, while the shearing stress on that plane is $\tau_{x'y'}$. In order to determine the new stress state, the equilibrium of the element in Fig. 2.3 must be considered. If forces in the x' direction are summed (taking the distance in the z direction, or normal to the paper, as unity), the following results:

$$\sigma_{x'}(AB) - \sigma_x \cos\theta(OA) - \sigma_y \sin\theta(OB) - \tau_{xy}\cos\theta(OB) - \tau_{xy}\sin\theta(OA) = 0$$

From Fig. 2.3, we see that $OA/AB = \cos\theta$ and $OB/AB = \sin\theta$. Dividing each term by AB and using these relationships, $\sigma_{x'}$ can be expressed in terms of σ_x, σ_y, τ_{xy}, and θ. Thus,

$$\sigma_{x'} = \sigma_x \cos^2\theta + \sigma_y \sin^2\theta + 2\tau_{xy}\sin\theta\cos\theta \qquad (2.3)$$

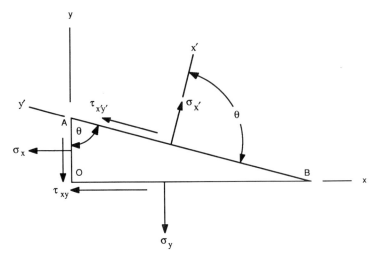

FIG. 2.3. Element cut by plane AB.

By summing forces in the y' direction, we obtain

$$\tau_{x'y'}(AB) + \sigma_x \sin\theta(OA) - \sigma_y \cos\theta(OB) - \tau_{xy}\cos\theta(OA) + \tau_{xy}\sin\theta(OB) = 0$$

Following the same procedure as before, the expression for $\tau_{x'y'}$ becomes

$$\tau_{x'y'} = -(\sigma_x - \sigma_y)\sin\theta\cos\theta + \tau_{xy}(\cos^2\theta - \sin^2\theta) \qquad (2.4)$$

Equations (2.3) and (2.4) can be expressed in terms of 2θ by using the following identities:

$$2\sin\theta\cos\theta = \sin 2\theta$$
$$\sin^2\theta = \tfrac{1}{2}(1 - \cos 2\theta)$$
$$\cos^2\theta = \tfrac{1}{2}(1 + \cos 2\theta)$$

The expressions for $\sigma_{x'}$ and $\tau_{x'y'}$ are rewritten as

$$\sigma_{x'} = \frac{\sigma_x + \sigma_y}{2} + \frac{\sigma_x - \sigma_y}{2}\cos 2\theta + \tau_{xy}\sin 2\theta \qquad (2.5)$$

$$\tau_{x'y'} = -\left(\frac{\sigma_x - \sigma_y}{2}\right)\sin 2\theta + \tau_{xy}\cos 2\theta \qquad (2.6)$$

Equations (2.5) and (2.6) allow the determination of $\sigma_{x'}$ and $\tau_{x'y'}$ at a point for any value of the cutting plane angle θ.

As the angle 2θ varies from $0°$ to $360°$, $\sigma_{x'}$ will change in value. The objective is to find its maximum and minimum values. This is accomplished by differentiating $\sigma_{x'}$ with respect to θ and setting the resulting equation equal to zero. From Eq. (2.5),

$$\frac{d\sigma_{x'}}{d\theta} = -(\sigma_x - \sigma_y)\sin 2\theta + 2\tau_{xy}\cos 2\theta = 0 \tag{a}$$

Dividing each term by $\cos 2\theta$ results in

$$\tan 2\theta = \frac{2\tau_{xy}}{\sigma_x - \sigma_y} \tag{2.7}$$

The directions of the principal stresses, and therefore the principal axes, are determined from Eq. (2.7). Thus, two values, $90°$ apart, for θ are determined. One value corresponds to the angle measured from the x axis to the first principal axis (counterclockwise is positive) along which the maximum principal stress acts. The other value corresponds to the angle measured from the x axis to the second principal axis along which the minimum principal stress acts. Equation (2.7) by itself does not allow us to distinguish between the two axes, and so we must call in another trigonometric relationship in order to distinguish between the two.

There are two quadrants in which $\tan 2\theta$ can have the value given by Eq. (2.7); these are the first and third quadrants. Considering first quadrant values for Eq. (2.7), we have

$$\sin 2\theta = \frac{2\tau_{xy}}{\sqrt{(\sigma_x - \sigma_y)^2 + (2\tau_{xy})^2}} \tag{2.8}$$

$$\cos 2\theta = \frac{\sigma_x - \sigma_y}{\sqrt{(\sigma_x - \sigma_y)^2 + (2\tau_{xy})^2}} \tag{2.9}$$

Substituting the values of $\sin 2\theta$ and $\cos 2\theta$ given by Eqs. (2.8) and (2.9), respectively, into Eq. (2.5) results in $\sigma_{x'} = \sigma_1$. Carrying out the required algebra,

$$\sigma_1 = \frac{\sigma_x + \sigma_y}{2} + \sqrt{\left(\frac{\sigma_x - \sigma_y}{2}\right)^2 + (\tau_{xy})^2}$$

If third-quadrant values are used, $\sin 2\theta$ and $\cos 2\theta$ are negative.

Substitution of these values into Eq. (2.5) gives the second principal stress as

$$\sigma_2 = \frac{\sigma_x + \sigma_y}{2} - \sqrt{\left(\frac{\sigma_x - \sigma_y}{2}\right)^2 + (\tau_{xy})^2}$$

We can now write the equation for the principal stresses as

$$\sigma_{1,2} = \frac{\sigma_x + \sigma_y}{2} \pm \sqrt{\left(\frac{\sigma_x - \sigma_y}{2}\right)^2 + (\tau_{xy})^2} \qquad (2.10)$$

Since Eqs. (2.8) and (2.9) give the values of $\sin 2\theta$ and $\cos 2\theta$ for the principal stresses, substitution of these values into Eq. (2.6) shows $\tau_{x'y'}$ to be zero. This tells us that there is no shearing stress on the planes containing the principal stresses. This is also apparent if Eqs. (2.6) and (a) are compared.

In order to determine the orientation of σ_1 with respect to the x axis, two of the three trigonometric relations given by Eqs. (2.7), (2.8), and (2.9) must be used.

The same procedure can be used in finding the maxmum value of $\tau_{x'y'}$ in the xy plane. This is achieved by differentiating $\tau_{x'y'}$ with respect to θ and setting the resulting equation equal to zero. From Eq. (2.6),

$$\frac{d\tau_{x'y'}}{d\theta} = -(\sigma_x - \sigma_y)\cos 2\theta - 2\tau_{xy}\sin 2\theta = 0$$

Dividing each term by $\cos 2\theta$,

$$\tan 2\theta = -\frac{(\sigma_x - \sigma_y)}{2\tau_{xy}} \qquad (2.11)$$

Note that Eq. (2.11) is the negative reciprocal of Eq. (2.7). In this case, the value of $\tan 2\theta$ given by Eq. (2.11) will be negative in either the second or fourth quadrant. Taking second-quadrant values,

$$\sin 2\theta = \frac{\sigma_x - \sigma_y}{\sqrt{(\sigma_x - \sigma_y)^2 + (2\tau_{xy})^2}} \qquad (2.12)$$

$$\cos 2\theta = \frac{-2\tau_{xy}}{\sqrt{(\sigma_x - \sigma_y)^2 + (2\tau_{xy})^2}} \qquad (2.13)$$

Substituting these values of $\sin 2\theta$ and $\cos 2\theta$ into Eq. (2.6) results in $\tau_{x'y'} = \tau_{max}$. Carrying out the operation,

$$\tau_{max} = -\sqrt{\left(\frac{\sigma_x - \sigma_y}{2}\right)^2 + (\tau_{xy})^2} \qquad (b)$$

If fourth-quadrant values are used, $\sin 2\theta$ is negative and $\cos 2\theta$ is positive. Substitution of these values into Eq. (2.6) yields

$$\tau_{max} = \sqrt{\left(\frac{\sigma_x - \sigma_y}{2}\right)^2 + (\tau_{xy})^2} \tag{c}$$

Thus, the maximum shearing stress can be written as

$$\tau_{max} = \pm\sqrt{\left(\frac{\sigma_x - \sigma_y}{2}\right)^2 + (\tau_{xy})^2} \tag{2.14}$$

If values of $\sin 2\theta$ and $\cos 2\theta$, given by Eqs. (2.12) and (2.13) and with appropriate signs for each of the two quadrants, are substituted into Eq. (2.5), we will find that each plane of the maximum shear stress element will be subjected to a normal stress that may be tensile, compressive, or zero. The value of the normal stress acting on these planes is

$$\sigma_H = \frac{\sigma_x + \sigma_y}{2} \tag{2.15}$$

It is best if the maximum shear stress is considered in terms of the principal stresses. In the plane stress state, Eq. (2.14) can be expressed in terms of σ_1 and σ_2 by using Eq. (2.10). If σ_2 is subtracted from σ_1, the result is

$$\sigma_1 - \sigma_2 = 2\sqrt{\left(\frac{\sigma_x - \sigma_y}{2}\right)^2 + (\tau_{xy})^2} = 2\tau_{max}$$

Thus,

$$\tau_{max} = \tfrac{1}{2}(\sigma_1 - \sigma_2) \tag{2.16}$$

The transformation equations have been developed for the biaxial stress state by taking $\sigma_z = \tau_{xz} = \tau_{yz} = 0$. The biaxial stress equations can be used even though σ_z is some value other than zero; that is, σ_z is the third principal stress, making $\sigma_z = \sigma_3$. The shearing stresses τ_{xz} and τ_{yz}, however, *must be zero*, otherwise we would be obliged to use the more complx stress equations for the triaxial stress state. Figure 2.4 shows such an element. If $\sigma_z = \sigma_3 = 0$, then the element of Fig. 2.4 reduces to the element in Fig. 2.2.

We see there will always be three mutually perpendicular principal stress axes, and corresponding to these directions there will be three principal stresses whose numerical values may be positive, negative, or zero. The

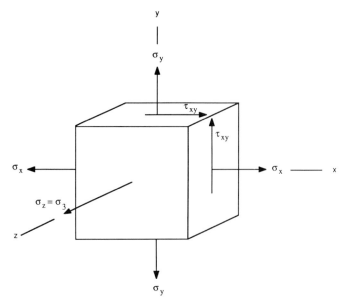

FIG. 2.4. Element with $\sigma_z = \sigma_3$ not equal to zero.

stresses are

$$\left.\begin{array}{l}\sigma_{maximum}\\ \sigma_{intermediate}\\ \sigma_{minimum}\end{array}\right\} \text{algebraically}$$

Figure 2.5 shows triaxial, biaxial, and uniaxial stress states. Note, however, that while tensile stresses are shown, some or all could also be compressive.

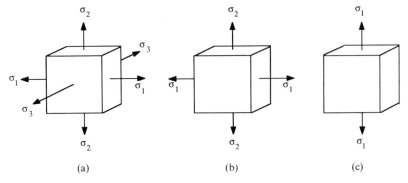

FIG. 2.5. Triaxial (a), biaxial (b), and uniaxial (c) stress states.

We turn our attention once again to the shearing stress. If the stress state is triaxial, then no matter what the values of the three principal stresses are, the maximum shear stress at the *point* will always be

$$\tau_{max} = \pm\tfrac{1}{2}(\sigma_{max} - \sigma_{min}) \qquad (2.17)$$

where σ_{max} and σ_{min} are principal stresses.

Since we are dealing with the biaxial stress state, the third principal stress, σ_3, will be zero. The maximum shearing stress at the point may or may not be in the xy plane, depending on the signs of σ_1 and σ_2. There are, then, three cases to consider.

(a) σ_1 and σ_2 have opposite signs; $\sigma_1 > 0$, $\sigma_3 = 0$, $\sigma_2 < 0$.

$$\sigma_1 = \sigma_{maximum}, \qquad 0 = \sigma_{intermediate}, \qquad \sigma_2 = \sigma_{minimum}$$

Thus,

$$\tau_{max} = \pm\tfrac{1}{2}(\sigma_{max} - \sigma_{min}) = \pm\tfrac{1}{2}(\sigma_1 - \sigma_2) \qquad (2.18)$$

(b) σ_1 and σ_2 greater than zero; $\sigma_1 > \sigma_2 > 0$, $\sigma_3 = 0$.

$$\sigma_1 = \sigma_{maximum}, \qquad \sigma_2 = \sigma_{intermediate}, \qquad 0 = \sigma_{minimum}$$

Thus

$$\tau_{max} = \pm\tfrac{1}{2}(\sigma_{max} - \sigma_{min}) = \pm\tfrac{1}{2}\sigma_1 \qquad (2.19)$$

(c) σ_1 and σ_2 less than zero; $0 > \sigma_1 > \sigma_2$, $\sigma_3 = 0$.

$$0 = \sigma_{maximum}, \qquad \sigma_1 = \sigma_{intermediate}, \qquad \sigma_2 = \sigma_{minimum}$$

Thus

$$\tau_{max} = \pm\tfrac{1}{2}(\sigma_{max} - \sigma_{min}) = \pm\tfrac{1}{2}\sigma_2 \qquad (2.20)$$

Figure 2.6 shows the three cases, with one of the shear planes marked for each case. The second shear plane for each case is at 90° to the one shown. Generally, in the case of the maximum shear stress, we are not concerned with the orientation of the element, but instead want to know its magnitude.

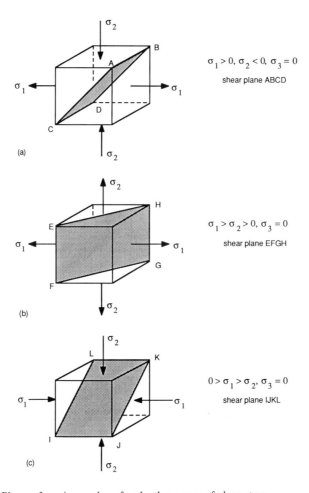

FIG. 2.6. Planes of maximum shear for the three cases of plane stress.

Summary of major equations

$$\sigma_{x'} = \frac{\sigma_x + \sigma_y}{2} + \frac{\sigma_x - \sigma_y}{2}\cos 2\theta + \tau_{xy}\sin 2\theta \tag{2.5}$$

$$\tau_{x'y'} = -\left(\frac{\sigma_x - \sigma_y}{2}\right)\sin 2\theta + \tau_{xy}\cos 2\theta \tag{2.6}$$

$$\tan 2\theta = \frac{2\tau_{xy}}{\sigma_x - \sigma_y} \tag{2.7}$$

$$\sin 2\theta = \frac{2\tau_{xy}}{\sqrt{(\sigma_x - \sigma_y)^2 + (2\tau_{xy})^2}} \tag{2.8}$$

STRESS–STRAIN ANALYSIS AND STRESS–STRAIN RELATIONS

$$\cos 2\theta = \frac{\sigma_x - \sigma_y}{\sqrt{(\sigma_x - \sigma_y)^2 + (2\tau_{xy})^2}} \tag{2.9}$$

$$\sigma_{1,2} = \frac{\sigma_x + \sigma_y}{2} \pm \sqrt{\left(\frac{\sigma_x - \sigma_y}{2}\right)^2 + (\tau_{xy})^2} \tag{2.10}$$

$$\tau_{max} = \pm \tfrac{1}{2}(\sigma_{max} - \sigma_{min}) \tag{2.17}$$

Example 2.1. A plane stress element, shown in Fig. 2.7, has the following stresses acting on it:

$$\sigma_x = -15\,000 \text{ psi}, \qquad \sigma_y = 6000 \text{ psi}, \qquad \tau_{xy} = -8000 \text{ psi}$$

Determine the principal stressss and their orientation relative to the x axis, then sketch the principal stress element. Compute the maximum shearing stress at the point.

Solution. The principal stresses are determined from Eq. (2.10).

$$\sigma_{1,2} = \frac{\sigma_x + \sigma_y}{2} \pm \sqrt{\left(\frac{\sigma_x - \sigma_y}{2}\right)^2 + (\tau_{xy})^2}$$

$$= \frac{-15\,000 + 6000}{2} \pm \sqrt{\left(\frac{-15\,000 - 6000}{2}\right)^2 + (-8000)^2}$$

$$= -4500 \pm 13\,200$$

$$\sigma_1 = 8700 \text{ psi}, \qquad \sigma_2 = -17\,700 \text{ psi}$$

Use Eqs. (2.7) and (2.8) to determine the orientation of σ_1 with respect to the x axis.

$$\tan 2\theta = \frac{2\tau_{xy}}{\sigma_x - \sigma_y} = \frac{2(-8000)}{-15\,000 - 6000} = 0.76191 \qquad \text{(1st or 3d quadrant)}$$

$$\sin 2\theta = \frac{2\tau_{xy}}{\sqrt{(\sigma_x - \sigma_y)^2 + (2\tau_{xy})^2}} = \frac{2(-8000)}{\sqrt{(-15\,000 - 6000)^2 + [2(-8000)]^2}}$$

$$= -0.606\,04 \qquad \text{(3d or 4th quadrant)}$$

Since the only match of tan 2θ and sin 2θ is in the third quadrant, the angle 2θ lies in the third quadrant. Thus, $2\theta = 217.3°$, or $\theta = 108.7°$, measured counterclockwise from the x axis. Figure 2.8 shows the orientation of σ_1 and σ_2 relative to the x axis.

The maximum principal stress is σ_1, and the minimum principal stress is σ_2. The intermediate principal stress is $\sigma_3 = 0$. Using Eq. (2.17), the maximum shear stress is

$$\tau_{max} = \pm \tfrac{1}{2}(\sigma_{max} - \sigma_{min}) = \pm \tfrac{1}{2}(\sigma_1 - \sigma_2)$$

$$= \pm \tfrac{1}{2}[8700 - (-17\,700)] = \pm 13\,200 \text{ psi}$$

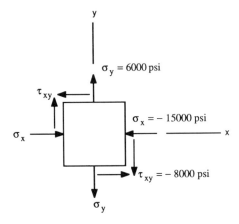

FIG. 2.7. Biaxial stress element for Example 2.1.

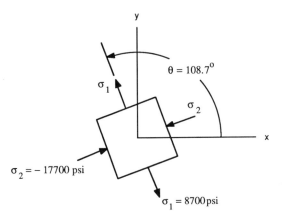

FIG. 2.8. Principal stress element for Example 2.1.

2.4. Mohr's circle for stress

Equations (2.5) and (2.6) are parametric equations for a circle, with the coordinate of any point on the circle being $(\sigma_{x'}, \tau_{x'y'})$. If these equations are plotted, the curve will advance in a clockwise direction rather than in the counterclockwise direction that is taken as positive. This condition can be alleviated by redefining the sign of the shearing stress. The graphical method we use is known as a Mohr's circle, and is named after Otto Mohr, a German engineer and professor, who proposed it in 1880.

The sign convention for normal stresses is the same as given in Section 2.3. The shearing stress, however, will be defined as follows: a shearing stress will be positive if the pair, acting on opposite and parallel faces of an element, form a clockwise couple.

STRESS–STRAIN ANALYSIS AND STRESS–STRAIN RELATIONS

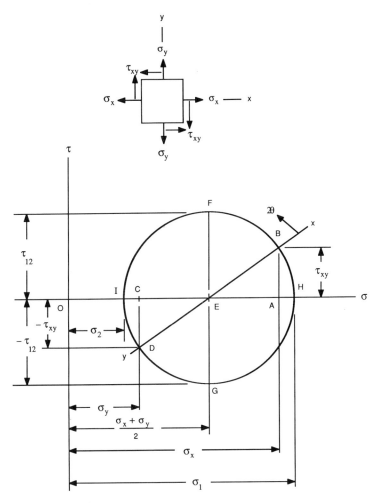

FIG. 2.9. Mohr's circle for stress.

Figure 2.9 shows an element and the corresponding Mohr's circle. To plot the circle, construct the orthogonal σ and τ axes, then start with the element face normal to the positive x axis. Here σ_x is a positive stress, so its value is laid off on the σ axis as OA. Next, the shear stress on this face is seen to form a clockwise couple with the shear stress on the face normal to the negative x axis. This is positive shear stress *for Mohr's circle*. The magnitude of τ_{xy} is plotted as AB parallel to the τ axis. The coordinates of point B are (σ_x, τ_{xy}). The stresses on the face normal to the positive y axis are plotted next. Here σ_y is positive and is laid off on the σ axis as OC. The shear stresses on the faces normal to the y axis form a counterclockwise couple and thus are negative. The magnitude of τ_{xy} on this face is plotted parallel to the τ axis in the negative direction as CD. The coordinates of point D are $(\sigma_y, -\tau_{xy})$. Points B and D are connected by a straight line whose length is the diameter of the circle with its center at point E. The diameter extended outward through point B represents the x axis, while the diameter extended outward through point D represents the y axis. On the circle, these points are 180° apart, which corresponds to 90° on the element. The angle 2θ is measured from BE (x axis) as positive in the counterclockwise direction. Points F and G are the maximum shear stress values (in the xy plane), designated here as τ_{12}, while points H and I are σ_1 and σ_2, respectively.

Mohr's circle is particularly useful with the hand calculator. A freehand sketch of Mohr's circle can be made and desired orientations taken from it. Stress magnitudes and angles can be computed with the calculator.

If we examine Mohr's circle shown in Fig. 2.9, the following can be seen:

1. The center of the circle corresponds to the isotropic (or hydrostatic) component of stress, σ_H. This is the stress defined by Eq. (2.15).

$$\sigma_H = \tfrac{1}{2}(\sigma_1 + \sigma_2) = \tfrac{1}{2}(\sigma_x + \sigma_y) \tag{2.21}$$

2. The radius of the circle is τ_{12}, given by

$$\tau_{12} = \tfrac{1}{2}(\sigma_1 - \sigma_2) = \sqrt{\left(\frac{\sigma_x - \sigma_y}{2}\right)^2 + (\tau_{xy})^2} \tag{2.22}$$

Note here that τ_{12} is the maximum shear stress in the xy plane, but if σ_1 and σ_2 are of opposite sign, then it is the maximum shear stress at the point.

3. From Mohr's circle we see that the principal stresses may be expressed as

$$\sigma_1 = \sigma_H + \tau_{12} \tag{2.23}$$
$$\sigma_2 = \sigma_H - \tau_{12} \tag{2.24}$$

Note that Eqs. (2.23) and (2.24) are another form of Eq. (2.10).

4. Mohr's circle is very helpful in determining the location of σ_1. We saw in Section 2.3 that two trigonometric relationships were required to locate σ_1, while on the circle we can locate it visually and compute the angle by using

$$2\theta = \tan^{-1} \left| \frac{2\tau_{xy}}{\sigma_x - \sigma_y} \right| \tag{2.25}$$

where 2θ is the acute angle between BE and the σ axis.

5. The signs of the stress components are easily determined from the circle. Normal stresses present no problem, but for the shear stress it is more convenient to calculate the magnitude analytically and then determine the sign (direction of the stress) by reference to the circle.

While we have developed transformation equations and Mohr's circle for plane stress, the fact that the third principal axis exists should be kept in mind, even though the stress in that direction is zero. As long as we have principal stresses σ_1, σ_2, and σ_3, three Mohr's circles may be drawn. Their radii will be given by

$$\tau_{12} = \tfrac{1}{2}(\sigma_1 - \sigma_2), \qquad \tau_{23} = \tfrac{1}{2}(\sigma_2 - \sigma_3), \qquad \tau_{13} = \tfrac{1}{2}(\sigma_1 - \sigma_3)$$

For our case of plane strain, $\sigma_3 = 0$, and so the radii become

$$\tau_{12} = \tfrac{1}{2}(\sigma_1 - \sigma_2) \tag{2.26}$$

$$\tau_{23} = \tfrac{1}{2}\sigma_2 \tag{2.27}$$

$$\tau_{13} = \tfrac{1}{2}\sigma_1 \tag{2.28}$$

Figure 2.10 shows the case where $\sigma_1 > \sigma_2 > 0$ and $\sigma_3 = 0$. It is obvious from the diagram that the maximum shear stress at the point is $\tau_{\max} = \tau_{13}$. If $\sigma_3 = 0$ had been ignored, we might have been fooled into thinking that τ_{12} was the maximum shear stress.

Example 2.2. The element in Fig. 2.11 has the following stresses acting on it:

$$\sigma_x = 10\,000 \text{ psi}, \qquad \sigma_y = 3000 \text{ psi}, \qquad \tau_{xy} = -8000 \text{ psi}$$

(the sign of τ_{xy} conforms to the convention established in Section 2.3).

Perform the listed tasks.
(a) Draw the corresponding Mohr's circle.
(b) Sketch an element showing the principal stresses and their orientation relative to the x axis.
(c) Sketch an element showing the maximum shear stress in the xy plane and its orientation relative to the x axis.

Solution. (a) Although a free-hand sketch of Mohr's circle could be made, it will be drawn to scale, but pertinent values will be calculated. In plotting Mohr's circle, the following steps are taken:

1. Plot $\sigma_x = 10\,000$ psi on the σ axis as point A.
2. Since τ_{xy} forms a clockwise couple on the parallel faces normal to the x axis, it is plotted as positive for Mohr's circle. Through point A, plot $\tau_{xy} = 8000$ psi in the positive τ direction. This gives point B, whose coordinates are (10 000, 8000).
3. Plot $\sigma_y = 3000$ psi on the σ axis as point C.

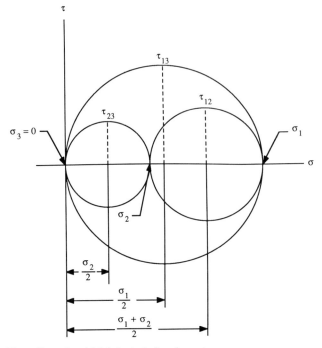

FIG. 2.10. Three-dimensional Mohr's circle for plane stress.

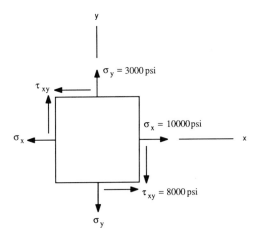

FIG. 2.11. Stress element for Example 2.2.

4. From point C, plot $\tau_{xy} = -8000$ psi in the negative τ direction. This gives point D, whose coordinates are $(3000, -8000)$.
5. Join points B and D to get the diameter of the circle. The intersection of the line BD with the σ axis is the center of the circle E.

STRESS–STRAIN ANALYSIS AND STRESS–STRAIN RELATIONS 59

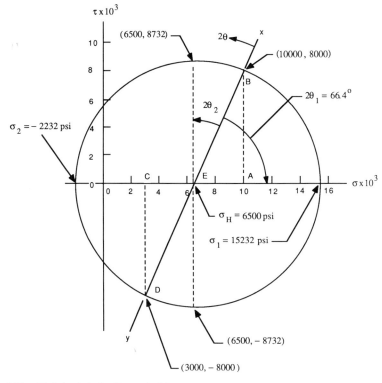

FIG. 2.12. Mohr's circle for Example 2.2.

6. Draw the circle and measure all angles from the x axis as shown, with the counterclockwise direction being positive. The completed Mohr's circle is shown in Fig. 2.12.

From Eq. (2.21),

$$\sigma_H = \tfrac{1}{2}(\sigma_x + \sigma_y) = \tfrac{1}{2}(10\,000 + 3000) = 6500 \text{ psi}$$

From Eq. (2.22),

$$\tau_{12} = \sqrt{\left(\frac{\sigma_x - \sigma_y}{2}\right)^2 + (\tau_{xy})^2}$$

$$= \sqrt{\left(\frac{10\,000 - 3000}{2}\right)^2 + (8000)^2} = 8732 \text{ psi}$$

Equations (2.23) and (2.24) give σ_1 and σ_2, respectively:

$$\sigma_1 = \sigma_H + \tau_{12} = 6500 + 8732 = 15\,232 \text{ psi}$$

$$\sigma_2 = \sigma_H - \tau_{12} = 6500 - 8732 = -2232 \text{ psi}$$

(b) In order to determine the orientation of σ_1 relative to the x axis, we go in a *clockwise* direction from the x axis on the circle through the angle $2\theta_1 = 66.4°$ to reach σ_1. The acute angle $2\theta_1$ can be computed using Eq. (2.25):

$$2\theta = \tan^{-1}\left|\frac{2\tau_{xy}}{\sigma_x - \sigma_y}\right| = \tan^{-1}\left|\frac{2(8000)}{10\,000 - 3000}\right| = 66.4°$$

Since we traveled in a clockwise direction on the circle to arrive at σ_1, we must go in the same direction when locating σ_1 relative to the x axis for the element; that is, $\theta_1 = 33.2°$ clockwise. The element is shown in Fig. 2.13. Note that we could also have gone in a counterclockwise direction on the circle to arrive at σ_1, thus traversing the circle through $2\theta_1 = 293.6°$. On the element, the angle between the x axis and σ_1 is 146.8°, which is the vector for σ_1 in the second quadrant of Fig. 2.13.

(c) The maximum shear stress in the xy plane is τ_{12}. For this problem, the maximum shear stress at the point is also $\tau_{\max} = \tau_{12} = 8732$ psi. To determine the orientation of the maximum shear stress element, go in a counterclockwise direction on the circle until τ_{12} is reached. This is the point whose coordinates are (6500, 8732). Since τ_{12} is positive here, the shear stresses on opposite faces of the element form a clockwise couple. Continue moving on the circle in a counterclockwise direction until $-\tau_{12}$ is reached. At this point, the coordinates of (6500, -8732) are the normal and shear stresses on the element face, which is 90° from the first face. The element is shown in Fig. 2.14. The angle $2\theta_2 = 90° - 2\theta_1$, and so $\theta_2 = 11.8°$.

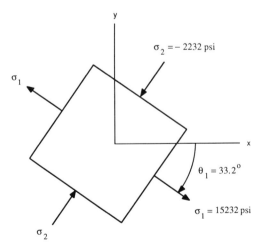

FIG. 2.13. Principal stress element for Example 2.2.

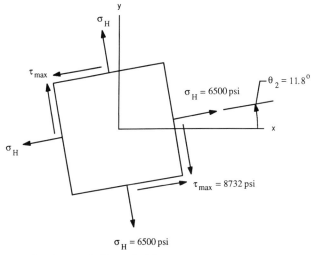

FIG. 2.14. Element showing τ_{max} in Example 2.2.

2.5. Basic concepts of strain

Accompanying stress there is usually some type of deformation which we regard as strain. As in the case of stress, we find there are two general kinds of strain; namely, linear strain and shear strain.

Linear strain is represented by the lengthening (+ for tension) or shortening (− for compression) of a straight line in the material. We assume that all longitudinal fibers of the bar elongate identically, and that the cross sections of the bar that are originally plane and perpendicular to the axis of the bar remain so during elongation. Such a bar is shown in Fig. 2.15, and its unit strain ε is given by the expression

$$\varepsilon = \frac{\delta}{L} \tag{2.29}$$

where δ = total elongation of the bar
L = original length of the bar

If the cross section of the bar is not constant, or if the load is not uniformly applied, all longitudinal fibers of the bar will not elongate uniformly, and so Eq. (2.29) represents average strain only. Thus, the unit strain varies from point to point along the bar. In this case, the unit strain is determined by considering the elongation $d\delta$ of a cross section of length

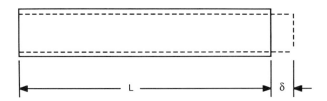

FIG. 2.15. Elongation of a bar.

dL. The unit strain at a point can be written as

$$\varepsilon = \frac{d\delta}{dL} \tag{2.30}$$

If axial compression is applied, Eqs. (2.29) and (2.30) apply, with the unit strain being negative.

When a bar, such as the one in Fig. 2.15, is loaded uniformly over the end faces, only those cross sections normal to the bar axis are subjected to stress. Observation of such tests shows that the extension of the bar in the axial direction is accompanied by a lateral contraction of the bar. Poisson, a French mathematician, demonstrated analytically that the axial and lateral strains are proportional to each other within the range of Hooke's law, and the ratio is constant for a given material. This ratio is known as Poisson's ratio and is expressed as

$$v = -\frac{\text{lateral strain}}{\text{axial strain}} \tag{2.31}$$

Figure 2.16 shows an element, given by *abcd* prior to loading, whose corners are square. The element is then loaded by the shearing stress shown and distorts into *ab'c'd*. Since the angle γ through which lines *ab* and *cd* rotate is very small, it is assumed that *ab'* is equal to *ab* and *dc'* is equal to *dc*. In this case, $\tan \gamma = bb'/ab$, and so for small angles $\tan \gamma$ may be replaced by γ. Thus, the shearing strain is given by the angle γ, whose value is in radians.

2.6. Plane strain

The transformation equations for plane stress were developed in Section 2.3. There the stresses in the z direction were zero; that is, $\sigma_z = \tau_{xz} = \tau_{yz} = 0$. We noted, however, that if $\sigma_z = \sigma_3(\tau_{xz} = \tau_{yz} = 0)$, then σ_3 could have a value other than zero and the biaxial transformation equations were still valid.

Since our work with strain gages will generally involve applying them to the surface of a machine element, we will need the transformation equations for plane strain. This implies that $\varepsilon_z = \gamma_{xz} = \gamma_{yz} = 0$, and we might

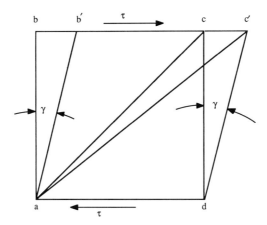

FIG. 2.16. Element subjected to pure shear.

assume that a plane stress state produces a plane strain state. This is not the case, however, for biaxial stresses produce a strain in the z direction because of the Poisson effect. The strain produced, ε_z, will be the principal strain ε_3, with $\gamma_{xz} = \gamma_{yz} = 0$. This will become apparent when stress–strain relationships are examined.

The stress transformation equations and the strain transformation equations have the same form, and so the strain transformation equations can be written directly by making the following substitutions into the stress transformation equations:

$$\varepsilon_x \text{ for } \sigma_x \qquad \varepsilon_y \text{ for } \sigma_y \qquad \gamma_{xy}/2 \text{ for } \tau_{xy}$$
$$\varepsilon_{x'} \text{ for } \sigma_{x'} \qquad \varepsilon_{y'} \text{ for } \sigma_{y'} \qquad \gamma_{x'y'}/2 \text{ for } \tau_{x'y'}$$

Making these substitutions into Eqs. (2.5) and (2.6) will yield the strain transformation equations. Thus,

$$\varepsilon_{x'} = \frac{\varepsilon_x + \varepsilon_y}{2} + \frac{\varepsilon_x - \varepsilon_y}{2} \cos 2\theta + \frac{\gamma_{xy}}{2} \sin 2\theta \qquad (2.32)$$

$$\gamma_{x'y'} = -(\varepsilon_x - \varepsilon_y) \sin 2\theta + \gamma_{xy} \cos 2\theta \qquad (2.33)$$

The sign of the shear strain must be compatible with shear stress. Figure 2.17 shows an element subjected to positive shear stress. Prior to loading, sides AB and AC are at right angles to each other. After the stresses are applied the right angle BAC will deform to angle B'AC', which is $\pi/2 - \gamma_{xy}$. Since this distortion is produced by positive shear stresses, the shear strain, γ_{xy}, will be defined as positive when the angle between two orthogonal lines decreases.

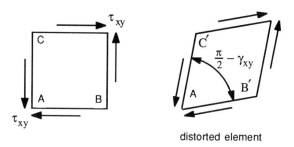

FIG. 2.17. Element subjected to positive shear stresses.

The orientation of σ_1 and ε_1 with respect to the x axis is the same. In order to determine the orientation of σ_1 with respect to the x axis, two of the three trigonometric relations given by Eqs. (2.7), (2.8), and (2.9) were used. These three expressions may be written in terms of strain, with the use of two of the three equations establishing the orientation of ε_1. The three required expressions in terms of strain are

$$\tan 2\theta = \frac{\gamma_{xy}}{\varepsilon_x - \varepsilon_y} \tag{2.34}$$

$$\sin 2\theta = \frac{\gamma_{xy}}{\sqrt{(\varepsilon_x - \varepsilon_y)^2 + (\gamma_{xy})^2}} \tag{2.35}$$

$$\cos 2\theta = \frac{\varepsilon_x - \varepsilon_y}{\sqrt{(\varepsilon_x - \varepsilon_y)^2 + (\gamma_{xy})^2}} \tag{2.36}$$

The principal strains, ε_1 and ε_2, follow directly from Eq. (2.10) when the appropriate values of strain are substituted for stress. This gives

$$\varepsilon_{1,2} = \frac{\varepsilon_x + \varepsilon_y}{2} \pm \sqrt{\left(\frac{\varepsilon_x - \varepsilon_y}{2}\right)^2 + \left(\frac{\gamma_{xy}}{2}\right)^2} \tag{2.37}$$

The second term on the right side of Eq. (2.37) is one-half of the maximum shear strain in the plane. Therefore,

$$\left(\frac{\gamma}{2}\right)_{\max} = \pm \sqrt{\left(\frac{\varepsilon_x - \varepsilon_y}{2}\right)^2 + \left(\frac{\gamma_{xy}}{2}\right)^2} \tag{2.38}$$

Unlike stresses, which can only be determined indirectly, linear strains are subject to direct measurement. If an xy reference system is chosen, then strain measurements are made in three known directions relative to the xy

STRESS–STRAIN ANALYSIS AND STRESS–STRAIN RELATIONS   65

coordinate system through the use of three-element strain rosettes. A delta rosette has three gages arranged at 60° (or 120°) intervals, while a rectangular rosette has three gages arranged at 45° intervals. Each measured strain is entered into Eq. (2.32) as $\varepsilon_{x'}$, and θ is the angle between the x axis and the measured strain. The three strain readings used in Eq. (2.32) produce three independent equations that are solved simultaneously for ε_x, ε_y, and γ_{xy}. Knowing the component strains ε_x, ε_y, and γ_{xy}, we can now compute the principal strains by using Eq. (2.37). The principal strain axes are located relative to the x axis by using any two of Eqs. (2.34), (2.35), and (2.36).

Summary of major equations

$$\varepsilon_{x'} = \frac{\varepsilon_x + \varepsilon_y}{2} + \frac{\varepsilon_x - \varepsilon_y}{2}\cos 2\theta + \frac{\gamma_{xy}}{2}\sin 2\theta \qquad (2.32)$$

$$\gamma_{x'y'} = -(\varepsilon_x - \varepsilon_y)\sin 2\theta + \gamma_{xy}\cos 2\theta \qquad (2.33)$$

$$\tan 2\theta = \frac{\gamma_{xy}}{\varepsilon_x - \varepsilon_y} \qquad (2.34)$$

$$\sin 2\theta = \frac{\gamma_{xy}}{\sqrt{(\varepsilon_x - \varepsilon_y)^2 + (\gamma_{xy})^2}} \qquad (2.35)$$

$$\cos 2\theta = \frac{\varepsilon_x - \varepsilon_y}{\sqrt{(\varepsilon_x - \varepsilon_y)^2 + (\gamma_{xy})^2}} \qquad (2.36)$$

$$\varepsilon_{1,2} = \frac{\varepsilon_x + \varepsilon_y}{2} \pm \sqrt{\left(\frac{\varepsilon_x - \varepsilon_y}{2}\right)^2 + \left(\frac{\gamma_{xy}}{2}\right)^2} \qquad (2.37)$$

$$\left(\frac{\gamma}{2}\right)_{max} = \pm\sqrt{\left(\frac{\varepsilon_x - \varepsilon_y}{2}\right)^2 + \left(\frac{\gamma_{xy}}{2}\right)^2} \qquad (2.38)$$

Example 2.3. The following strains and their orientation relative to an xy coordinate system on a machine element are given. (Note: The symbol μ stands for 1×10^{-6}).

$$\varepsilon_a = 745 \text{ }\mu\text{in/in at } \theta_a = 50°$$
$$\varepsilon_b = -396 \text{ }\mu\text{in/in at } \theta_b = 95°$$
$$\varepsilon_c = -245 \text{ }\mu\text{in/in at } \theta_c = 140°$$

The arrangement of the gages giving these readings is shown in Fig. 2.18.

(a) Determine ε_x, ε_y, and γ_{xy}.
(b) Determine ε_1 and ε_2 and the orientation of ε_1 realtive to the x axis.
(c) Determine $(\gamma/2)_{max}$ in the plane.

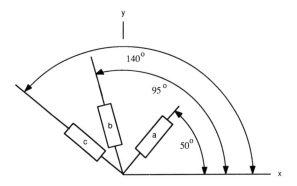

FIG. 2.18. Gage arrangement for Example 2.3.

Solution. In order to determine ε_1 and ε_2, we must first determine ε_x, ε_y, and γ_{xy} in the xy coordinate system. This requires three independent equations, which we obtain by using Eq. (2.32). Note that $\varepsilon_{x'}$ becomes in turn ε_a, ε_b, and ε_c, while θ takes on the corresponding values of θ_a, θ_b, and θ_c.

$$\varepsilon_{x'} = \varepsilon_a = 745 = \frac{\varepsilon_x + \varepsilon_y}{2} + \frac{\varepsilon_x - \varepsilon_y}{2}\cos 2(50) + \frac{\gamma_{xy}}{2}\sin 2(50)$$

$$\varepsilon_{x'} = \varepsilon_b = -396 = \frac{\varepsilon_x + \varepsilon_y}{2} + \frac{\varepsilon_x - \varepsilon_y}{2}\cos 2(95) + \frac{\gamma_{xy}}{2}\sin 2(95)$$

$$\varepsilon_{x'} = \varepsilon_c = -245 = \frac{\varepsilon_x + \varepsilon_y}{2} + \frac{\varepsilon_x - \varepsilon_y}{2}\cos 2(140) + \frac{\gamma_{xy}}{2}\sin 2(140)$$

The three equations reduce to

$$745 = 0.4132\varepsilon_x + 0.5868\varepsilon_y + 0.4924\gamma_{xy}$$
$$-396 = 0.0076\varepsilon_x + 0.9924\varepsilon_y - 0.0868\gamma_{xy}$$
$$-245 = 0.5868\varepsilon_x + 0.4132\varepsilon_y - 0.4924\gamma_{xy}$$

Solving the equations simultaneously, we obtain $\varepsilon_x = 800$ μin/in; $\varepsilon_y = -300$ μin/in; $\gamma_{xy} = 1200$ μradians. The principal stresses are computed using Eq. (2.37).

(a)

$$\varepsilon_{1,2} = \frac{\varepsilon_x + \varepsilon_y}{2} \pm \sqrt{\left(\frac{\varepsilon_x - \varepsilon_y}{2}\right)^2 + \left(\frac{\gamma_{xy}}{2}\right)^2}$$

$$= \frac{800 - 300}{2} \pm \sqrt{\left(\frac{800 + 300}{2}\right)^2 + \left(\frac{1200}{2}\right)^2} = 250 \pm 814$$

$$\varepsilon_1 = 1064 \text{ μin/in}; \qquad \varepsilon_2 = -564 \text{ μin/in}$$

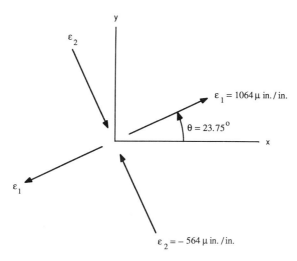

FIG. 2.19. Orientation of the principal stresses in Example 2.3.

(b) In order to determine the orientation of ε_1 with respect to the x axis, Eqs. (2.34) and (2.35) will be used. From Eq. (2.34),

$$\tan 2\theta = \frac{\gamma_{xy}}{\varepsilon_x - \varepsilon_y} = \frac{1200}{800 + 300} = 1.09091 \quad \text{(1st or 3d quadrant)}$$

From Eq. (2.35),

$$\sin 2\theta = \frac{\gamma_{xy}}{\sqrt{(\varepsilon_x - \varepsilon_y)^2 + (\gamma_{xy})^2}} = \frac{1200}{\sqrt{(800 + 300)^2 + (1200)^2}}$$
$$= 0.73715 \quad \text{(1st or 2d quadrant)}$$

The common quadrant is the first, and so $2\theta = 47.5°$. Thus, ε_1 lies at an angle of $\theta = 23.75°$ in a counterclockwise direction from the x axis. The orientation is shown in Fig. 2.19.

(c) The maximum shearing strain *in the plane* is obtained by using Eq. (2.38).

$$\left(\frac{\gamma}{2}\right)_{max} = \pm\sqrt{\left(\frac{\varepsilon_x - \varepsilon_y}{2}\right)^2 + \left(\frac{\gamma_{xy}}{2}\right)^2} = \pm\sqrt{\left(\frac{800 + 300}{2}\right)^2 + \left(\frac{1200}{2}\right)^2}$$

$$\gamma_{max} = 1628 \text{ μradians}$$

Note: The problem could also be solved by aligning the x' and y' axes along ε_a and ε_c, respectively, then through the use of Eq. (2.32), determining the component strains in the $x'y'$ system. Finally, the transformation equations could be used to get the desired values in the xy coordinate system.

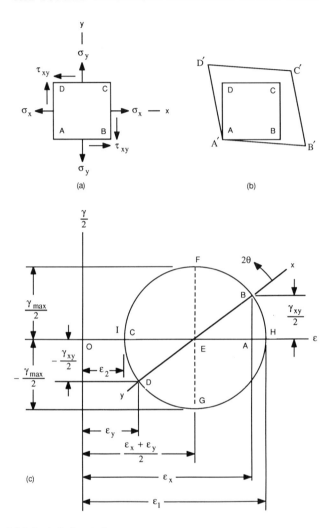

FIG. 2.20. Mohr's circle for strain.

2.7. Mohr's circle for strain

Mohr's circle for strain is constructed in a manner similar to that of Mohr's circle for stress. Figure 2.20 shows the Mohr's circle for strain and its attendant element. Figure 2.20a is the element used for the construction of Mohr's circle for stress, while Fig. 2.20b shows the element in its distorted position (greatly exaggerated).

To plot the diagram, construct the orthogonal ε and $\gamma/2$ axes, then start with the x axis and plot ε_x as OA on the ε axis. The shearing stresses acting on the element faces open angle DAB to $D'A'B'$, and so it is taken as *positive for Mohr's circle*. This is the shearing strain γ_{xy}, and since it is positive, $\gamma_{xy}/2$, is plotted as AB parallel to the $\gamma/2$ axis. The coordinates of point B are $(\varepsilon_x, \gamma_{xy}/2)$. Next, plot ε_y as OC on the ε axis and $\gamma_{xy}/2$ as CD parallel to the $\gamma/2$ axis. The coordinates of point D are $(\varepsilon_y, -\gamma_{xy}/2)$. Connect points B and D with a straight line; length BD is the diameter of the circle with its center at point E. The diameter extended outward through point B represents the x axis, while the diameter extended outward through point D represents the y axis. These two points, 180° apart on the circle, correspond to 90° on the element. The angle 2θ is measured positive in a counterclockwise direction from the x axis. Point F is the maximum value of $\gamma/2$ in the xy plane, and its coordinates are $[(\varepsilon_x + \varepsilon_y)/2, \gamma_{max}/2]$. Point G is the other maximum value of $\gamma/2$, and its coordinates are $[(\varepsilon_x + \varepsilon_y)/2, -\gamma_{max}/2]$. Points H and I are the values of the principal strains, ε_1 and ε_2, respectively. The values of 2θ on the circles for stress and strain correspond. For instance, the angle between σ_1 and the x axis on the circle for stress is identical to the angle between ε_1 and the x axis on the circle for strain.

If we examine Mohr's circle shown in Fig. 2.20, the following can be seen:

1. The center of the circle corresponds to an isotropic (hydrostatic) component of strain, ε_H.

$$\varepsilon_H = \frac{\varepsilon_1 + \varepsilon_2}{2} = \frac{\varepsilon_x + \varepsilon_y}{2} \tag{2.39}$$

2. The radius of the circle is

$$\frac{\gamma_{max}}{2} = \frac{\varepsilon_1 - \varepsilon_2}{2} = \sqrt{\left(\frac{\varepsilon_x - \varepsilon_y}{2}\right)^2 + \left(\frac{\gamma_{xy}}{2}\right)^2} \tag{2.40}$$

3. From Mohr's circle we see that the principal strains are

$$\varepsilon_1 = \varepsilon_H + \frac{\gamma_{max}}{2} \tag{2.41}$$

$$\varepsilon_2 = \varepsilon_H - \frac{\gamma_{max}}{2} \tag{2.42}$$

Note that Eqs. (2.41) and (2.42) are another form of Eq. (2.37).

4. Mohr's circle is very helpful in determining the location of ε_1. We saw in Section 2.6 that two trigonometric relationships were required to locate ε_1, while on the circle we can locate it visually and compute the angle by using

$$2\theta = \tan^{-1}\left|\frac{\gamma_{xy}}{\varepsilon_x - \varepsilon_y}\right| \tag{2.43}$$

where 2θ is the acute angle between BE and the ε axis.

70   THE BONDED ELECTRICAL RESISTANCE STRAIN GAGE

5. The signs of the strain components are easily determined from the circle. Normal strains present no problem, but for the shear strain, it is more convenient to calculate its magnitude and then determine its sign by reference to the circle.

Example 2.4. The strains on the surface of a machine element are the following:

$$\varepsilon_x = 1000 \text{ μin/in}, \qquad \varepsilon_y = -400 \text{ μin/in}, \qquad \gamma_{xy} = 800 \text{ μradians}$$

(a) Plot Mohr's circle and determine the principal strains ε_1 and ε_2.
(b) Determine the orientation of ε_1 relative to the x axis.
(c) Determine $\varepsilon_{x'}$, $\varepsilon_{y'}$, and $\gamma_{x'y'}$ for $\theta = 35°$.

Note: γ_{xy} is positive by the sign convention established in Section 2.6, and so must be plotted as *negative* for Mohr's circle.

Solution. (a) Mohr's circle for strain will be plotted to scale, but the pertinent values will be calculated. The following steps are taken in plotting the diagram shown in Fig. 2.21.

1. Plot $\varepsilon_x = 1000$ μin/in on the ε axis as point A.
2. Since $\gamma_{xy} = 800$ μradians, it must be taken as *negative* for Mohr's circle. Thus, we plot $\gamma_{xy}/2 = -400$ μradians from ε_x parallel to the negative $\gamma/2$ axis. This is point B, whose coordinates are $(1000, -400)$.
3. Plot $\varepsilon_y = -400$ μin/in on the ε axis as point C.
4. From ε_y, plot $\gamma_{xy}/2 = 400$ μradians parallel to the positive $\gamma/2$ axis. This is point D, whose coordinates are $(-400, 400)$.
5. Join points B and D to get the diameter of the circle. The intersection of line BD with the ε axis is the center of the circle E.
6. Draw the circle and measure all angles from the x axis as shown, with the counterclockwise direction being positive.

From Eq. (2.39),

$$\varepsilon_H = \frac{\varepsilon_x + \varepsilon_y}{2} = \frac{1000 - 400}{2} = 300 \text{ μin/in}$$

From Eq. (2.40),

$$\frac{\gamma_{max}}{2} = \sqrt{\left(\frac{\varepsilon_x - \varepsilon_y}{2}\right)^2 + \left(\frac{\gamma_{xy}}{2}\right)^2} = \sqrt{\left(\frac{1000 + 400}{2}\right)^2 + (-400)^2}$$

$$= 806 \text{ μradians}$$

From Eq. (2.41),

$$\varepsilon_1 = \varepsilon_H + \frac{\gamma_{max}}{2} = 300 + 806 = 1106 \text{ μin/in}$$

STRESS–STRAIN ANALYSIS AND STRESS–STRAIN RELATIONS

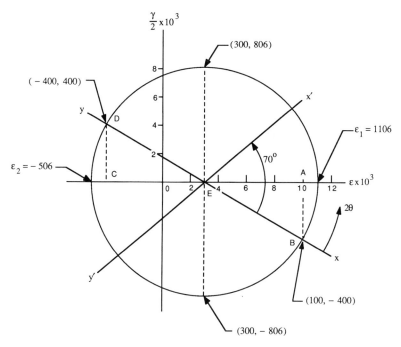

FIG. 2.21. Mohr's circle for Example 2.4.

From Eq. (2.42),

$$\varepsilon_2 = \varepsilon_H - \frac{\gamma_{max}}{2} = 300 - 806 = -504 \ \mu\text{in/in}$$

(b) The orientation of ε_1 relative to the x axis can be measured directly from the circle. It lies in a counterclockwise direction from the x axis. We can also calculate its value using Eq. (2.43):

$$2\theta = \tan^{-1}\left|\frac{\gamma_{xy}}{\varepsilon_x - \varepsilon_y}\right| = \tan^{-1}\left|\frac{-800}{1000 + 400}\right| = 29.74°$$

Therefore, $\theta = 14.87°$. The orientations of ε_1 and ε_2 are shown in Fig. 2.22.

(c) The three values can be scaled directly from the circle if so desired. We can also use Eq. (2.32) to determine the normal strains, and Eq. (2.33) to determine the shearing strain. Using Eq. (2.32) and $\theta = 35°$, $\varepsilon_{x'}$ can be determined.

$$\varepsilon_{x'} = \frac{\varepsilon_x + \varepsilon_y}{2} + \frac{\varepsilon_x - \varepsilon_y}{2}\cos 2\theta + \frac{\gamma_{xy}}{2}\sin 2\theta$$

$$= \frac{1000 - 400}{2} + \frac{1000 + 400}{2}\cos 2(35) + \frac{800}{2}\sin 2(35)$$

$$= 915 \ \mu\text{in/in}$$

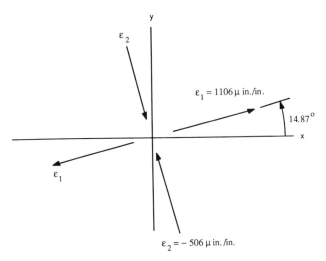

FIG. 2.22. Principal strains for Example 2.4.

For $\varepsilon_{y'}$, the angle is $\theta + 90°$.

$$\varepsilon_{y'} = \frac{\varepsilon_x + \varepsilon_y}{2} + \frac{\varepsilon_x - \varepsilon_y}{2} \cos 2(\theta + 90) + \frac{\gamma_{xy}}{2} \sin 2(\theta + 90)$$

$$= \frac{1000 - 400}{2} + \frac{1000 + 400}{2} \cos 2(35 + 90) + \frac{800}{2} \sin 2(35 + 90)$$

$$= -315 \; \mu\text{in/in}$$

Using Eq. (2.33) and $\theta = 35°$,

$$\gamma_{x'y'} = -(\varepsilon_x - \varepsilon_y) \sin 2\theta + \gamma_{xy} \cos 2\theta$$

$$= -(1000 + 400) \sin 2(35) + 800 \cos 2(35) = -1042 \; \mu\text{radians}$$

Observe here that $\gamma_{x'y'}$ is negative when computed using the transformation equation, which is in keeping with the value of $\gamma_{x'y'}$ from the Mohr's diagram. All of the computed values can be checked by using Mohr's diagram, shown in Fig. 2.21.

2.8. Stress–strain relationships

We have considered stress and strain separately at a point. In developing the transformation equations for stress, the static equilibrium of the element was examined and the resulting equations were not dependent on material properties. Although the strain transformation equations were written directly from those of stress, one should be aware of the fact that they may be developed from the geometry of small deformations, and therefore

STRESS–STRAIN ANALYSIS AND STRESS–STRAIN RELATIONS

material properties do not appear. The objective, now, is to relate stress and strain, and so material properties will be the link.

Robert Hooke was the first to state the relationship between stress and strain. For a tensile test it can be stated as

$$\sigma_x = E\varepsilon_x \tag{2.44}$$

where σ_x = the longitudinal stress

ε_x = the longitudinal strain

E = a constant of proportionality called the modulus of elasticity

In the most general form of Hooke's law, it is assumed each stress component has a linear relationship with the six strain components, resulting in 36 elastic constants. For an isotropic material, however, it can be shown that the 36 constants are not all independent and that only two independent constants exist (1).

By testing, three elastic constants can be determined for a given material. The elastic constants are the modulus of elasticity, E, the shear modulus (modulus of rigidity), G, and Poisson's ratio, v. If a tensile test is conducted on a specimen with a rectangular cross section, the stress σ_x will be uniformly distributed over two parallel sections normal to the x axis, with the x axis taken in the direction of loading; the faces normal to the y and z axes would be stress free. Observation of such tests show the extension in the x direction is accompanied by a lateral contraction in the y and z directions. The strain in these two directions is negative and proportional to the strain in the x direction. Thus, as shown by Eq. (2.31),

$$\varepsilon_y = \varepsilon_z = -v\varepsilon_x = -v\frac{\sigma_x}{E} \tag{2.45}$$

where v is Poisson's ratio.

Since there are only two independent constants for an isotropic material, a relationship must exist between E, v, and G. The shear modulus G can be expressed in terms of E and v as

$$G = \frac{E}{2(1+v)} \tag{2.46}$$

The six equations relating strain in terms of stress are

$$\varepsilon_x = \frac{1}{E}[\sigma_x - v(\sigma_y + \sigma_z)] \tag{2.47a}$$

$$\varepsilon_y = \frac{1}{E}[\sigma_y - \nu(\sigma_z + \sigma_x)] \tag{2.47b}$$

$$\varepsilon_z = \frac{1}{E}[\sigma_z - \nu(\sigma_x + \sigma_y)] \tag{2.47c}$$

$$\gamma_{xy} = \frac{\tau_{xy}}{G} \tag{2.47d}$$

$$\gamma_{xz} = \frac{\tau_{xz}}{G} \tag{2.47e}$$

$$\gamma_{yz} = \frac{\tau_{yz}}{G} \tag{2.47f}$$

If Eqs. (2.47) are solved for stress in terms of strain, we have

$$\sigma_x = \frac{E}{(1+\nu)(1-2\nu)}[(1-\nu)\varepsilon_x + \nu(\varepsilon_y + \varepsilon_z)] \tag{2.48a}$$

$$\sigma_y = \frac{E}{(1+\nu)(1-2\nu)}[(1-\nu)\varepsilon_y + \nu(\varepsilon_z + \varepsilon_x)] \tag{2.48b}$$

$$\sigma_z = \frac{E}{(1+\nu)(1+2\nu)}[(1-\nu)\varepsilon_z + \nu(\varepsilon_x + \varepsilon_y)] \tag{2.48c}$$

$$\tau_{xy} = G\gamma_{xy} \tag{2.48d}$$

$$\tau_{xz} = G\gamma_{xz} \tag{2.48e}$$

$$\tau_{yz} = G\gamma_{yz} \tag{2.48f}$$

Equations (2.47) and (2.48) represent the triaxial stress and triaxial strain case. Special stress and strain states may be determined from these equations.

Plane stress state: $\sigma_z = \gamma_{xz} = \gamma_{yz} = 0$. The plane, or biaxial, stress case was developed in Section 2.3. Since $\sigma_z = 0$, Eq. (2.48c) can be used to determine ε_z in terms of ε_x and ε_y. Therefore

$$\varepsilon_z = -\frac{\nu(\varepsilon_x + \varepsilon_y)}{(1-\nu)} \tag{2.49}$$

By substituting the expression for ε_z given by Eq. (2.49) into Eqs. (2.48a)

STRESS–STRAIN ANALYSIS AND STRESS–STRAIN RELATIONS   75

and (2.48b), we arrive at

$$\sigma_x = \frac{E}{1-v^2}(\varepsilon_x + v\varepsilon_y) \tag{2.50a}$$

$$\sigma_y = \frac{E}{1-v^2}(\varepsilon_y + v\varepsilon_x) \tag{2.50b}$$

$$\tau_{xy} = G\gamma_{xy} \tag{2.50c}$$

The corresponding strain equations can be obtained from Eqs. (2.47):

$$\varepsilon_x = \frac{1}{E}[\sigma_x - v\sigma_y] \tag{2.51a}$$

$$\varepsilon_y = \frac{1}{E}[\sigma_y - v\sigma_x] \tag{2.51b}$$

$$\varepsilon_z = -\frac{v}{E}[\sigma_x + \sigma_y] \tag{2.51c}$$

$$\gamma_{xy} = \frac{\tau_{xy}}{G} \tag{2.51d}$$

The expressions for ε_z given by Eqs. (2.49) and (2.51c) give identical results, of course. As pointed out earlier, even though a plane stress state exists, the strain state is triaxial.

Plane strain state: $\varepsilon_z = \gamma_{xz} = \gamma_{yz} = 0$. Since $\varepsilon_z = 0$, σ_z can be written in terms of σ_x and σ_y by using Eq. (2.47c). This gives

$$\sigma_z = v(\sigma_x + \sigma_y) \tag{2.52}$$

The value of σ_z given by Eq. (2.52) can be substituted into the expressions for ε_x and ε_y, given by Eqs. (2.47a) and (2.47b), respectively. The expressions for ε_x and ε_y in terms of stress then become

$$\varepsilon_x = \frac{1+v}{E}[(1-v)\sigma_x - v\sigma_y] \tag{2.53a}$$

$$\varepsilon_y = \frac{1+v}{E}[(1-v)\sigma_y - v\sigma_x] \tag{2.53b}$$

$$\gamma_{xy} = \frac{\tau_{xy}}{G} \tag{2.53c}$$

The corresponding equations for stress in terms of strain are, from Eqs. (2.48),

$$\sigma_x = \frac{E}{(1+v)(1-2v)}[(1-v)\varepsilon_x + v\varepsilon_y] \qquad (2.54a)$$

$$\sigma_y = \frac{E}{(1+v)(1-2v)}[(1-v)\varepsilon_y + v\varepsilon_x] \qquad (2.54b)$$

$$\sigma_z = \frac{vE}{(1+v)(1-2v)}[\varepsilon_x + \varepsilon_y] \qquad (2.54c)$$

$$\tau_{xy} = G\gamma_{xy} \qquad (2.54d)$$

Once again, note that when we have plane strain we do not have plane stress. Furthermore, the values of σ_z given by Eqs. (2.52) and (2.54c) produce identical results.

Uniaxial stress state: $\sigma_y = \sigma_z = \tau_{xy} = \tau_{xz} = \tau_{yz} = 0$. In the case of a uniaxial stress state, Eqs. (2.47) reduce to

$$\varepsilon_x = \frac{\sigma_x}{E} \qquad (2.55a)$$

$$\varepsilon_y = \varepsilon_z = -v\frac{\sigma_x}{E} \qquad (2.55b)$$

We see here that even though a uniaxial stress state exists, the strain state is triaxial.

Our equations have been written in terms of the xyz coordinate system, but if we are dealing with the principal stresses and strains, then the subscripts x, y, and z can be replaced by subscripts 1, 2, and 3 to put the equations in terms of principal stresses and strains. In this case, there would be no shearing strains and hence no shearing stresses. Equations (2.47) become

$$\varepsilon_1 = \frac{1}{E}[\sigma_1 - v(\sigma_2 + \sigma_3)] \qquad (2.56a)$$

$$\varepsilon_2 = \frac{1}{E}[\sigma_2 - v(\sigma_3 + \sigma_1)] \qquad (2.56b)$$

$$\varepsilon_3 = \frac{1}{E}[\sigma_3 - v(\sigma_1 + \sigma_2)] \qquad (2.56c)$$

STRESS–STRAIN ANALYSIS AND STRESS–STRAIN RELATIONS 77

Equations (2.48) become

$$\sigma_1 = \frac{E}{(1 + v)(1 - 2v)}[(1 - v)\varepsilon_1 + v(\varepsilon_2 + \varepsilon_3)] \qquad (2.57a)$$

$$\sigma_2 = \frac{E}{(1 + v)(1 - 2v)}[(1 - v)\varepsilon_2 + v(\varepsilon_3 + \varepsilon_1)] \qquad (2.57b)$$

$$\sigma_3 = \frac{E}{(1 + v)(1 - 2v)}[(1 - v)\varepsilon_3 + v(\varepsilon_1 + \varepsilon_2)] \qquad (2.57c)$$

Corresponding changes can be made for the other stress and strain states.

2.9. Application of equations

The material developed can now be used in an application. Suppose we have an existing steel machine element of such a shape that the stresses cannot be determined analytically. At a point in question a three-element rectangular strain rosette is applied in order to determine the strains. The gages are applied so that one gage is aligned along the chosen x axis, as shown in Fig. 2.23. The gages are designated as a, b, and c. As the member is loaded, each gage will be strained. Our goal is to obtain the principal stresses and their orientation relative to the x axis.

After testing has been completed, the following information is presented for analysis:

$$\varepsilon_a = -800 \text{ μin/in at } \theta_a = 0°$$
$$\varepsilon_b = -300 \text{ μin/in at } \theta_b = 45°$$
$$\varepsilon_c = 1200 \text{ μin/in at } \theta_c = 90°$$
$$E = 30 \times 10^6 \text{ psi and } v = 0.3$$

Before the principal strains can be calculated, ε_x, ε_y, and γ_{xy} must be determined. The desired values can be computed using the strain readings in conjunction with Eq. (2.32). It is obvious that $\varepsilon_x = \varepsilon_a$ and $\varepsilon_y = \varepsilon_c$. Thus, we can use the reading given by ε_b and Eq. (2.32) in order to obtain $\gamma_{xy}/2$:

$$\varepsilon_b = \frac{\varepsilon_x + \varepsilon_y}{2} + \frac{\varepsilon_x - \varepsilon_y}{2}\cos 2\theta + \frac{\gamma_{xy}}{2}\sin 2\theta$$

where $\theta = 45°$. This gives

$$-300 = \frac{-800 + 1200}{2} + \frac{-800 - 1200}{2}\cos 2(45) + \frac{\gamma_{xy}}{2}\sin 2(45)$$

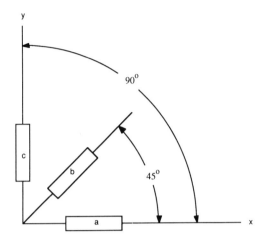

FIG. 2.23. Three-element rectangular rosette.

Thus,

$$\frac{\gamma_{xy}}{2} = -300 - \left(\frac{-800 + 1200}{2}\right) = -500 \text{ μradians}$$

The principal strains can be computed using Eq. (2.37):

$$\varepsilon_{1,2} = \frac{\varepsilon_x + \varepsilon_y}{2} \pm \sqrt{\left(\frac{\varepsilon_x - \varepsilon_y}{2}\right)^2 + \left(\frac{\gamma_{xy}}{2}\right)^2}$$

$$= \frac{-800 + 1200}{2} \pm \sqrt{\left(\frac{-800 - 1200}{2}\right)^2 + (-500)^2} = 200 \pm 1118$$

$$\varepsilon_1 = 1318 \text{ μin/in}; \qquad \varepsilon_2 = -918 \text{ μin/in}$$

The orientation of ε_1 can be determined by using any two of Eqs. (2.34), (2.35), and (2.36). From Eq. (2.34),

$$\tan 2\theta = \frac{\gamma_{xy}}{\varepsilon_x - \varepsilon_y} = \frac{-1000}{-800 - 1200} = 0.500\,00 \text{ (1st or 3d quadrant)}$$

From Eq. (2.35),

$$\sin 2\theta = \frac{\gamma_{xy}}{\sqrt{(\varepsilon_x - \varepsilon_y)^2 + (\gamma_{xy})^2}} = \frac{-1000}{\sqrt{(-800 - 1200)^2 + (-1000)^2}}$$

$$= -0.447\,21 \text{ (3d or 4th quadrant)}$$

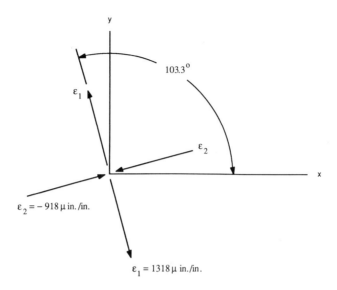

FIG. 2.24. Orientation of principal strains.

Since the only match is third quadrant, 2θ must be in that quadrant; thus, $2\theta = 206.6°$, or $\theta = 103.3°$. Figure 2.24 shows the orientation of the principal strains relative to the x axis.

A Mohr's circle could also have been used, and so it will be drawn in order to check the values of the principal strains and their orientation. Note, however, that while $\gamma_{xy}/2$ for the transformation equation is negative, its sign must be changed to positive when plotting Mohr's circle. Figure 2.25 shows the diagram for strain. Mohr's diagram shows quite clearly the orientation of ε_1 relative to the x axis. From the circle, we see that we could have gone in a negative (clockwise) direction from the x axis to ε_1 through an angle of $2\theta = 153.4°$. In Fig. 2.24, this would be the clockwise angle of $\theta = 76.7°$ from the x axis to ε_1 shown in the fourth quadrant.

Since this is a plane stress problem, σ_1 and σ_2 may be determined by using Eqs. (2.50a) and (2.50b). Here the subscripts x and y are changed to 1 and 2, respectively. Thus,

$$\sigma_1 = \frac{E}{1-v^2}(\varepsilon_1 + v\varepsilon_2) = \frac{30 \times 10^6}{1-(0.3)^2}[1318 + 0.3(-918)]10^{-6}$$
$$= 34\,371 \text{ psi}$$

$$\sigma_2 = \frac{E}{1-v^2}(\varepsilon_2 + v\varepsilon_1) = \frac{30 \times 10^6}{1-(0.3)^2}[-918 + 0.3(1318)]10^{-6}$$
$$= -17\,229 \text{ psi}$$

The orientation of σ_1 and σ_2 will, of course, be the same as ε_1 and ε_2.

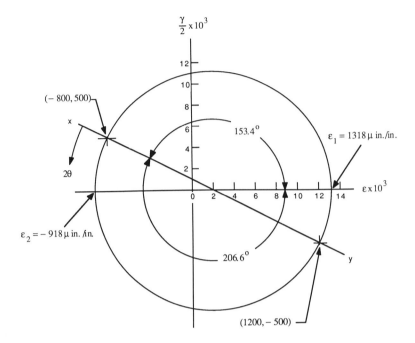

FIG. 2.25. Mohr's circle for strain.

The two principal strains ε_1 and ε_2 have been calculated. Because of the Poisson effect, there will also be a strain along the z, or 3, axis and so $\varepsilon_3 = \varepsilon_z$. This strain can be computed by using either Eq. (2.49) or Eq. (2.51c). Choosing Eq. (2.51c),

$$\varepsilon_3 = \varepsilon_z = -\frac{\nu}{E}(\sigma_1 + \sigma_2) = -\frac{0.3}{30 \times 10^6}(34\,371 - 17\,229)$$

$$= -171 \; \mu\text{in/in}$$

Since $\sigma_3 = 0$ and σ_1 and σ_2 are of opposite sign, the maximum value of the shear stress at the point is

$$\tau_{\max} = \tfrac{1}{2}(\sigma_1 - \sigma_2) = \tfrac{1}{2}[34\,371 - (-17\,229)] = 25\,800 \text{ psi}$$

The problem could have been approached in a different manner. Once ε_x, ε_y, and γ_{xy} were determined, σ_x, σ_y, and τ_{xy} could have been computed by using the stress–strain relationships, and the problem completed by using the stresses.

2.10. Stress and strain invariants

The development of the transformation equations for stress has been limited to the biaxial stress state. The derivation for the triaxial stress state, as well as the determination of the principal stresses and their orientation relative to the original coordinate system, is more complex. During this process a cubic equation is developed whose roots are real and are the values of the principal stresses, σ_1, σ_2, and σ_3(1). The cubic equation is

$$\sigma^3 - I_1\sigma^2 + I_2\sigma - I_3 = 0 \tag{2.58}$$

where
$$I_1 = \sigma_x + \sigma_y + \sigma_z \tag{2.59}$$

$$I_2 = \sigma_x\sigma_y + \sigma_y\sigma_z + \sigma_z\sigma_x - \tau_{xy}^2 - \tau_{yz}^2 - \tau_{zx}^2 \tag{2.60}$$

$$I_3 = \sigma_x\sigma_y\sigma_z - \sigma_x\tau_{yz}^2 - \sigma_y\tau_{zx}^2 - \sigma_z\tau_{xy}^2 + 2\tau_{xy}\tau_{yz}\tau_{zx} \tag{2.61}$$

The terms I_1, I_2, and I_3 are called stress invariants, since they are constants for any axis transformation. Considering I_1 as an example,

$$I_1 = \sigma_x + \sigma_y + \sigma_z = \sigma_{x'} + \sigma_{y'} + \sigma_{z'} = \sigma_1 + \sigma_2 + \sigma_3 = \text{a constant}$$

Thus, the sum of the normal stresses for any transformed axes will always have the same value. One can check Ex. 2.1, where $\sigma_3 = 0$, and find that

$$\sigma_x + \sigma_y = \sigma_1 + \sigma_2 = -9000 \text{ psi}$$

A cubic equation similar to Eq. (2.58) can also be developed for the determination of the principal strains (1). It is

$$\varepsilon^3 - I_1\varepsilon^2 + I_2\varepsilon - I_3 = 0 \tag{2.62}$$

where

$$I_1 = \varepsilon_x + \varepsilon_y + \varepsilon_z \tag{2.63}$$

$$I_2 = \varepsilon_x\varepsilon_y + \varepsilon_y\varepsilon_z + \varepsilon_z\varepsilon_x - \left(\frac{\gamma_{xy}}{2}\right)^2 - \left(\frac{\gamma_{yz}}{2}\right)^2 - \left(\frac{\gamma_{zx}}{2}\right)^2 \tag{2.64}$$

$$I_3 = \varepsilon_x\varepsilon_y\varepsilon_z - \varepsilon_x\left(\frac{\gamma_{yz}}{2}\right)^2 - \varepsilon_y\left(\frac{\gamma_{zx}}{2}\right)^2 - \varepsilon_z\left(\frac{\gamma_{xy}}{2}\right)^2 + \frac{\gamma_{xy}\gamma_{yz}\gamma_{zx}}{4} \tag{2.65}$$

Again, the roots of Eq. (2.62) are the values of the principal strains, ε_1, ε_2, and ε_3. The terms I_1, I_2, and I_3 are called strain invariants.

The stress states that we deal with are generally biaxial. The strain state will be triaxial, but since the strains that will be measured are on the surface of a machine element, the strain normal to the surface at that point will be due to the Poisson effect. In this case the strain normal to the surface at the point will be constant regardless of the orientation of the axes. Therefore, for plane strain, I_1 can be written as

$$I_1 = \varepsilon_x + \varepsilon_y = \varepsilon_{x'} + \varepsilon_{y'} = \varepsilon_1 + \varepsilon_2 = \text{a constant}$$

since $\varepsilon_z = \varepsilon_{z'} = \varepsilon_3$.

It will be pointed out later in the text how I_1 can be used when strain rosettes are considered.

Problems

In problems 2.1 through 2.8, determine the principal stresses and show, by sketching, their orientation relative to the xy coordinate system. Draw Mohr's circle for each and determine the maximum shear stress at the point.

For all problems in this chapter, use $v = 0.3$ and $E = 30 \times 10^6$ psi.

Prob. No.	σ_x, psi	σ_y, psi	τ_{xy}, psi
2.1	10 000	3 500	−5 000
2.2	12 500	12 500	8 000
2.3	−8 000	4 700	5 500
2.4	9 500	9 500	0
2.5	−15 000	−8 500	−3 000
2.6	0	0	7 500
2.7	−15 000	−15 000	−8 000
2.8	12 000	−4 000	0

2.9. For the cantilever beam and the loading shown in Fig. 2.26, determine the following:

   (a) The principal stresses at point A and B and their orientation relative to the x axis.
   (b) The maximum shear stress at points A and B.

2.10. A closed-end tube has an inside diameter of 2.000 in and an outside diameter of 2.125 in. The internal pressure is 750 psi and the tube is subjected to a torsional moment of 3000 in-lb. Determine the principal stresses.

2.11. Two gears are keyed to a rotating shaft as shown in Fig. 2.27. Force F_D is applied to gear D in the yz plane. Determine the reactive force F_C (also in the yz plane), and the bearing reactions, assuming frictionless bearings. Draw a free-body diagram of the assembly and determine the maximum shearing stress in the shaft between the gears.

2.12. A mechanic uses a torque wrench and an extension bar to tighten a nut (Fig. 2.8). If the torque wrench reads 100 ft-lb, determine the principal tensile stress and the maximum shear stress at the section shown on the extension bar.

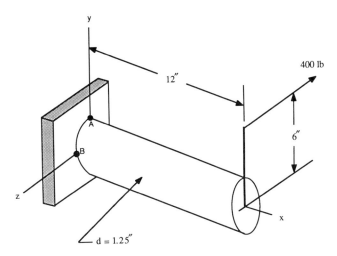

FIG. 2.26.

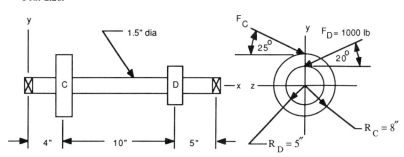

FIG. 2.27.

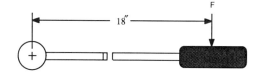

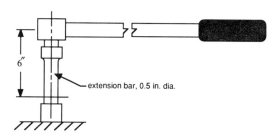

FIG. 2.28.

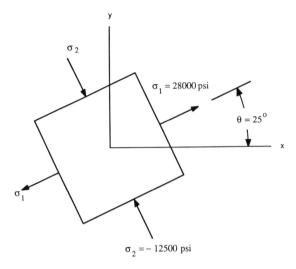

FIG. 2.29.

2.13. Figure 2.29 shows a principal stress element and its orientation relative to the x axis. Determine σ_x, σ_y, and τ_{xy}, then sketch the element showing the stresses in their proper directions.

2.14. Figure 2.30 shows a stress element. Using Mohr's circle, determine the following:

(a) The principal stresses. Sketch the principal stress element and show its orientation relative to the x axis.
(b) The maximum shear stress. Sketch the maximum shear stress element and show its orientation relative to the x axis.
(c) Verify your answers by analytical methods.

In Problems 2.15 through 2.24, determine the principal strains, ε_1 and ε_2 and, by a sketching, their orientation relative to the xy coordinate system. Draw a Mohr's circle for each problem. All values are in µin/in.

Prob. No.	ε_x	ε_y	γ_{xy}
2.15	1 800	−800	1 520
2.16	1 660	355	−960
2.17	−1 035	−260	770
2.18	−140	710	−390
2.19	0	0	2 000
2.20	0	0	−500
2.21	1 400	400	800
2.22	−800	400	0
2.23	1 150	1 150	0
2.24	640	430	−1 430

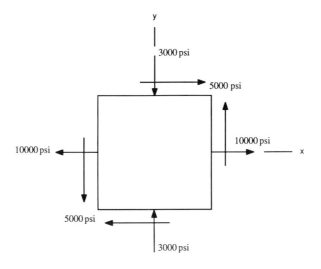

FIG. 2.30.

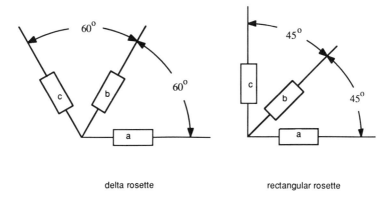

delta rosette   rectangular rosette

FIG. 2.31.

2.25. The following strains are given:

$$\varepsilon_x = 2500 \ \mu\text{in/in}, \qquad \varepsilon_y = -1000 \ \mu\text{in/in}, \qquad \gamma_{xy} = 1500 \ \mu\text{radians}$$

(a) Determine ε_1, ε_2, and γ_{max}.
(b) Determine the gage readings for (i) a three-element rectangular rosette and (ii) a three-element delta rosette, assuming gage a is aligned along the x axis, as shown in Fig. 2.31.

2.26. The three-element rectangular rosette shown in Fig. 2.32 gives the following strains:

$$\varepsilon_a = -800 \ \mu\text{in/in}, \qquad \varepsilon_b = 500 \ \mu\text{in/in}, \qquad \varepsilon_c = 1200 \ \mu\text{in/in}$$

Determine the principal strains and their orientation relative to the x axis.

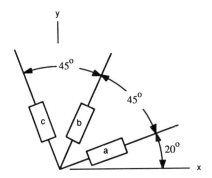

FIG. 2.32.

2.27. A three-element rectangular rosette gives the following readings:

$$\varepsilon_a = 0 \text{ μin/in at } \theta_a = 0°, \quad \varepsilon_b = -1500 \text{ μin/in at } \theta_b = 45°,$$
$$\varepsilon_c = 0 \text{ μin/in at } \theta_c = 90°$$

Determine the principal strains and their orientation relative to gage a, which is aligned along the x axis.

2.28. A three-element delta rosette gives the following readings:

$$\varepsilon_a = -1000 \text{ μin/in at } \theta_a = 0°, \quad \varepsilon_b = -1500 \text{ μin/in at } \theta_b = 60°,$$
$$\varepsilon_c = 1000 \text{ μin/in at } \theta_c = 120°$$

Determine the principal strains and their orientation relative to the x axis if gage b is aligned along the x axis.

2.29. The rosette in Fig. 2.33 is attached to a machine member. Determine the principal strains and their orientation if the following strain readings have been recorded:

$$\varepsilon_a = 1000 \text{ μin/in}, \quad \varepsilon_b = 1000 \text{ μin/in}, \quad \varepsilon_c = -1000 \text{ μin/in}$$

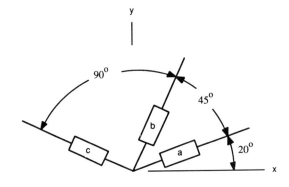

FIG. 2.33.

2.30. Determine σ_x, σ_y, τ_{xy}, σ_1, and σ_2 for Problem 2.26.
2.31. Determine σ_x, σ_y, τ_{xy}, σ_1, and σ_2 for Problem 2.27.
2.32. Determine σ_x, σ_y, τ_{xy}, σ_1, and σ_2 for Problem 2.28.
2.33. Given the following strains:

$$\varepsilon_x = 750 \text{ μin/in} \qquad \varepsilon_y = -800 \text{ μin/in}$$
$$\varepsilon_z = 450 \text{ μin/in} \qquad \gamma_{xy} = 200 \text{ μradians}$$
$$\gamma_{yz} = -5000 \text{ μradians} \quad \gamma_{xz} = 3000 \text{ μradians}$$

Determine σ_x, σ_y, σ_z, τ_{xy}, τ_{yz}, and τ_{xz}.

2.34. If $\varepsilon_1 = 800$ μin/in and $\varepsilon_2 = -200$ μin/in, determine the stress necessary to make $\varepsilon_3 = 0$.

2.35. Given the following stresses:

$$\sigma_x = 25\,000 \text{ psi} \qquad \sigma_y = 8000 \text{ psi}$$
$$\tau_{xy} = 7000 \text{ psi} \qquad \sigma_z = \tau_{yz} = \tau_{xz} = 0$$

Determine the principal strains ε_1, ε_2, and ε_3.

2.36. The strain gage rosette, Fig. 2.34, is mounted on the surface of a machine member. Previous calculations have yielded the following strains in the xy plane at point 0:

$$\varepsilon_x = 1570 \text{ μin/in}, \qquad \varepsilon_y = -472 \text{ μin/in}, \qquad \gamma_{xy} = 1416 \text{ μradians}$$

(a) Determine the expected strain rosette readings.
(b) Determine all of the principal strains.
(c) Determine σ_x, σ_y, and τ_{xy} at point 0.
(d) Sketch the principal stress element and its orientation relative to the x axis. Show the values of σ_1 and σ_2.

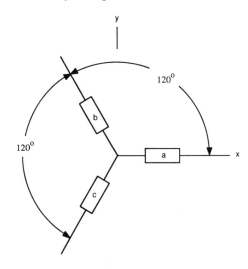

Fig. 2.34.

2.37. For the stress element shown in Fig. 2.35, determine the following:
   (a) The stress acting along the z axis that will make $\varepsilon_z = 0$.
   (b) With the stress from part (a) acting on the element, compute ε_x, ε_y, and γ_{xy}.

2.38. The 1.0-in diameter steel shaft is loaded as shown in Fig. 2.36 and has a strain gage rosette attached at point A. Determine the expected values of ε_a, ε_b, and ε_c.

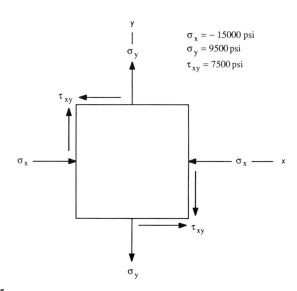

FIG. 2.35.

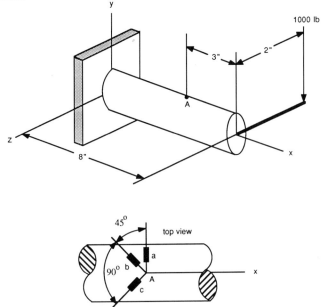

FIG. 2.36.

REFERENCES

1. Durelli, A. J., E. A. Phillips, and C. H. Tsao, *Introduction to the Theoretical and Experimental Analysis of Stress and Strain*, New York, McGraw-Hill, 1958, Chaps. 1, 2, 4.

3

ELEMENTARY CIRCUITS

3.1. Introduction

Since the change in resistance of a strain gage is measured by its effect upon either the current passing through the gage or the voltage drop across it, the gage must form part of some kind of electrical circuit. Figure 3.1 shows such a strain gage.

For initial conditions we can write

$$E = IR_g$$

where E = voltage drop across the gage

I = current passing through the gage

R_g = gage resistance

When the gage resistance changes from R_g to $R_g + \Delta R_g$, either the current, I, or the voltage, E, or both, will be changed.

It is our purpose now to explore two simple circuits and to investigate the corresponding effects of unit changes, $\Delta R_g/R_g$, in gage resistance upon voltage and current. Schematic diagrams of the two elementary circuits, each containing a single strain gage, are shown in Figs. 3.2 and 3.3. The first of these indicates a constant-voltage source connected to the gage, while the second represents a constant-current circuit.

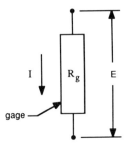

FIG. 3.1. Strain gage.

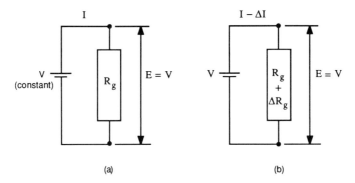

FIG. 3.2. Constant voltage applied to gage.

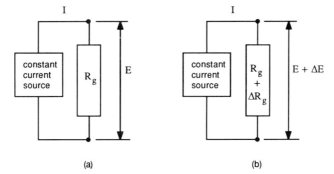

FIG. 3.3. Constant-current circuit.

3.2. Constant-voltage circuit

For the case shown in Fig. 3.2, the applied voltage, V, will be the same as E, the potential drop across the gage. Because this circuit contains a constant-voltage source (for example, a battery of sufficient size), there will be no change in potential drop across the gage even though there may be variations in resistance.

Due to the application of constant potential, V, the only thing that happens when the gage resistance changes is a change in the current. Thus, when the gage resistance changes from R_g to $R_g + \Delta R_g$, the corresponding change in current is from I to $I - \Delta I$.

We can now examine how the change in current, ΔI, is related to strain, or to the unit change in gage resistance. Initially,

$$E = IR_g \tag{3.1}$$

After a change in gage resistance of ΔR_g, the current changes by $-\Delta I$, and so

$$E = (I - \Delta I)(R_g + \Delta R_g)$$

Expanding the right side,

$$E = IR_g - \Delta I\,\Delta R_g - \Delta I\, R_g + I\,\Delta R_g \tag{3.2}$$

Since $E = IR_g$, Eq. (3.2) becomes

$$0 = -\Delta I\,\Delta R_g - \Delta I\, R_g + I\,\Delta R_g$$

or

$$0 = -\Delta I(R_g + \Delta R_g) + I\,\Delta R_g \tag{3.3}$$

Equation (3.3) may now be expressed in terms of unit changes in resistance by dividing each term by R_g. This gives

$$0 = -\Delta I\left(1 + \frac{\Delta R_g}{R_g}\right) + I\frac{\Delta R_g}{R_g}$$

Solving for ΔI produces

$$\Delta I = \left(\frac{1}{1 + \Delta R_g/R_g}\right) I \frac{\Delta R_g}{R_g} \tag{3.4}$$

We know, however, that

$$\frac{\Delta R_g}{R_g} = \varepsilon G_F \tag{3.5}$$

where ε = strain, in/in
   G_F = gage factor

Substituting the value of $\Delta R_g/R_g$ given by Eq. (3.5) into Eq. (3.4) gives

$$\Delta I = \left(\frac{1}{1 + \Delta R_g/R_g}\right) I G_F \tag{3.6}$$

We can now write the expression for the change in current per unit of strain as

$$\frac{\Delta I}{\varepsilon} = \left(\frac{1}{1 + \Delta R_g/R_g}\right) I G_F \tag{3.7}$$

The current can be written in terms of the applied voltage as

$$I = \frac{E}{R_g} = \frac{V}{R_g}$$

Using this value of I, Eq. (3.7) becomes

$$\frac{\Delta I}{\varepsilon} = \left(\frac{1}{1 + \Delta R_g/R_g}\right)\frac{VG_F}{R_g} \tag{3.8}$$

Equations (3.4) and (3.6) indicate that the change in current, ΔI, is a nonlinear function of the unit change in gage resistance, or the strain. On this account it is sometimes more convenient to express Eq. (3.7) in the following modified form:

$$\frac{\Delta I}{\varepsilon} = IG_F(1 - n) \tag{3.9}$$

where n, the nonlinearity factor for this case, is

$$n = \frac{\Delta R_g/R_g}{1 + \Delta R_g/R_g} \tag{3.10}$$

Figure 3.4 shows a curve representing the values of the nonlinearity factor, n, as given in Eq. (3.10), in terms of the change in gage resistance, $\Delta R_g/R_g$.

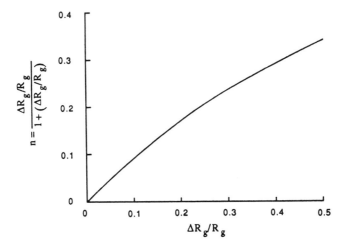

FIG. 3.4. Nonlinearity factor vs. unit change in gage resistance.

From the relationship expressed in Eq. (3.9), we can draw two interesting conclusions:

1. The circuit indication, or output, per unit strain, $\Delta I/\varepsilon$, is a nonlinear function of the current I and the gage factor G_F.
2. The maximum indication, or output, in terms of current change per unit of strain will occur for very small (theoretically zero) strains. In quantitative terms this will be represented by the product of the initial current and the gage factor, IG_F.

Since the maximum current (roughly 30 milliamperes) that can be carried by the gage depends upon the ability to dissipate heat, there will definitely be an upper limit for the output per unit strain, $\Delta I/\varepsilon$, that can be achieved for any particular installation. For example, if the maximum current I_m is 30 milliamperes and the gage factor G_F is 3.5, then the maximum value of $\Delta I/\varepsilon$ will be

$$\left(\frac{\Delta I}{\varepsilon}\right)_{max} = IG_F = (0.030)(3.5) = 0.105 \text{ amps/unit strain}$$

or

$$\left(\frac{\Delta I}{\varepsilon}\right)_{max} = 0.105 \text{ microamps/microstrain}$$

It should be noted that, for the indication of strain, this is a current-sensitive circuit. As such, it possesses certain characteristics that it shares with other types of current-sensitive circuits used with strain gages.

Since $\Delta I/\varepsilon$, the output per unit strain, varies directly with the gage current, the highest possible current consistent with the limitations imposed by heating effects should be employed. Equation (3.8) indicates that this objective can be achieved either by the use of low-resistance gages or by employing high values for the applied voltage. However, because considerations of safety and convenience impose an upper limit on the applied voltage V, it will be desirable to select strain gages of low resistance and high gage factor, for this type of circuit, in order to achieve the maximum possible indication for a given strain.

3.3. Constant-current circuit

An alternative to the circuit which applies a constant voltage to the gage is the circuit delivering a constant current to the gage. Figure 3.3 shows the circuit. In this case $\Delta I = 0$ at all times. However, due to the fact that the current is constant, there will be changes in the voltage drop across the gage as its resistance changes. We are able, therefore, to determine the resistance change by measuring the change in voltage drop across the gage.

ELEMENTARY CIRCUITS

We now investigate the relationship between the unit change in gage resistance and the corresponding change in voltage drop across the gage for the constant current circuit. In making the analysis, we refer to Fig. 3.3. The initial conditions are again given by Eq. (3.1).

$$E = IR_g \qquad (3.1)$$

When the resistance of the gage changes from R_g to $R_g + \Delta R_g$, we can write the corresponding expression for the voltage drop across the gage as

$$E + \Delta E = I(R_g + \Delta R_g) = IR_g + I\,\Delta R_g \qquad (3.11)$$

Substituting the value of E from Eq. (3.1) into Eq. (3.11), we have

$$\Delta E = I\,\Delta R_g = IR_g \frac{\Delta R_g}{R_g} \qquad (3.12)$$

Equation (3.12) can be written in terms of ε and G_F by using Eq. (3.5). Thus,

$$\Delta E = IR_g G_F \varepsilon \qquad (3.13)$$

The potential drop across the gage per unit of strain may be written as

$$\frac{\Delta E}{\varepsilon} = IR_g G_F \qquad (3.14)$$

From these equations we can draw the following conclusions for the constant current circuit:

1. The change in potential drop across the gage, ΔE, will be a linear function of the strain (or the unit change in gage resistance, $\Delta R_g/R_g$).
2. The indication, or output, per unit strain is a linear function of each of the three quantities: (a) gage current I, (b) gage resistance R_g, and (c) gage factor G_F, as well as their product.
3. The maximum output per unit strain, $(\Delta E/\varepsilon)_{max}$, will occur when the product $IR_g G_F$ reaches a maximum.
4. For this type of circuit, which is voltage-sensitive, the preceding indicates that the maximum output will be achieved with *high-resistance* gages with high gage factors. This is in direct contrast to the constant-voltage (current-sensitive) circuit for which the maximum output is achieved with low-resistance gages possessing high gage factors.
5. Since the gage resistance R_g and the gage factor G_F are both properties of the gage, and because the maximum current I_m is determined by the gage's ability to dissipate heat, the maximum attainable output per unit strain, $(\Delta E/\varepsilon)_{max}$, depends entirely upon the characteristics of the gage.

Item (5) is rather important because it tells us that the *maximum possible* output, per unit of strain, is dependent only upon the properties of the gage and not upon the characteristics of the circuit to which it is attached. If the efficiency of the electric circuit is 100 percent, then the maximum possible output can be achieved.

The efficiency of the electric circuit is expressed as the ratio of the maximum output from the circuit per unit of strain divided by the maximum possible output from the gage per unit of strain. According to this definition, the constant-current circuit is 100 percent efficient since it delivers the maximum possible output from the gage.

3.4. Advantages of the constant-current circuit

The two circuits just discussed represent different approaches to the determination of the same thing, namely, strain. They are both special cases of the potentiometric circuit, which is of a more general nature, and include certain of the advantages of each of these two elementary forms.

The linear relation between strain and the output, ΔE, of the constant current circuit is a tremendous advantage, if not a necessity. For metallic sensors this characteristic is desirable but not so important since the resistance changes are small. However, for smeiconductor gages subjected to any appreciable amount of strain this is practically a necessity, particularly since the degree of nonlinearity varies with changes of reference or initial reading.

Since the constant-current circuit gives its indication of strain in terms of a change in voltage, it is ideally suited for use with numerous well-known techniques and standard instruments already developed (for other purposes) to measure small voltage changes precisely. This represents both convenience and economy.

About the only real advantage of the constant-voltage (across the gage) circuit lies in its ability to use a simple, inexpensive battery as a power supply. However, this advantage is also possessed by the potentiometric circuit, which can always approximate, and sometimes achieve, constant-current conditions, with the corresponding advantages of the constant-current circuit.

There was no constant-current power supply commercially available for strain gage use until the early 1960s (1). Although the constant-current power supply costs considerably more than a battery, nevertheless, it is not an expensive instrument. Reference 1 contains a list of its specifications and characteristics, while Fig. 3.5 is a schematic diagram of a constant-current circuit.

Some additional advantages of a constant current circuit are as follows:

1. Within the power capability of the constant-current source, there will be no effect from long leads of appreciable resistance, since the lead

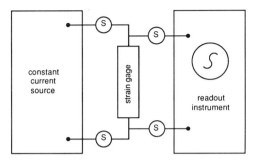

FIG. 3.5. Schematic diagram of constant-current circuit.

resistance will not alter the current flow. Since a high-impedance readout instrument, which draws essentially no current, must be employed, appreciable resistance in the leads to it will cause no trouble, for practically no current flows in these leads.
2. For the same reasons given in Item (1), if the arrangement shown in Fig. 3.5 is employed, variations in contact resistance at switches or slip rings will have no effect on the indicated output from the gage as long as the current source can respond rapidly enough to the resistance changes to maintain the constant current.

3.5. Fundamental laws of measurement

In the foregoing circuit analysis it has been assumed that the readout instrument would draw no current. This brings us to the consideration of two fundamental concepts which apply not only to strain gages and their associated electrical instruments but in general to measurements of all kinds (2).

These concepts are frequently referred to as the *fundamental laws of measurement* and can be briefly stated as follows:

1. The instrument, or device, used to make a measurement should have no (or negligible) effect upon the quantity being measured.
2. The quantity being measured should have no (or negligible) effect upon the instrument, or device, used to make the measurement.

Numerous examples of violations of these laws may be cited. However, the following examples taken from strain gage studies will serve as illustrations:

1. A large strain gage on a stiff carrier is used to measure strain on a slender specimen of low-modulus material. This violates the first law because the stiffening effect of the gage masks the true value of the strain that is to be measured.

98   THE BONDED ELECTRICAL RESISTANCE STRAIN GAGE

2. A high gage factor strain gage, which has been developed to measure very small strains, and which will suffer a loss in gage factor if overstrained, is inadvertently used to measure strains well into a condition of yielding. This violates the second law because a characteristic of the gage, namely the gage factor, has been changed by the strain that the instrument is endeavoring to measure.

In estimating conformity with the laws of measurement, since most measuring devices have some influence (although perhaps very small) on the quantity being measured, it is necessary to determine to what degree this influence is taking place and whether or not this can be considered negligible for the particular set of conditions at hand. Under one set of conditions a given effect might be quite negligible, whereas in other circumstances the same thing might be very important. For example, for stress analysis one can often safely neglect errors which are exceedingly important when relating to load cells and other weighing devices.

Problems

3.1. Plot $\Delta I/I$ vs. $\Delta R_g/R_g$ for values of $\Delta R_g/R_g$ between 0 and 1.0.

3.2. A strain gage with $R_g = 120$ ohms and $G_F = 2.5$ is bonded to the simply supported beam shown in Fig. 3.6. A constant voltage of $V = 2.4$ volts is applied across the gage. The beam is restrained in such a manner that it is free to bend but not to buckle. Determine (a) the gage current after loading and (b) the nonlinearity factor.

3.3. Redo Problem 3.2 for $R_g = 350$ ohms but all other factors remaining the same.

3.4. In Fig. 3.2, a resistor is shunted across the strain gage, R_g, in order to simulate a high strain. If $R_g = 120$ ohms, $G_F = 2.15$, $V = 3$ volts, and the shunt resistor is $R_p = 1000$ ohms, determine the final current I_f, and the nonlinearity factor.

3.5. In the constant-current circuit shown in Fig. 3.3, a resistor, R_p, is shunted across R_g. If $I = 0.025$ amperes, $R_g = 120$ ohms, $G_F = 2.0$, and $\Delta E = -0.06$ volts, determine the value of R_p.

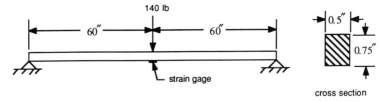

FIG. 3.6.

REFERENCES

1. Stein, Peter K., "The Constant Current Concept for Dynamic Strain Measurement," *Strain Gage Readings*, Vol. VI, No. 3, Aug–Sept. 1963, pp. 53–72. Also *BLH Measurement Topics*, Vol. 6, No. 2, Spring 1968, pp. 1–2, and *Instruments & Control Systems*, Vol. 38, No. 5, May 1965, pp. 145–155.

2. Stein, Peter K., *Measurement Engineering*, Stein Engineering Services, Inc., 5603 East Monte Rosa, Phoenix, AZ 85018-4646, Vol. II, 1962, Chap. 24. Vol. II is: *The Strain Gage Encylopaedia*. Chap. 24 is on *Circuits for Non-Self-Generating Transducers*.

4
THE POTENTIOMETRIC CIRCUIT

4.1. Introduction

The potentiometric circuit is also known as the ballast circuit, or series circuit. Because, in effect, it corresponds to half a Wheatstone bridge, it is sometimes referred to as the half bridge. The circuit is represented schematically in Fig. 4.1.

In its elementary form, as applied to strain gages, the potentiometric circuit contains the following three major components:

1. A power supply, usually a battery, which will furnish constant voltage V to the circuit.
2. A strain gage of initial resistance R_g.
3. A ballast resistance, of initial value R_b, to control the current in the circuit. Sometimes the ballast resistance consists of a second strain gage which, depending upon the particular conditions prevailing, may or may not be identical to R_g.

In addition to the above components, there must also be some means of obtaining a measure, or readout, of the change in voltage drop across the gage (or ballast resistance). This provides an indication of the change in gage resistance, ΔR_g, which, in turn, represents a measure of the strain. The exact nature of the readout device, or system, will depend upon the magnitude of the signal, ΔE, and the precision with which it is desired to make the observation.

A study of Fig. 4.1 reveals that the potentiometric circuit is really a compromise between the two simple arrangements described in Chapter 3 on elementary circuits. Both of the elementary circuits are actually special cases of this somewhat more generalized form. The following concepts will assist in clarification:

1. $R_b = 0$: If the ballast resistance is reduced to zero, we have the case of the strain gage directly connected to a constant-voltage power supply.
2. $R_b \to \infty$: In this case R_b is very large relative to R_g. Let us consider what happens as the ballast resistance is increased and the applied voltage, V, is correspondingly stepped up to maintain some desired initial value of gage current before strain takes place at the gage (i.e., when $\Delta R_g = 0$).

   As R_b becomes progressively larger, the gage resistance R_g assumes

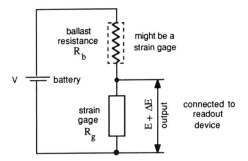

FIG. 4.1. The potentiometric circuit.

a smaller proportion of the total resistance in the circuit. In consequence, any changes in gage resistance, ΔR_g, will have a progressively smaller influence on the flow of current, until ultimately, when R_b is large enough, the effect of changes in gage resistance will have an insignificant effect on the current. When this condition has been reached, for practical purposes, we have essentially achieved a constant-current circuit.

Subject to the limitation of keeping the applied voltage within safe working limits, the potentiometric circuit may be made to approximate a constant-current circuit to any degree of precision. Under these conditions we might think of the power supply and the ballast resistor as being combined into a single unit providing, within specified limits, a constant-current source connected to the strain gage.

3. $\Delta R_b = -\Delta R_g$: Sometimes it is possible to vary the ballast resistance (for example, when it consists of a second strain gage) in such a manner that its change, ΔR_b, is equal in magnitude, but of opposite sign, to the change in gage resistance ΔR_g. Under these conditions the total resistance in the circuit remains constant. Thus, for a constant applied voltage, we have a true constant-current circuit possessing all the advantages indicated previously.

4.2. Circuit equations

In order to obtain an immediate insight into the properties of the potentiometric circuit, the circuit equations and some discussion of them are presented here. The complete derivations will be developed later in the chapter.

For convenience, the relationship between the ballast resistor, R_b, and the gage resistance, R_g, is expressed as a dimensionless ratio as follows:

$$\frac{R_b}{R_g} = a$$

The incremental output from the circuit, when written in terms of unit changes in ballast and gage resistances, and the ratio a, is expressed as

$$\Delta E = V \frac{a}{(1+a)^2} \left[\frac{\Delta R_g}{R_g} - \frac{\Delta R_b}{R_b} \right] (1-n) \tag{4.1}$$

where n is a nonlinearity factor given as

$$n = \frac{1}{1 + \dfrac{1+a}{\dfrac{\Delta R_g}{R_g} + a \dfrac{\Delta R_b}{R_b}}} \tag{4.2}$$

For a single active gage, R_b is a constant and $\Delta R_b = 0$, so Eqs. (4.1) and (4.2) reduce to the following:

$$\Delta E = V \frac{a}{(1+a)^2} \left[\frac{\Delta R_g}{R_g} \right] (1-n) \tag{4.3}$$

$$n = \frac{1}{1 + \dfrac{1+a}{\Delta R_g / R_g}} \tag{4.4}$$

Since $\Delta R_g / R_g$ = strain times gage factor = εG_F, Eqs. (4.3) and (4.4) can be rewritten as

$$\Delta E = V \frac{a}{(1+a)^2} [\varepsilon G_F](1-n) \tag{4.5}$$

$$n = \frac{1}{1 + \dfrac{1+a}{\varepsilon G_F}} \tag{4.6}$$

Characteristics of the circuit

Some discussion of the equations for the potentiometric circuit is now in order.

1. *Difference of two strains*—Eq. (4.1). When the ballast resistance is variable, the change in output voltage, ΔE, is directly proportional to the algebraic difference between the unit changes in the gage and ballast resistances, providing the nonlinearity factor, n, can be neglected. This

means the circuit is capable of providing a reading directly proportional to the algebraic difference between the strains at two gage locations. If the gages have positive and negative gage factors, then the reading will be the algebraic sum.
2. *Magnification of the strain gage signal*—Eq. (4.1). When strains of known ratio but of opposite sign prevail at two locations, the signal $\Delta R_g/R_g$ can be increased by using a second active gage for the ballast resistor, R_b. For example, if the strain at the gage comprising the ballast resistor, R_b, is equal but of opposite sign to R_g, the output will be doubled. In this particular case we have a constant-current circuit whose output will be linear with strain, and of the maximum obtainable value per unit of strain.
3. *Linearity*—Eqs. (4.2) and (4.4). Basically, the incremental output of the circuit, ΔE, is a nonlinear function of the strain.

> *Single gage.* For a single gage with a fixed ballast resistance, R_b, nonlinearity is always the case. Nevertheless, the nonlinearity factor, n, can be made negligibly small by having R_b large relative to R_g.
>
> *Two gages.* When the ballast resistance, R_b, consists of a second strain gage (which is not necessarily required to have the same resistance or gage factor as R_g), the optimum condition is achieved when
>
> $$\Delta R_b = -\Delta R_g$$
>
> When this situation prevails we have a constant-current circuit that gives a linear output of maximum attainable value per unit strain; that is,
>
> $$\frac{\Delta E}{\varepsilon} = I_m R_g G_F \qquad (4.7)$$
>
> where I_m is the maximum permissible gage current.

When the ballast resistance is fixed and several *like* gages are connected in series in the adjacent arm, shown in Fig. 4.2, *the change in voltage drop across all the gages will correspond to the average of the strains experienced by the gages.* In other words, ΔE represents the average strain, ε, for this arrangement.

For like gages in series, the previous equations can be used with the following modifications:

$$a = \frac{R_b}{\sum R_g} \qquad (4.8)$$

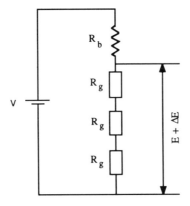

FIG. 4.2. Gages in series in the same arm.

What was previously expressed as the unit change in gage resistance is now

$$\text{unit change} = \frac{\sum \Delta R_g}{\sum R_g} \quad (4.9)$$

Substituting the values of a given by Eq. (4.8) and the unit change in gage resistance given by Eq. (4.9) into Eq. (4.3), we arrive at

$$\Delta E = V \frac{R_b/\sum R_g}{[1 + R_b/\sum R_g]^2} \left(\frac{\sum \Delta R_g}{\sum R_g}\right)(1 - n) \quad (4.10)$$

Similarly, the value of n is

$$n = \frac{1}{1 + \dfrac{1 + (R_b/\sum R_g)}{\sum \Delta R_g/\sum R_g}} \quad (4.11)$$

Applications

The potentiometric circuit shown in Fig. 4.1 will work equally well for static or dynamic strains, or combinations thereof. However, the means employed to measure ΔE impose certain limitations which determine its applicability for static strain as well as dynamic strain observations.

If the resistance change in the strain gage is large enough so that an instrument employed to measure E, the initial voltage drop across the strain gage (for zero strain), is also capable of measuring the change, ΔE, to the desired degree of precision, then both static and dynamic observations can be made (as long as the dynamic response of the instrument is suited to the frequencies of the strain signals).

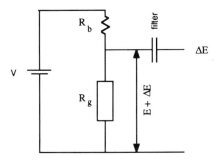

FIG. 4.3. Potentiometric circuit with filter to eliminate steady component, E, from the output.

In general, however, ΔE will be too small to be measured conveniently, and with the desired precision, on the same scale as that used for the measurement of E. In order to overcome this difficulty, it is customary to use a filter that will eliminate the steady voltage, E, so that ΔE can be amplified and measured by itself. The method works very well for the determination of dynamic strains, but the filter which eliminates the steady (zero-frequency) component, E, also eliminates any other zero-frequency signals and, in consequence, precludes the possibility of making static strain observations. The potentiometric circuit with a filter to eliminate the steady component, E, is shown schematically in Fig. 4.3.

Due to the relatively small signals produced by metallic strain gages, and the consequent use of the filter to eliminate E, the use of the potentiometric circuit has, in the past, been limited to dynamic strain measurements. As a result of the development of semiconductor strain gages with high gage factors, and the availability of four- or five-place digital voltmeters, it is likely that this circuit will also be used for numerous static applications.

Advantages and limitations of the potentiometric circuit

Among the advantages of the potentiometric circuit the following may be included:

1. Extreme simplicity.
2. Ability to approach, and in some cases to reach, the linearity and optimum output of the constant-current circuit.
3. The circuit is able to use a simple constant-voltage, ripple-free, power supply from a battery (dry cell) and, at the same time, to provide an output in the form of a voltage change that can be measured relatively easily.
4. The circuit, the readout instrument, and associated amplifier (if one is used), can all be connected to a common ground.

Among the limitations of the potentiometric circuit we find the following:

1. Inability to measure static strains with metallic or other strain gages producing very low-level signals. This is not really a limitation of the circuit, but of the readout equipment associated with it.
2. The strain signal, ΔE, is directly proportional to the battery voltage, V. If the battery runs down, the strain signal will be influenced. For dynamic measurements which can be completed in a short space of time, this will probably cause no difficulty, but if observations are to be made over a considerable time interval, then periodic checks of the battery condition should be made.

4.3. Analysis of the circuit

For the purpose of analyzing the potentiometric circuit, refer again to Fig. 4.1 and consider the case where the resistances of both the strain gage, R_g, and the ballast resistor, R_b, undergo changes.

Under initial conditions, before any changes take place, the expression for the voltage drop across the gage is

$$E = V \frac{R_g}{R_g + R_b} \tag{4.12}$$

If R_b and R_g change to $R_b + \Delta R_b$ and $R_g + \Delta R_g$, respectively, the voltage drop across the gage becomes $E + \Delta E$ and is expressed as

$$E + \Delta E = V \frac{R_g + \Delta R_g}{R_g + \Delta R_g + R_b + \Delta R_b} \tag{4.13}$$

By subtracting the value of E given by Eq. (4.12) from both sides of Eq. (4.13), the change in voltage drop across the gage, ΔE, is

$$\Delta E = V \left[\frac{R_g + \Delta R_g}{R_g + \Delta R_g + R_b + \Delta R_b} - \frac{R_g}{R_g + R_b} \right] \tag{4.14}$$

Now divide the numerator and denominator of the right-hand side of Eq. (4.14) by R_g. This puts all of the quantities into terms of dimensionless ratios and unit changes. Thus,

$$\Delta E = V \left[\frac{1 + \dfrac{\Delta R_g}{R_g}}{1 + \dfrac{\Delta R_g}{R_g} + \dfrac{R_b}{R_g} + \dfrac{\Delta R_b}{R_g}} - \frac{1}{1 + \dfrac{R_b}{R_g}} \right] \tag{4.15}$$

Again we can write the ratio of the ballast to gage resistance as

$$a = \frac{R_b}{R_g} \tag{4.16}$$

and so

$$\frac{1}{R_g} = \frac{a}{R_b} \tag{4.17}$$

Insertion of the values of a and $1/R_g$, given by Eqs. (4.16) and (4.17), respectively, into Eq. (4.15) gives

$$\Delta E = V \left[\frac{1 + \dfrac{\Delta R_g}{R_g}}{1 + \dfrac{\Delta R_g}{R_g} + a + a \dfrac{\Delta R_b}{R_b}} - \frac{1}{1 + a} \right] \tag{4.18}$$

We observe that, in Eq. (4.18), all of the quantities involving resistance are expressed either in terms of the dimensionless ratio, a, or as unit changes in gage and ballast resistances. If we assume the ballast resistance to be a strain gage, then the unit changes in the resistances are directly related to strain, since

$$\frac{\Delta R}{R} = \varepsilon G_F$$

Equation (4.18) can be reduced to a simpler and more convenient form. We start by slightly rearranging the terms and putting everything over a common denominator. This gives

$$\Delta E = V \left[\frac{\left(1 + \dfrac{\Delta R_g}{R_g}\right)(1 + a) - \left(1 + a + \dfrac{\Delta R_g}{R_g} + a \dfrac{\Delta R_b}{R_b}\right)}{\left(1 + a + \dfrac{\Delta R_g}{R_g} + a \dfrac{\Delta R_b}{R_b}\right)(1 + a)} \right] \tag{4.19}$$

Clearing the parentheses in the numerator, Eq. (4.19) becomes

$$\Delta E = \frac{V}{1 + a} \left[\frac{1 + a + \dfrac{\Delta R_g}{R_g} + a \dfrac{\Delta R_g}{R_g} - 1 - a - \dfrac{\Delta R_g}{R_g} - a \dfrac{\Delta R_b}{R_b}}{1 + a + \dfrac{\Delta R_g}{R_g} + a \dfrac{\Delta R_b}{R_b}} \right]$$

This reduces to

$$\Delta E = V \frac{a}{1+a} \left[\frac{\Delta R_g}{R_g} - \frac{\Delta R_b}{R_b} \right] \left[\frac{1}{1 + a + \frac{\Delta R_g}{R_g} + a \frac{\Delta R_b}{R_b}} \right] \quad (4.20)$$

Multiplying the numerator and denominator of Eq. (4.20) by $(1 + a)$ results in

$$\Delta E = V \frac{a}{(1+a)^2} \left[\frac{\Delta R_g}{R_g} - \frac{\Delta R_b}{R_b} \right] \left[\frac{1 + a}{1 + a + \frac{\Delta R_g}{R_g} + a \frac{\Delta R_b}{R_b}} \right] \quad (4.21)$$

Equation (4.21) indicates that ΔE will not be a linear function of the difference between the unit changes in gage and ballast resistances unless

$$\left[\frac{1 + a}{1 + a + \frac{\Delta R_g}{R_g} + a \frac{\Delta R_b}{R_b}} \right] = 1 \quad (4.22)$$

This will take place when

$$\frac{\Delta R_g}{R_g} + a \frac{\Delta R_b}{R_b} = 0 \quad (4.23)$$

which gives us

$$\frac{\Delta R_g}{R_g} + \frac{R_b}{R_g} \frac{\Delta R_b}{R_b} = \frac{\Delta R_g}{R_g} + \frac{\Delta R_b}{R_g} = 0 \quad (4.24)$$

This means

$$\Delta R_g = -\Delta R_b \quad (4.25)$$

Since ΔR_g is usually not equal to $-\Delta R_b$, in general ΔE will be a nonlinear function of the difference between the unit changes in gage and ballast resistances.

We now examine the deviation from linearity, represented by the symbol n, in a modified version of Eq. (4.21). It may be rewritten as

$$\Delta E = V \frac{a}{(1+a)^2} \left[\frac{\Delta R_g}{R_g} - \frac{\Delta R_b}{R_b} \right] (1 - n) \quad (4.26)$$

where n is the nonlinearity factor.

Since Eqs. (4.21) and (4.26) represent the same quantity, ΔE, we see that

$$1 - n = \frac{1 + a}{1 + a + \dfrac{\Delta R_g}{R_g} + a \dfrac{\Delta R_b}{R_b}} \tag{4.27}$$

We now solve for the nonlinearity factor, n, in terms of the unit changes in the ballast and gage resistances, $\Delta R_g/R_g$, $\Delta R_b/R_b$, and the ratio of the ballast to the gage resistance, $a = R_b/R_g$. Rearrangement of Eq. (4.27) shows that

$$n = 1 - \frac{1 + a}{1 + a + \dfrac{\Delta R_g}{R_g} + a \dfrac{\Delta R_b}{R_b}} \tag{4.28}$$

Putting the right-hand side over a common denominator gives

$$n = \frac{1 + a + \dfrac{\Delta R_g}{R_g} + a \dfrac{\Delta R_b}{R_b} - 1 - a}{1 + a + \dfrac{\Delta R_g}{R_g} + a \dfrac{\Delta R_b}{R_b}}$$

This reduces to

$$n = \frac{\dfrac{\Delta R_g}{R_g} + a \dfrac{\Delta R_b}{R_b}}{1 + a + \dfrac{\Delta R_g}{R_g} + a \dfrac{\Delta R_b}{R_b}}$$

Dividing the numerator and denominator of the right-hand side by $\Delta R_g/R_g + a \Delta R_b/R_b$, we arrive at

$$n = \frac{1}{1 + \dfrac{1 + a}{\dfrac{\Delta R_g}{R_g} + a \dfrac{\Delta R_b}{R_b}}} \tag{4.29}$$

Thus, for a given set of conditions, the deviation from linearity can be determined from Eq. (4.29). Note that the nonlinearity is a function of a, $\Delta R_g/R_g$, and $\Delta R_b/R_b$.

110   THE BONDED ELECTRICAL RESISTANCE STRAIN GAGE

Limitation on applied voltage V

Since the signal, ΔE, is directly proportional to the applied voltage, the maximum output will be achieved for the largest value of V. This will be limited by the following two practical considerations:

1. The maximum current, I_m, that the gage can carry. Frequently the maximum current will be limited to about 30 milliamperes, but it may be less, depending on the particular conditions prevailing.
2. The maximum voltage that can be safely handled for a given application. Up to 300 volts have been used in some cases, but it is preferred to keep V down to 90 volts. If it were not for this restriction, any potentiometric circuit, for practical purposes, could be made linear by making the ratio $a = R_b/R_g$ indefinitely large.

From the first limitation, when I_m is the maximum permissible gage current, the maximum allowable voltage will be

$$V_{\max} = I_m(R_b + R_g) \tag{4.30}$$

However, the second restriction of safety may require the use of a somewhat lower value.

Use with a single gage

When only one strain gage is used in the circuit, the ballast resistance will be fixed. Under this condition, R_b is constant and $\Delta R_b = 0$. Equations (4.26) and (4.29) then reduce to Eqs. (4.3) and (4.4). The latter two equations will be renumbered in this section for convenience; they are

$$\Delta E = V \frac{a}{(1+a)^2} \left[\frac{\Delta R_g}{R_g} \right] (1 - n) \tag{4.31}$$

and

$$n = \frac{1}{1 + \dfrac{1+a}{\Delta R_g/R_g}} \tag{4.32}$$

If it is more convenient to deal in terms of strain rather than in terms of unit changes in gage resistance, then Eqs. (4.5) and (4.6) may be used. Again, renumbering gives

$$\Delta E = V \frac{a}{(1+a)^2} [\varepsilon G_F](1 - n) \tag{4.33}$$

and

$$n = \frac{1}{1 + \dfrac{1 + a}{\varepsilon G_F}} \qquad (4.34)$$

Circuit efficiency

The circuit efficiency, η, of a particular circuit may be expressed as the ratio of its maximum output, per unit of strain, to the corresponding value for the constant-current circuit that produces the maximum obtainable output. Thus,

$$\text{Circuit efficiency} = \frac{(\Delta E/\varepsilon)_{max} \text{ for a given circuit}}{(\Delta E/\varepsilon) \text{ for a constant current circuit}}$$

By expressing Eq. (4.33) in terms of current, we can readily determine the efficiency of a given circuit with a single gage. Since

$$V = I(R_g + R_b) = IR_g(1 + a)$$

we can substitute this value of V into Eq. (4.33) to obtain

$$\Delta E = IR_g \varepsilon G_F \left[\frac{a}{1 + a} \right] (1 - n) \qquad (4.35)$$

From Eq. (4.35),

$$\left(\frac{\Delta E}{\varepsilon} \right)_{max} = IR_g G_F \left[\frac{a}{1 + a} \right]$$

Equation (3.14) gives the potential drop across the gage per unit of strain, for a constant-current circuit, as

$$\frac{\Delta E}{\varepsilon} = IR_g G_F \qquad (3.14)$$

and so

$$\eta = \frac{(\Delta E/\varepsilon)_{max}}{(\Delta E/\varepsilon)_{I = const}} = \frac{IR_g G_F \left(\dfrac{a}{1 + a} \right)}{IR_g G_F} = \frac{a}{1 + a} \qquad (4.36)$$

It is interesting to note that, for the potentiometric circuit, the maximum value of $\Delta E/\varepsilon$ occurs when $n = 0$, which corresponds to zero strain. For the constant-current circuit, whose output is linear with strain, $n = 0$ for all values of strain.

Since n is a function of strain, all comparisons should be made on the same basis of strain, or for the same value of n. However, the only value of n common to the constant-current circuit and all potentiometric circuits is $n = 0$. Hence this value must be employed for the previous analysis.

Gages in series

Sometimes it is desirable to obtain the average value of the strains at several different locations. This can always be done by measuring the individual strains at each location and subsequently calculating the average value. For static observations there is no problem because we merely employ a switching device to connect each gage, in turn, to the strain-indicating instrument. But for dynamic observations, in order to determine the strains at all locations simultaneously, it is necessary to have a complete channel of instrumentation for each strain gage, or to have a high-speed scanning device.

Unless we need to know the individual values of the strain at each gage, time, equipment, and effort can be saved if a reading of the average value can be obtained directly. Fortunately, we are able to do this by connecting a number of like gages in series, as shown in Fig. 4.4.

Equations (4.31) through (4.34) still apply for gages in series, but the values of the symbols will be somewhat different. In the arm of the circuit containing the strain gages, the resistance will now be made up of the sum of the resistances of the individual gages. Thus,

$$\sum R_g = xR_g$$

where $x =$ the number of like gages. The total resistance change in this arm

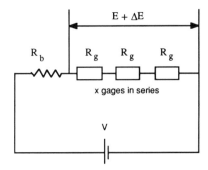

FIG. 4.4. Potentiometric circuit with strain gages in series.

consists of the sum of the changes in the individual gages. Thus,

$$\sum \Delta R_g = \Delta R_{g1} + \Delta R_{g2} + \Delta R_{g3} + \cdots + \Delta R_{gx}$$

The ratio a is expressed as

$$a = \frac{R_b}{\sum R_g} = \frac{R_b}{xR_g}$$

Equation (4.31) now becomes

$$\Delta E = V \frac{R_b/\sum R_g}{[1 + (R_b/\sum R_g)]^2} \left[\frac{\sum \Delta R_g}{\sum R_g}\right](1 - n) \tag{4.37}$$

and Eq. (4.32) is

$$n = \frac{1}{1 + \dfrac{1 + (R_b/\sum R_g)}{\sum \Delta R_g/\sum R_g}} \tag{4.38}$$

For x like gages of resistance R_g, Eq. (4.37) may be modified as follows:

$$\Delta E = V \frac{R_b/xR_g}{[1 + (R_b/xR_g)]^2} \left[\frac{\sum \Delta R_g}{xR_g}\right](1 - n)$$

or

$$\Delta E = V \frac{R_b/xR_g}{[1 + (R_b/xR_g)]^2} \left[\frac{1}{x}\left(\frac{\Delta R_{g1}}{R_g} + \frac{\Delta R_{g2}}{R_g} + \cdots + \frac{\Delta R_{gx}}{R_g}\right)\right](1 - n)$$

Using $\Delta R_g/R_g = \varepsilon G_F$, ΔE can be written in terms of strain:

$$\Delta E = V \frac{R_b/xR_g}{[1 + (R_b/xR_g)]^2} \left[\frac{1}{x}(\varepsilon_1 G_F + \varepsilon_2 G_F + \cdots + \varepsilon_x G_F)\right](1 - n)$$

or

$$\Delta E = V \frac{R_b/xR_g}{[1 + (R_b/xR_g)]^2} \left[\frac{G_F}{x}(\varepsilon_1 + \varepsilon_2 + \cdots + \varepsilon_x)\right](1 - n)$$

We see that the average strain is

$$\varepsilon_{av} = \frac{\varepsilon_1 + \varepsilon_2 + \cdots + \varepsilon_x}{x}$$

and so

$$\Delta E = V \frac{R_b/xR_g}{[1 + (R_b/xR_g)]^2} G_F \varepsilon_{av}(1 - n) \quad (4.39)$$

Equation (4.39) tells us that if n is small enough to be neglected, then $\Delta E \propto \varepsilon_{av}$.

Equation (4.38) becomes

$$n = \frac{1}{1 + \dfrac{1 + (R_b/XR_g)}{\sum \Delta R_g/xR_g}} = \frac{1}{1 + \dfrac{1 + (R_b/xR_g)}{G_F \varepsilon_{av}}} \quad (4.40)$$

Maximum applied voltage

When gages are added in series, the ratio a must be kept fixed in order to maintain a given condition of linearity. This means the ballast resistance must be stepped up proportionately. However, to obtain the greatest output per unit of strain, the gage current must be maintained at its maximum value, I_m. The applied voltage must then be increased in proportion to the total resistance in the circuit. Therefore, subject to the limitations of safety, the maximum applied voltage, V_{max}, for gages in series will be given by

$$V_{max} = I_m(R_b + \sum R_g)$$

or

$$V_{max} = I_m(R_b + xR_g) \quad (4.41)$$

Static vs. dynamic measurements

An examination of the schematic diagram Fig. 4.1 for the potentiometric circuit, and Eqs. (4.1), (4.3), and (4.5) for the incremental output ΔE, indicates that all we need for a strain measurement is to observe the change in voltage drop across the gage. This applies to either static or dynamic conditions.

This is perfectly correct. However, when we begin to look into the practical aspects of selecting a suitable measuring instrument, we run into the difficulty that ΔE may be very small relative to E. In this case, if the instrument has a readout scale suitable for measuring E, it may be entirely unsuited for the measurement of ΔE, or vice versa. We should therefore make some preliminary estimate of the approximate values of E and ΔE in order to decide upon an instrument, or readout system, which will determine ΔE (or the strain) to the desired degree of precision.

To illustrate this point, we take up two examples. The first considers a semiconductor strain gage, the second a metallic strain gage. For simplicity, consider that $a = R_b/R_g$ will be large enough so the output and linearity of the constant-current circuit are closely enough approximated for all practical purposes.

Example 4.1. The following values are given for a potentiometric circuit using a semiconductor strain gage:

Gage resistance, R_g 120 ohms
Gage factor, G_F 104
Gage current, I 20 milliamps

Solution. The voltage, E, across the gage is

$$E = IR_g = (0.020)(120) = 2.4 \text{ volts}$$

For optimum conditions (constant current) and assuming the gage has linear response,

$$\Delta E = I(R_g + \Delta R_g) - IR_g = I \Delta R_g$$

or

$$\Delta E = IR_g \frac{\Delta R_g}{R_g} = IR_g G_F \varepsilon = (0.020)(120)(104)\varepsilon = 250\varepsilon$$

When $\varepsilon = 4000$ microstrain $= 4000 \, \mu\text{in/in}$, then

$$\Delta E = 250\varepsilon = (250)(4000 \times 10^{-6}) = 1 \text{ volt}$$

We can look at how this might be represented on a DC voltmeter (or recorder) with a linear scale 5 inches long marked off in inches and subdivided in tenths of inches. This is represented graphically in Fig. 4.5. Note that the voltmeter should have high impedance to avoid loading the circuit.

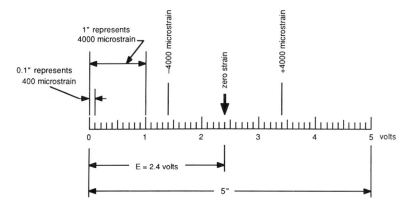

FIG. 4.5. Voltmeter readings on linear scale, semiconductor gage.

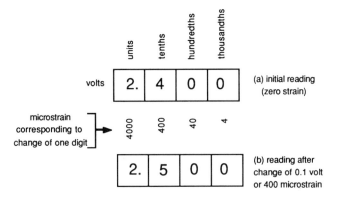

FIG. 4.6. Readings on digital voltmeter, semiconductor gage.

For zero strain the pointer will come to the position for 2.4 volts (2.4 inches along the scale). If the gage factor is positive, the pointer will move to the right when strain is applied (for positive strain) or to the left (for negative strain) by an amount of 1 inch for every 4000 microstrain, or 0.1 in for every 400 microstrain. If we can make observations of the position of the pointer to the nearest half-division on the scale, the readings will be good to the nearest 200 microstrain.

For strains of the order of 4000 to 5000 microstrain, a reading to the nearest 200 microstrain represents the nearest 4 or 5 percent, which in many cases is good enough. However, if we are dealing with magnitudes of 400 or 500 microstrain, then a reading to the nearest 200 microstrain (± 50 percent) is not nearly good enough, and so a different type of voltmeter is required.

Let us see how the same situation appears on a high-impedance, four-place digital voltmeter capable of measuring from 0 to 9.999 volts. Initially, the meter will read 2.400 volts for zero strain, as shown in Fig. 4.6. For a strain of $+400$ microstrain, ΔE will be 0.100, so the meter will read 2.500 volts, or a change of 0.100 to the nearest 1 in 100 or the nearest 1 percent.

This example indicates that for this particular semiconductor gage, operating under the stated conditions, the potentiometric circuit can be used for static or dynamic (up to the frequency limits of the instruments) strain measurements as follows:

1. With the simple meter for strains of 4000 microstrain and above.
2. With the four-place digital voltmeter for strains above 200 microstrain.

For dynamic observations at frequencies higher than those to which these meters will faithfully respond, a different system will have to be used.

Example 4.2. The following values are given for a metallic strain gage used in a potentiometric circuit:

Gage resistance, R_g 120 ohms
Gage factor, G_F 2.08
Gage current, I 20 milliamps

Solution. The voltage, E, across the gage is

$$E = IR_g = (0.020)(120) = 2.4 \text{ volts}$$

For optimum conditions (constant current),

$$\Delta E = IR_g G_F \varepsilon = (0.020)(120)(2.08)\varepsilon = 5\varepsilon$$

Thus, when $\varepsilon = 4000$ microstrain $= 4000$ µin/in,

$$\Delta E = 5\varepsilon = 5(4000 \times 10^{-6}) = 0.02 \text{ volts}$$

or

$$\Delta E = 0.01 \text{ volts for 2000 microstrain}$$

On the meter scale illustrated in Fig. 4.5, this would be equivalent to 20 000 microstrain for one minor division, as shown in Fig. 4.7. For 4000 microstrain the pointer would move 1/5 of a minor division, and for 400 microstrain there would hardly be any perceptible motion at all. Obviously, this kind of meter cannot be used with metallic gages because the ratio $\Delta E/E$ is too small.

We now consider what will happen with a four-place digital voltmeter. This is indicated in Fig. 4.8, where we observe that a reading to the nearest 200 microstrain is possible. If the instrument had five places, however, we could obtain a reading to the nearest ± 20 microstrain. This would be adequate for essentially all requirements. Therefore, for the metallic gage,

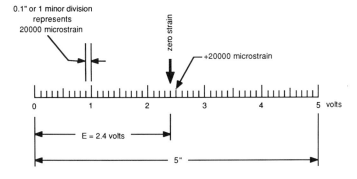

FIG. 4.7. Voltmeter readings on linear scale, metallic gage.

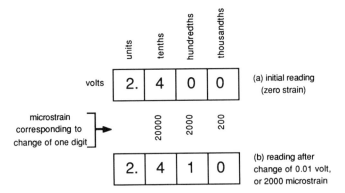

FIG. 4.8. Readings on digital voltmeter, metallic gage.

the simple voltmeter is unsuitable, but a four-place digital voltmeter might be used for relatively rough readings (± 5 percent) of high strains. A five-place digital voltmeter would be capable of indications down to ± 20 microstrain, and for all readings above 200 microstrain capable of achieving a precision of 1 percent or better.

Static strains

The preceding examples indicate that static strain measurements can be made with the potentiometric circuit provided that we have a suitable readout instrument and that the strains are sufficiently large.

When large strains are to be measured with a semiconductor strain gage possessing a high gage factor, the change in the voltage drop across the gage will be large enough, with respect to the initial value, to permit satisfactory observations with an inexpensive meter. However, when semiconductor gages are subjected to small strains, or for metallic gages, the change in potential drop across the gages will be so small, relative to the ambient value, that a comparatively expensive digital voltmeter will be required to obtain a reasonably precise strain indication.

Since static strain measurements requiring the use of an expensive digital voltmeter can be obtained equally well by other methods with less expensive instruments, the use of the potentiometric circuit for static readings is not very attractive.

Dynamic strains

For dynamic strain measurement, the simplicity of the potentiometric circuit and the convenience of using a common ground for the circuit and related components, make it very attractive, in spite of the fact that there are other cicuits that can also be used to determine time-varying strain. Where static

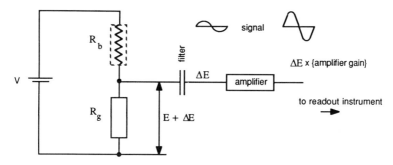

FIG. 4.9. Potentiometric circuit as applied to dynamic strain measurements.

observations are not required, the potentiometric circuit is very popular and widely used.

The usual arrangement for using a potentiometric circuit to measure dynamic strains is shown in Fig. 4.9. This includes a filter (condenser) that will eliminate the steady component, E, but will pass the dynamic part, ΔE (representing the strain) of the potential drop across the gage. When the signal, ΔE, has been isolated, it can be amplified and shown as a function of time on some readily available instrument such as a cathode-ray oscilloscope.

The result of eliminating E in order to observe ΔE is to impose the frequency limitations (both upper and lower) of the filter and the amplifier upon the final output signal. Since the filter was put into the system for the specific purpose of eliminating the steady component, E, it will also eliminate any steady strain signals.

We should note that it is the filter, which is a part of the readout apparatus, that makes the system unsuited for static strain measurement. The potentiometric circuit itself responds to both static and dynamic strains, although we can only observe dynamic strains with this particular method of obtaining the indication.

4.4. Linearity considerations

We can now examine the deviation from linearity of the signal, ΔE, with respect to strain, or $\Delta R_g/R_g$, in a potentiometric circuit with a single strain gage and a fixed ballast resistance R_b. For this purpose, it will be best to express the signal ΔE as a fraction of E, the initial potential drop across the gage. We know that

$$\Delta E = V \frac{a}{(1 + a)^2} \left[\frac{\Delta R_g}{R_g} \right] (1 - n) \qquad (4.31)$$

where

$$n = \cfrac{1}{1 + \cfrac{1+a}{\Delta R_g/R_g}} \qquad (4.32)$$

The initial potential drop across the gage is

$$E = IR_g \qquad (4.42)$$

The applied voltage across the circuit is

$$V = I(R_g + R_b) = IR_g(1 + a) = E(1 + a) \qquad (4.43)$$

Substituting the value of V given by Eq. (4.43) into Eq. (4.31) results in

$$\frac{\Delta E}{E} = \frac{a}{(1+a)}\left[\frac{\Delta R_g}{R_g}\right](1 - n) \qquad (4.44)$$

Since the circuit efficiency given by Eq. (4.36) is $\eta = a/(1 + a)$, Eq. (4.44) can be rewritten as

$$\frac{\Delta E}{E} = \eta\left[\frac{\Delta R_g}{R_g}\right](1 - n) \qquad (4.45)$$

Figure 4.10 shows $\Delta E/E$ from Eq. (4.44) plotted versus $\Delta R_g/R_g$ for various values of $a = R_b/R_g$. For comparative purposes, the linear relation for the constant-current circuit, representing the optimum, is also shown.

From Fig. 4.10, we observe the following characteristics:

1. The deviation from linearity becomes larger as $\Delta R_g/R_g$ increases.
2. For a given value of $\Delta R_g/R_g$, the deviation from linearity is less for larger values of a and approaches zero as a becomes very large.

Since it is necessary to know the applied voltage, V, that is required for a given potentiometric circuit, we can choose the gage resistance and current from which E, the voltage drop across the gage, can be computed by using Eq. (4.42). Using Eqs. (4.42) and (4.43), the ratio V/E is

$$\frac{V}{E} = \frac{E(1 + a)}{E} = 1 + a \qquad (4.46)$$

The applied voltage V can now be calculated using Eq. (4.46).

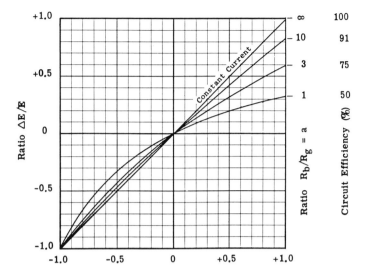

FIG. 4.10. Ratio $\Delta E/E$ as a function of $\Delta R/R$.

Linearization with variable ballast resistance

The general case. Let us now look into what may be achieved when the ratio $\varepsilon_g/\varepsilon_b$, at two locations, is known from the physical conditions which prevail, especially when either ε_g or ε_b is tension and the other is compression. By the use of one, or more, strain gages for the ballast resistance, R_b, it may be possible to produce a constant-current circuit with a constant applied voltage. To accomplish this, the total resistance in the circuit must remain constant at all times, so that, in symbols, one can write

$$R_g + \Delta R_g + R_b + \Delta R_b = \text{a constant} \quad (4.47)$$

This will take place when $\Delta R_b = -\Delta R_g$, so that the total resistance of the circuit, neglecting lead resistance, is given by the expression

$$R_g + \Delta R_g + R_b + (-\Delta R_g) = R_g + R_b = \text{constant} \quad (4.48)$$

Let us now examine the relationship between ΔR_b and $-\Delta R_g$ that will permit us to have a constant-current circuit with constant applied voltage; that is, when $\Delta R_b = -\Delta R_g$. From the basic strain gage relation given in Chapter 1, we can write

$$\frac{\Delta R/R}{\varepsilon} = \text{gage factor}$$

so that

$$\Delta R_g = R_g \varepsilon_g (G_F)_g$$

and

$$\Delta R_b = R_b \varepsilon_b (G_F)_b$$

Therefore, the ratio $-\Delta R_b/\Delta R_g$ may be written as

$$\frac{-\Delta R_b}{\Delta R_g} = -\left(\frac{R_b}{R_g}\right)\left(\frac{\varepsilon_b}{\varepsilon_g}\right)\left[\frac{(G_F)_b}{(G_F)_g}\right] \qquad (4.49)$$

Since the ratio R_b/R_g will always be positive, Eq. (4.49) indicates that either $\varepsilon_b/\varepsilon_g$ or $(G_F)_b/(G_F)_g$ must be negative in order to achieve constant-current conditions.

It should be noted that if R_g and the ballast gage, or gages, are not operating under identical lateral strain conditions, the term $(G_F)_b/(G_F)_g$ may have to be modified slightly to take into account the differences in the ratios of lateral strain to axial strain on each gage (lateral effects are discussed in a later chapter). However, this problem can be completely eliminated by selecting gages, for both R_g and R_b, which have *transverse sensitivity factors* equal to zero.

For semiconductor gages, which can be manufactured with either positive or negative gage factors, this means that these gages can be used in locations of strain of either the same or opposite sign. However, for gages with metal sensors, for which the gage factors are only positive, one is limited to the requirement that R_g must be located in the region of tensile strain while R_b must be subjected to compressive strain, or vice versa.

Let us assume for the moment that $(G_F)_b = (G_F)_g = G_F$. For the constant-current circuits the nonlinearity factor, n, becomes zero. This can be shown by referring to either Eq. (4.2) or Eq. (4.29). Thus,

$$n = \frac{1}{1 + \dfrac{1 + a}{\dfrac{\Delta R_g}{R_g} + a\dfrac{\Delta R_b}{R_b}}} = \frac{1}{1 + \dfrac{1 + a}{\dfrac{\Delta R_g}{R_g} + \dfrac{(-\Delta R_g)}{R_g}}}$$

From this,

$$n = \frac{1}{1 + \dfrac{1 + a}{0}} = \frac{1}{1 + \infty} = 0$$

If the ballast resistance consists of a strain gage, which is subjected to the appropriate amount of strain with respect to that occurring at R_g, then, even though R_b and R_g may not be equal, it is still possible, by suitable choice of relative strain, gage factor, and resistance, to produce a situation such that $\Delta R_b = -\Delta R_g$, and thereby to achieve linearity between incremental output, ΔE, and the strain, as well as the maximum signal per unit of strain.

The analysis of this is done by using Eq. (4.26), then taking $\Delta R_b = -\Delta R_g$ and $n = 0$. Thus,

$$\Delta E = V \frac{a}{(1+a)^2} \left[\frac{\Delta R_g}{R_g} - \frac{\Delta R_b}{R_b} \right] = V \frac{a}{(1+a)^2} \left[\frac{\Delta R_g}{R_g} + \frac{\Delta R_g}{aR_g} \right]$$

This reduces to

$$\Delta E = V \left(\frac{1}{1+a} \right) \frac{\Delta R_g}{R_g}$$

Since

$$V = I(R_g + R_b) = IR_g(1+a)$$

then

$$\Delta E = IR_g(1+a) \left(\frac{1}{1+a} \right) \frac{\Delta R_g}{R_g} = IR_g \frac{\Delta R_g}{R_g}$$

Using $\Delta R_g / R_g = G_F \varepsilon$, ΔE becomes

$$\Delta E = IR_g G_F \varepsilon \qquad (4.50)$$

Equation (4.50) is valid for all values of $R_b / R_g = a$.

The relation expressed in Eq. (4.50) indicates that when one has selected the gage resistance, R_g, the gage current, I, and determined the gage factor, G_F, one has established the value of output per unit strain, since

$$\frac{\Delta E}{\varepsilon} = IR_g G_F \qquad (4.51)$$

For a given gage current and resistance, I and R_g, the choice of the ballast resistance, R_b, will determine the necessary applied voltage, V, or vice versa.

Comparison with fixed ballast resistance

It is of interest to compare the output per unit strain for this variable-ballast constant-current circuit, given by Eq. (4.51), with the corresponding circuit containing a fixed ballast, R_b. For the fixed ballast, ΔE is given by Eq. (4.35):

$$\Delta E = IR_g \varepsilon G_F \left[\frac{a}{1+a} \right] (1-n) \tag{4.35}$$

From this, the output per unit strain is

$$\frac{\Delta E}{\varepsilon} = IR_g G_F \left[\frac{a}{1+a} \right] (1-n) \tag{4.52}$$

The relative output of the fixed-ballast circuit to the constant-current circuit is obtained by taking the ratio of Eq. (4.52) to Eq. (4.51). Calling this ratio R_{rel}, we obtain

$$R_{rel} = \frac{IR_g G_F \left(\dfrac{a}{1+a} \right) (1-n)}{IR_g G_F}$$

or

$$R_{rel} = \left(\frac{a}{1+a} \right) (1-n) = \eta(1-n) \tag{4.53}$$

where η = circuit efficiency.

Figure 4.10 also shows the same information in terms of the ratio of the ordinates of the curves to the corresponding ordinates of the straight line for the constant-current circuit. In this figure we see two points common to all the curves, including the straight line for constant current. The first point corresponds to the origin, or zero value for $\Delta R_g/R_g$. The second point corresponds to $\Delta R_g/R_g = -1$. This latter point represents a somewhat theoretical concept, since it corresponds to a reduction in gage resistance equal to the original value, R_g. This means that the gage resistance has been reduced to zero, which cannot be achieved in actual practice with conventional strain gages, although it might be possible with a slide-wire device under short-circuit conditions. If $\Delta R_g/R_g = -1$ could be achieved, this would mean the voltage drop across the gage had been reduced to zero and consequently $\Delta E/E = -1$.

Let us now examine two different situations involving the linearization of the potentiometric circuit with variable ballast resistance. In the first case the gage and ballast resistances will be equal, $R_b = R_g$, and in the second case they will be unequal, $R_b \neq R_g$.

Example with equal ballast and gage resistances

A usual case of this nature is represented by the use of two like gages mounted back to back on a uniform beam of rectangular cross section and subjected to simple bending, as shown in Fig. 4.11. In this particular case, $R_b = R_g$, $(G_F)_b = (G_F)_g$, and, due to the characteristics of the beam, $\varepsilon_b = -\varepsilon_g$, so that the general equation for the ratio of changes in the resistance of the ballast to that of the gage for this special case reduces to

$$\frac{\Delta R_b}{\Delta R_g} = \left(\frac{R_b}{R_g}\right)\left(\frac{\varepsilon_b}{\varepsilon_g}\right)\left[\frac{(G_F)_b}{(G_F)_g}\right] = (1)(-1)(1) = -1$$

Therefore, this ratio fulfills the requirements for a constant-current circuit (when supplied with constant voltage). Thus, the equation for ΔE, given by the general expression of Eq. (4.26), is

$$\Delta E = V\frac{a}{(1+a)^2}\left[\frac{\Delta R_g}{R_g} - \left(\frac{-\Delta R_b}{R_b}\right)\right]$$

$$= V\frac{a}{(1+a)^2}\left[\frac{\Delta R_g}{R_g} - \left(\frac{-\Delta R_g}{R_g}\right)\right] = V\left(\frac{1}{4}\right)\left(2\frac{\Delta R_g}{R_g}\right) = \frac{V}{2}\frac{\Delta R_g}{R_g}$$

Since, for this case, $V = 2IR_g$, the value of ΔE is

$$\Delta E = IR_g\left(\frac{\Delta R_g}{R_g}\right) = IR_g\varepsilon G_F$$

From this,

$$\frac{\Delta E}{\varepsilon} = IR_g G_F$$

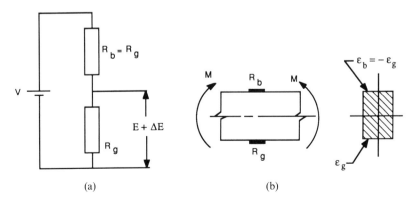

FIG. 4.11. Beam in bending with potentiometric circuit. (a) Wiring diagram. (b) Space diagram.

and

$$\left(\frac{\Delta E}{\varepsilon}\right)_{max} = I_m R_g G_F$$

This circuit is automatically temperature-compensated, as an active-dummy system, as long as the two gages can be maintained at equal temperatures.

Gages with positive and negative gage factors

Another method of achieving the same electrical characteristics is to use two gages of equal resistance having gage factors of equal magnitude but opposite sign. The gages are then installed side by side, either independently or on a common carrier. Thus, for any strain, positive or negative, the increase in resistance of one gage is just equal to the decrease in resistance of the other, and constant-current (optimum) conditions will therefore prevail. This system has the advantage that both gages are subjected to the same strain. The circuit is shown in Fig. 4.12.

For gages with metallic sensors, the concept of employing elements with positive and negative gage factors is somewhat academic, as few metals possess negative strain sensitivity. Most of those that do possess this characteristic have other properties that make them undesirable for strain gages. However, since the advent of semiconductor gages, which can be produced with an infinite variety of gage factors running from about -100 to about $+200$, this concept of linearizing the circuit has become very important. Sanchez and Wright (1) give excellent quantitative information.

Example with unequal ballast and gage resistances

The approach to linearization under the special conditions of $R_b = R_g$, while convenient, is not an essential condition. We now look into the general case to determine the relationship actually required between R_b and R_g. We know

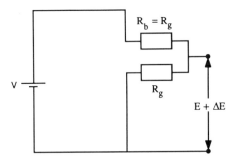

FIG. 4.12. Gages with (+) and (−) gage factors placed side by side.

THE POTENTIOMETRIC CIRCUIT

that for conditions of constant current (V assumed fixed) the total resistance in the circuit must remain constant. To achieve this, it is necessary that

$$\Delta R_g = -\Delta R_b \tag{4.25}$$

If strain gages are used for both R_b and R_g, then, since $\varepsilon = (\Delta R/R)/G_F$ for a strain gage, we can write

$$\Delta R_b = R_b \varepsilon_b (G_F)_b \tag{4.54}$$

and

$$\Delta R_g = R_g \varepsilon_g (G_F)_g \tag{4.55}$$

The subscripts b and g refer to quantities related to the ballast and the gage, across which ΔE is being measured, respectively.

In order to satisfy Eq. (4.25), we must have

$$R_b \varepsilon_b (G_F)_b = -R_g \varepsilon_g (G_F)_g \tag{4.56}$$

This means we can have any physically possible values for the six quantities in Eq. (4.56) as long as we satisfy the equation. The ratio R_b/R_g may now be expressed as

$$\frac{R_b}{R_g} = -\frac{\varepsilon_g (G_F)_g}{\varepsilon_b (G_F)_b} \tag{4.57}$$

Since R_b and R_g must always be positive for strain gages, Eq. (4.57) tells us that if the two gage factors have the same sign, then the strains must have opposite sign, or vice versa.

Theoretically, there is a wide choice of values for the quantities in Eq. (4.57). From a practical point of view, however, there are some limitations. For example, when two gages have been chosen, R_b/R_g is fixed as well as the ratio of the gage factors, which do not have to be the same. This means that the gages must be installed at locations such as $\varepsilon_g/\varepsilon_b$ will satisfy Eq. (4.57). If this can be done conveniently, we have a means of adjustment for difference between gage factors.

When $(G_F)_g = (G_F)_b$, Eq. (4.57) reduces to

$$\frac{R_b}{R_g} = -\frac{\varepsilon_g}{\varepsilon_b} \tag{4.58}$$

Equations (4.57) and (4.58) indicate the possibility of linearizing the circuit with a pair of unlike strain gages when $\varepsilon_g/\varepsilon_b \neq -1$.

Example 4.3. Two gages of unequal resistance, but of equal gage factor, are to be used on a cantilever beam, as shown in Fig. 4.13. The gages are arranged along the longitudinal axis, top and bottom, and the purpose is to design the beam cross section so the potentiometric circuit is linearized.

Solution. Since $a = R_b/R_g$, we choose $\varepsilon_b = -\varepsilon_g/a$. From Eq. (4.54) we have

$$\Delta R_b = R_b \varepsilon_b (G_F)_b = aR_g(-\varepsilon_g/a)(G_F)_b$$
$$= -R_g \varepsilon_g (G_F)_b$$

From Eq. (4.55),

$$\Delta R_g = R_g \varepsilon_g (G_F)_g$$

With $(G_F)_b = (G_F)_g$, then $\Delta R_b = -\Delta R_g$, and so the nonlinearity factor, n, is zero. The signal from the circuit, ΔE, is given by Eq. (4.1):

$$\Delta E = V \frac{a}{(1+a)^2} \left[\frac{\Delta R_g}{R_g} - \frac{\Delta R_b}{R_b} \right](1-n)$$

or

$$\Delta E = V \frac{a}{(1+a)^2} \left[\frac{\Delta R_g}{R_g} - \left(-\frac{\Delta R_g}{aR_g} \right) \right](1-n)$$

and so

$$\Delta E = \frac{V}{(1+a)} \left[\frac{\Delta R_g}{R_g} \right]$$

We also know that

$$V = I(R_b + R_g) = IR_g(1+a)$$

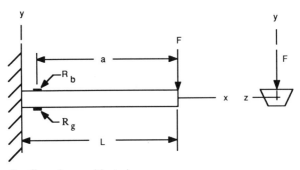

FIG. 4.13. Cantilever beam with strain gages.

Substituting this value of V into the expression for ΔE, we obtain

$$\Delta E = IR_g \frac{\Delta R_g}{R_g}$$

or

$$\Delta E = IR_g \varepsilon_g (G_F)_g$$

Thus, under these particular conditions, ΔE is a linear function of strain.

To complete the problem, R_b and R_g must be chosen, and then an appropriate beam cross section determined so that the necessary requirements will be met. Two gages readily available with the same gage factor have resistances of 120 ohms and 350 ohms. Minor differences between the gages can be expected, of course, but they will probably be less than 1 percent. For this problem, choose $R_b = 350$ ohms and place it on the top of the beam, then $R_g = 120$ ohms is placed on the bottom of the beam directly underneath R_b. This produces the value of a and the relationship between ε_b and ε_g. Determining the beam cross section is left as a homework problem.

4.5. Temperature effects

Whenever the mechanical strain varies rapidly in relation to change of temperature, it is perfectly permissible to employ a single strain gage, as shown in Fig. 4.14, and to neglect the effect of the temperature change upon the signal for the time-varying part of the strain, even though the sensing element of the strain gage may have a high response to changes in temperature. This procedure is quite appropriate when the mechanical effect takes place in such a relatively short interval of time that the accompanying change in temperature is too small to cause an appreciable error in the indication of the dynamic component of the strain. However, it is always desirable to make an estimate of the approximate error anticipated from this procedure as applied to a particular set of conditions.

Let us now look into what may be expected from a strain gage of known temperature response, as mounted on a particular material, when a given amount of strain is to be measured at some particular frequency in the presence of a varying temperature.

To illustrate the point, consider the following:

Strain magnitude	500 microstrain
Frequency	60 cycles/sec
Gage response to temperature change	150 microstrain/°F
Rate of temperature change	12°F/min

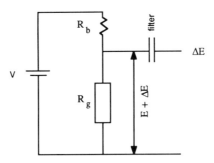

FIG. 4.14.

What will be the error, during one cycle, on the trace of a recording oscillograph?

Time for one cycle = 1/60 sec
Temperature change in 1/60 sec = (1/60)(12/60) = 1/300°F
Gage response for 1/300°F = (1/300)(150) = 0.5 microstrain
Percent error in signal = 0.5/500 = 0.1 percent

For stress analysis, in general, an error of 1 percent can safely be neglected. However, we must realize that the preceding calculation applies only to the dynamic strain signal for a single cycle, which is completed in a very short interval of time. It gives no information in regard to the gradual change in reference or zero shift.

Let us now look into the question of the length of time for recording a transient strain without exceeding a specified amount of error due to change in temperature. We assume the same numerical values used in the preceding example, and determine the time to develop a 1 percent error due to zero shift or reference change. Figure 4.15 will help to illustrate what is taking place.

Limiting error in microstrain = 1 percent of 500 = 5 microstrain
Change in temperature to develop this error = 5/150 = 1/30°F
Time for temperature change of 1/30°F to take place = (1/30)/12
$$= 1/360 \text{ min}$$
$$= 1/6 \text{ sec}$$

This neglects any errors produced by temperature changes in the lead wires and soldered joints. It also assumes that the amplifiers transmit the strain signals faithfully at these frequencies.

Slowly varying strains vs. temperature change

We have just discussed the measurement of dynamic strain without regard to temperature. We now consider the measurement of dynamic strain when the influences of temperature change cannot be neglected. Thermal effects

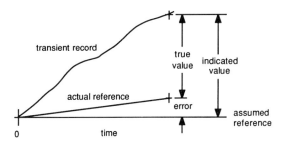

Fig. 4.15.

can produce intolerably large errors not only when the mechanical effect varies slowly in the presence of moderate variation in temperature, but also for certain combinations of high-frequency strains and violently fluctuating temperature, such as might occur in certain parts of gas turbines. However, since the important factor to consider is represented by the relative rates of change of strain and temperature with respect to time, the same methods of combating the temperature effects may be applied to either of these two conditions. Some approaches to this problem, with comments on the relative merits, will now be noted.

Self-temperature-compensated strain gage

When applicable, one of the most effective ways of minimizing the influence of temperature change is to employ a self-temperature-compensated strain gage. However, it is necessary that the environmental conditions be suitable and, in general, we have to consider the following points.

1. One may have to be satisfied with a gage of lower factor, since the usual self-temperature-compensated gages have gage factors of about 2.0 as contrasted with isoelastic gages with a gage factor of about 3.5 (or platinum–tungsten alloys and pure platinum, whose strain sensitivities are about 4 and 6, respectively).
2. The maximum temperature (or minimum temperature for cryogenic applications) at which the strain gage is expected to operate will determine whether a conventional self-temperature-compensated gage can be employed at all, or if it will be desirable to use a gage that permits adjustment to suit a particular set of conditions.
3. The average operating temperature will have to be considered so that one may select a gage with the best compensation for the operating conditions. This is due to the fact that the gage's temperature response per unit change in temperature varies with temperature. A gage possessing the best compensation on a particular material at room temperature may not be nearly as well-compensated at higher or lower

132   THE BONDED ELECTRICAL RESISTANCE STRAIN GAGE

   temperature as some other gage with a much poorer performance at
   room temperature.
4. The range of temperature variation is naturally of utmost importance,
   since it is the magnitude of the temperature change which determines
   the error from this source. If temperature variation could be reduced
   to zero, there would be no error from this cause. Zero shift might still
   occur if the operating temperature is different from ambient temperature, but this will not appear in the dynamic signal from the potentiometric circuit unless the temperature change from ambient to operating
   conditions takes place very rapidly.
5. Lead wire errors must also be considered. Even though a self-temperature-compensated strain gage may be employed with great success to
   minimize the effects of temperature changes within the gage itself, there
   still may be appreciable errors arising from the temperature changes
   occurring in the lead wires, especially if the range of temperature
   variation is large. The length of lead wire subjected to temperature
   change will, of course, be important. The common method for eliminating the error caused by changes of temperature in the lead wires
   is to use the three-wire system shown in Fig. 4.16.

   The junction, C in Fig. 4.16, between the gage and the two leads
indicated as ballast and gage must be made right at the gage. The ballast
and gage leads must be brought out to the rest of the circuit in close contact
with each other so they will be subjected to the same temperature effects.
For convenience, the common lead is usually brought out in contact with
the other two, but this is not essential, since it carries no current (high-
impedance readout instrument assumed) and does not form a part of the
actual strain gage circuit. We now look into the required relationships
between the resistances of the three leads: (a) the common lead, and (b) the
ballast and gage leads.

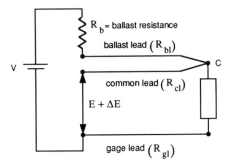

FIG. 4.16. Three-wire system for potentiometric circuit.

The common lead

The sole purpose of the common lead is to sense the change in voltage drop across the gage. Because it carries no current and is not really a part of the strain gage circuit, its resistance, and any changes thereof, will not influence the indicated output from the circuit. For convenience, this lead is frequently identical to one or both of the other two, but this is not essential, nor is it necessary that the common lead should be subjected to the same temperature conditions as the other two.

The ballast and gage leads

We now consider means of eliminating (or minimizing) errors caused by temperature changes in the other two leads. Since the ballast lead and the gage lead are in series with the ballast and gage resistances, any change of resistance produced in either of the leads by a change in temperature will appear to the readout instrument as a change in gage resistance (strain). This results in an error in the strain indication. However, by appropriate proportioning of the resistances, it is possible to make the temperature-induced errors cancel each other so that the output signal from the circuit is independent of this temperature effect.

Because the total indicated output from the circuit may be obtained by superposition of the effects in the leads on the indication from the strain gage, it is in order to consider the resistance changes in the leads by themselves and to determine under what conditions the indicated output from them can be reduced to zero.

The output of the potentiometric circuit, ΔE, is given by Eq. (4.1). It is

$$\Delta E = V \frac{a}{(1+a)^2} \left[\frac{\Delta R_g}{R_g} - \frac{\Delta R_b}{R_b} \right] (1 - n) \qquad (4.1)$$

Here we observe that the output is proportional to the difference between the unit changes in gage and ballast resistances. Thus,

$$\Delta E \propto \left[\frac{\Delta R_g}{R_g} - \frac{\Delta R_b}{R_b} \right] \qquad (4.59)$$

Equation (4.59) indicates that there will be no output from the circuit when the unit changes in gage resistance and ballast resistance are equal. Therefore, if this condition can be fulfilled when the lead resistance changes, the effect will not be seen by the readout instrument, and the circuit output will be independent of temperature effects in the leads. Thus, we see by inspection that, if the unit changes in lead resistance are equal, the temperature effects will balance out. The leads, then, should be selected so that

$$\frac{R_{bL}}{R_{gL}} = \frac{R_b}{R_g} = a$$

where R_{bL} = ballast lead resistance

R_{gL} = gage lead resistance

To prove the statement mathematically, we make two assumptions:

1. For the time being the ballast and gage resistances will remain fixed, while the resistances of the leads undergo changes. The resistance changes in the circuit will then be ΔR_{bL} on the ballast side and ΔR_{gL} on the gage side.
2. Both leads will have the same temperature coefficient, so that each will exhibit the same percentage change in resistance per unit change in temperature. Thus,

$$\Delta R_{bL} = \Delta T\, K R_{bL} \tag{4.60}$$

$$\Delta R_{gL} = \Delta T\, K R_{gL} \tag{4.61}$$

From Eq. (4.59) we see that for zero output, $\Delta E = 0$, and so

$$\frac{\Delta R_g}{R_g} - \frac{\Delta R_b}{R_b} = 0 \tag{4.62}$$

For the particular situation at hand, when lead resistance is taken into account, we can write

$$\frac{\Delta R_{gL}}{R_g + R_{gL}} - \frac{\Delta R_{bL}}{R_b + R_{bL}} = 0 \tag{4.63}$$

If we substitute the values of ΔR_{bL} and ΔR_{gL}, given by Eqs. (4.60) and (4.61), respectively, into Eq. (4.63), we have

$$\frac{\Delta T\, K R_{gL}}{R_g + R_{gL}} - \frac{\Delta T\, K R_{bL}}{R_b + R_{bL}} = 0$$

This reduces to

$$\frac{R_{gL}}{R_g + R_{gL}} = \frac{R_{bL}}{R_b + R_{bL}}$$

By inversion, we have

$$\frac{R_g}{R_{gL}} + 1 = \frac{R_b}{R_{bL}} + 1$$

which gives us

$$\frac{R_{bL}}{R_{gL}} = \frac{R_b}{R_g} = a \qquad (4.64)$$

Equation (4.64) is the necessary relation between the resistances of the leads in order to permit canceling out of the effects of the temperature changes upon them. This equation also tells us that, for the special case in which the ballast and gage resistances are equal, the two leads should be alike.

The main point of the analysis is to draw attention to the fact that it is necessary to consider the circuit parameters in order to achieve complete elimination of the errors arising from changes in temperature of the leads. For example, if we were to use identical leads with a ratio of $R_b/R_g = 5$, only 20 percent of the error would be eliminated.

Temperature compensation with two active strain gages

Whenever there is a known fixed ratio between the strains at two nearby locations on the same member (or between strains in two directions at a single location), it may be possible to achieve temperature compensation by using a second active strain gage for the ballast resistance. Successful application of this method of temperature compensation requires the following:

1. The two gages must always be maintained at equal temperatures in spite of fluctuations in the temperature of the member upon which they are mounted.
2. The temperature characteristics of the gages must be matched as closely as possible over the operating range of temperature.
3. The signs of the gage factors and the signs and relative magnitudes of the strains must be compatible.

Equal strains of opposite sign

The method of equal strains of opposite sign is best suited, although certainly not limited, to conditions involving two strains of equal magnitude but of opposite sign, such as encountered on opposite sides of a beam of rectangular cross section under the influence of bending. Under these conditions, two identical strain gages are used. The gages may be connected to the circuit by two identical pairs of leads, or by the three-lead system where identical ballast and gage leads are used. The circuits are shown in Fig. 4.17.

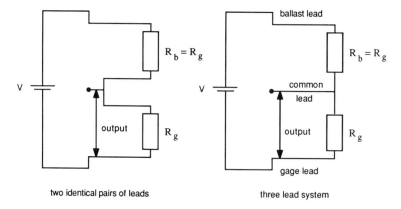

FIG. 4.17. Two identical gages.

With this arrangement, changes in temperature will cause equal resistance changes in both the ballast and gage sides of the circuit. Due to the temperature change, this results in $\Delta R_b = \Delta R_g$. Since the gages are identical, $R_b = R_g$, and so, due to temperature change,

$$\frac{\Delta R_g}{R_g} - \frac{\Delta R_b}{R_b} = 0$$

Thus, the changes in temperature will not affect the output from the circuit, as can be seen from Eq. (4.1), and it will respond only to the influence of the mechanical strains on the gages.

The properties of the circuit for this particular situation may be summarized as follows:

1. Temperature compensation, as just shown.
2. Linearity between output, ΔE, and the mechanical strain, ε. For the mechanical effect, $\Delta R_g = -\Delta R_b$, hence $n = 0$.
3. Maximum obtainable output per unit of strain. This is equal to the output per unit strain from a constant current circuit.

This comes about because $\Delta R_g = -\Delta R_b$, $n = 0$, $R_g = R_b$, and $a = 1$. Therefore, from Eq. (4.1),

$$\Delta E = V \frac{a}{(1+a)^2} \left[\frac{\Delta R_g}{R_g} - \frac{\Delta R_b}{R_b} \right] (1-n)$$

$$= V \left(\frac{1}{4}\right) \left[2 \frac{\Delta R_g}{R_g} \right] = \frac{V}{2} \left(\frac{\Delta R_g}{R_g} \right) = \frac{V}{2} \varepsilon G_F$$

For the conditions at hand,

$$V = IR_g(1 + a) = 2IR_g$$

Substituting this value of V into the preceding equation for ΔE, and then dividing both sides by ε, gives us

$$\frac{\Delta E}{\varepsilon} = IR_g G_F$$

This neglects lead resistance.

It is interesting to observe that, for any fixed temperature, this is a constant-current circuit because the resistance changes in the two gages just balance each other and the total resistance in the circuit remains constant. However, when the temperature changes, the total resistance in the circuit changes and consequently, for constant applied voltage, V, the current changes. Due to the constant voltage, though, the current change just compensates for the overall resistance change, so that even when the temperature is changing (and the current is varying) the behavior of the circuit is the same as that of a constant-current circuit at constant temperature.

Unequal strains of opposite sign

When the strains are of opposite sign, it is always possible to obtain temperature compensation by using a second active strain gage for the ballast resistance provided that

1. The gage factors of both strain gages have the same sign.
2. The two gages have identical temperature characteristics.

When *two identical gages* are used, equal changes in temperature will produce equal changes in resistance and therefore equal unit changes in resistance (because the gages are identical). This means there will be no influence on the output from the circuit. In other words, temperature compensation has been achieved. However, the output and linearity of the circuit will depend upon the ratio of the strains.

In the analysis dealing with strains of equal magnitude, but opposite sign, except for the sign of the circuit output, it was unimportant to distinguish between the ballast gage corresponding to R_b and the active gage corresponding to R_g, since both were equally active. However, when the strains to which the gages are subjected are unequal, it is necessary to specify whether R_g corresponds to the numerically larger or smaller strain.

In accordance with this requirement, we will consider that the active gage, R_g, is subjected to the numerically larger strain, and the ballast gage,

R_b, experiences the numerically smaller strain. According to this specification, the ratio of the strain on the ballast gage, R_b, to the strain on the active gage, R_g, will lie between 0, when there is no strain on the ballast gage, and -1, when the strain on the ballast gage is just equal (but opposite in sign) to that on the active gage. Under these conditions, the circuit with two identical gages will exhibit the following characteristics:

1. Temperature compensation will be achieved.
2. The sign of the output will correspond to the sign of the strain on R_g.
3. The magnitude of the output will always be larger than that available from a single gage with a fixed ballast of equal initial resistance. As the strain ratio approaches -1, the output will approach a value twice as large as this latter figure.
4. The output will always be at least 50 percent of the maximum obtainable (with a constant-current circuit), and, as the strain ratio approaches -1, will actually approach this optimum value.
5. The nonlinearity of the output, with regard to strain, will disappear as the strain ratio approaches -1.

It is interesting to note that, since none of the preceding properties depends upon any specific value of the ratio of the strains on the two gages, the circuit will work equally well for all strain ratios between 0 and -1. The actual value of the strain ratio prevailing under a particular set of conditions will, of course, be reflected, either directly or indirectly, in the calibration in terms of the strain on R_g.

When the strain ratio approaches -1, the output and linearity may be sufficiently close to the optimum (constant-current conditions) for the particular requirements at hand. If, however, the strain ratio is nearer to 0, it may be preferable to consider an alternative method of temperature compensation that will yield higher output and better linearity.

Use of more than two identical gages or two similar gages of unequal resistance

When $\varepsilon_b/\varepsilon_g$, the ratio of the strains on R_b and R_g, is small, it will be possible to improve the linearity, and to increase the output from the circuit, by making the ballast resistance, R_b, larger than the gage resistance, R_g. For best results with this approach, the ratio $\varepsilon_b/\varepsilon_g$ must be known, and its value must remain fixed.

As shown previously, the optimum conditions will prevail, for gages of equal gage factor, when the ratio

$$\frac{R_b}{R_g} = -\frac{\varepsilon_g}{\varepsilon_b} \tag{4.58}$$

This is a necessary requirement when dealing with gages whose temperature response and other characteristics must be matched as nearly as possible.

Thus, under the conditions expressed in Eq. (4.58), we will have a constant-current circuit, with all the advantages, for a *negative* strain ratio when the ratio R_b/R_g is *numerically* equal to the inverse ratio of the strains.

How will the difference in resistance between R_b and R_g influence the temperature compensation? Fortunately, this difference between R_b and R_g will not alter the temperature-compensating characteristics as long as the gages used for R_b and R_g have matched temperature characteristics, because unit changes in resistances are involved; whereas, to establish the constant-current circuit, we had to consider total resistance changes in the two arms of the circuit. We can best illustrate this by considering the situation in which we have a single gage for R_g and a number of gages, x, all identical with R_g, connected in series to form R_b which will thus be x times as large as R_g. If there is a change in temperature, then we have the following:

	For R_g	For $R_b = xR_g$
Change in resistance	ΔR_g	$x\,\Delta R_g$
Unit change in resistance	$\dfrac{\Delta R_g}{R_g}$	$\dfrac{x\,\Delta R_g}{xR_g} = \dfrac{\Delta R_g}{R_g}$

Hence, for a change in temperature, the output from the circuit, ΔE, will be zero because

$$\frac{\Delta R_g}{R_g} - \frac{\Delta R_b}{R_b} = \frac{\Delta R_g}{R_g} - \frac{\Delta R_g}{R_g} = 0$$

This means we still have temperature compensation even through $R_b > R_g$. Furthermore, since x is not required to be an integral number, the compensation may be affected either by using integral numbers of identical gages or by employing any two gages having the appropriate resistance ratio, as long as the gage factor and temperature characteristics are the same.

We may therefore summarize the characteristics of this particular arrangement of the potentiometric circuit by saying that

1. Temperature compensation can always be achieved as long as the gages corresponding to R_b and R_g have the same temperature characteristics, even though they have different resistances.

Furthermore, when $R_b/R_g = -\varepsilon_g/\varepsilon_b$, the following additional properties will be exhibited:

2. The output, ΔE, will be linear with strain.
3. The magnitude of the output will correspond to that obtainable from a constant-current circuit, i.e.,

$$\Delta E = IR_g G_F \varepsilon_g$$

It should be noted that if the resistance of the ballast gage, R_b, is made greater than that required to produce constant-current conditions, the output can be increased somewhat more, but at the expense of linearity. The writers feel it would be better to have R_g larger in the first place and to keep R_b in the proper proportion to produce constant-current conditions, and hence linearity, between strain and circuit output.

Strains of one sign only

When the strains at all locations where gages can be mounted are of one sign only (either all tension or all compression), it is still possible to achieve temperature compensation by using an active ballast. However, unless the strain on the ballast gages is relatively small, this is not an attractive method for eliminating the temperature effect, for the following reasons:

1. When the gages corresponding to R_b and R_g both have the same sign for the gage factor, the result of strains of like sign acting on them will be to reduce the circuit output below that available from R_g acting alone in conjunction with a fixed or inactive ballast. In certain types of transducers, however, the advantage of temperature compensation more than offsets a slight loss in sensitivity.
2. With changes of resistance of the same sign in both sides of the circuit, it is impossible to achieve constant-current conditions, and the corresponding linearity between strain and the output. Actually, the deviation from linearity will be greater than that for fixed R_b and variable R_g.
3. Although the use of gages with positive and negative gage factors may be very attractive for increasing the circuit output at constant temperature, there may be considerable difficulty if the temperature changes. The magnitude of the difficulty will depend upon the precision desired and the magnitude of the temperature fluctuation.

    Since the temperature response of a gage depends upon the effect of temperature upon the gage factor, the temperature coefficient of resistance of the material of the sensing element, and the difference in coefficients of expansion of the sensing element and the material upon which it is mounted, it is very difficult to make all these effects balance out, except at one or two temperature levels, because they are actually nonlinear functions of temperature.

    As an example, let us imagine that the coefficients of expansion of the sensing elements of both gages are the same but different from that of the material upon which they are mounted. If there is a change in temperature, both gages will feel an expansion or contraction. However, since this effect will be indistinguishable from similar strains produced by direct mechanical action, it will appear in the form of an output from the circuit unless the temperature change also produces compensating changes in the resistances of the sensing elements.

We might think of a special case in which temperature compensation might be achieved as follows:

1. Imagine that the coefficient of expansion of both gages is matched with the material upon which they are mounted. Under these conditions, when there is a change in temperature, the sensing elements will move freely with the base material and no resistance change takes place as a result of differential expansion or contraction.
2. If the temperature coefficients of the two sensing elements are the same, then equal unit changes in resistance will appear in both sides of the circuit and there will be no effect on the output.
3. The effect of temperature on the value of the gage factors should be negligible (or compensating), since otherwise a change in temperature, although producing no direct output from the circuit, may have the inconvenience of changing the calibration.

Further details in regard to temperature effects and methods of allowing for them are given by Hines and Weymouth (2), and Wnuk (3).

4.6. Calibration

In order to determine what the signal from the circuit represents in terms of strain, some type of calibration is required (4). There are a number of different ways in which this can be done, and each method will have some special advantages with respect to some particular application. For the purpose here, however, one usual method will be discussed in some detail.

We will consider the shunt calibration method as applied to a single gage with a fixed ballast. This involves the simulation of a change in gage resistance by the introduction of a large known resistance in parallel with the gage, and calculation of the equivalent strain which corresponds to the circuit output. Theoretically, we should be able to employ a series resistance, but in general this will be so small that variations in contact resistance at switches are likely to impair the accuracy. From a practical point of view, it is better to use a large parallel resistance because the variations in contact resistance at switches will then be reltively insignificant.

Figure 4.18 represents a potentiometric circuit with a fixed ballast resistance, R_b, a strain gage, R_g, a calibrating resistor, R_c, and a switch, S, to bring R_c into the circuit. Although not shown in the diagram, there should be some means (chopper) of opening and closing the switch, S, at approximately the same frequency as the strain gage signal.

Let us consider that, for the moment, the strain gage is at rest under zero strain. When the switch, S, is closed, the readout device will sense a change in resistance, ΔR_c, which produces a change in voltage, ΔE_c, at the output terminals. This change in resistance, ΔR_c, corresponds to the difference between the gage resistance, R_g, and the combined effect, R_{cg}, of

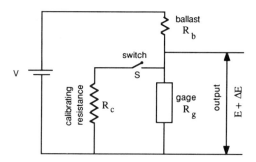

FIG. 4.18. Potentiometric circuit with calibration resistor.

R_g and R_c in parallel. For the parallel resistances,

$$\frac{1}{R_{cg}} = \frac{1}{R_g} + \frac{1}{R_c} = \frac{R_c + R_g}{R_g R_c}$$

This gives R_{cg} as

$$R_{cg} = \frac{R_g R_c}{R_c + R_g} \tag{4.65}$$

The change in resistance, ΔR_c, is

$$\Delta R_c = R_g - R_{cg} = R_g - \frac{R_g R_c}{R_c + R_g} \tag{4.66}$$

Since strain is represented by unit change in resistance, we now divide both sides of Eq. (4.66) by R_g, so that

$$\frac{\Delta R_c}{R_g} = 1 - \frac{R_c}{R_c + R_g} = \frac{R_g}{R_c + R_g} \tag{4.67}$$

Solving for R_c, the calibrating resistance, gives

$$R_c = R_g \left[\frac{1}{\Delta R_c / R_g} - 1 \right] \tag{4.68}$$

Since the readout device cannot determine whether the change in resistance that it senses comes from strain in the gage or the introduction of the parallel calibrating resistance, as far as it is concerned $\Delta R_c/R_g$

represents

$$\frac{\Delta R_g}{R_g} = \varepsilon G_F$$

or

$$\varepsilon = \left(\frac{\Delta R_g}{R_g}\right)\bigg/ G_F$$

Substituting this equivalent value of strain into Eq. (4.68), we may now write the expression for the size of the calibrating resistance to represent a given strain as

$$R_c = R_g\left[\frac{1}{\varepsilon G_F} - 1\right] \qquad (4.69)$$

In some cases, however, it will be necessary to determine the strain simulated by a calibrating resistance of some arbitrary or predetermined value. Under these conditions Eq. (4.69) is used to solve for ε, which gives

$$\varepsilon = \frac{R_g}{(R_g + R_c)G_F} \qquad (4.70)$$

From Eq. (4.70) we can compute the strain simulated by a calibrating resistance of some particular magnitude.

Special case for uniaxial stress

For uniaxial stress conditions, when the gage axis is lined up with the stress axis, Eqs. (4.69) and (4.70) can conveniently be expressed directly in terms of stress. This is due to the fact that for uniaxial stress

$$\varepsilon = \frac{\sigma}{E}$$

where σ = stress

E = modulus of elasticity

Substituting this value of ε into Eq. (4.69) produces

$$R_c = R_g\left[\frac{E}{\sigma G_F} - 1\right] \qquad (4.71)$$

Thus,

$$\sigma = \frac{ER_g}{(R_g + R_c)G_F} \tag{4.72}$$

Problems

4.1. Verify Eq. (4.21).

4.2. Verify Eq. (4.29).

4.3. In Eq. (4.29), let $\Delta R_b = 0$ and $\Delta R_g/R_g = G_F \varepsilon$, where $G_F = 2.0$. Plot the nonlinearity factor, n, vs. the strain, ε, on log–log paper for values of $a = 1, 2, 5$, and 9 in order to show the dependency of n on the strain level, ε.

4.4. Using the data in Problem 4.3, on semilog paper plot n vs. the ratio a in order to show the dependency of n on the ratio a.

4.5. The following data are available for the potentiometric circuit: $V = 35$ volts, $R_g = 120$ ohms, $R_b = 1080$ ohms, and $G_F = 2.0$. Determine the circuit efficiency and the strain that will result in a 2 percent error in ΔE. Will the current in the circuit exceed 0.03 amperes?

4.6. Redo Problem 4.5 if the strain gage is changed to $R_g = 350$ ohms, all other factors remaining the same.

4.7. If the voltage in Problem 4.6 is increased so that the circuit current is 0.03 amperes, will the error be affected If the error is not to exceed 2 percent, compute the change in ΔE.

4.8. A steel tension link of rectangular cross section is subjected to an axial load that varies between 0 and 33 750 lb. The load is offset from the longitudinal axis of the bar, as shown in Fig. 4.19. Four gages, arranged along the longitudinal axis, are bonded at the center of each face of the bar and wired in series to form the potentiometric circuit of Fig. 4.4. If $R_g = 350$ ohms, $G_F = 2.5$, $R_b = 7000$ ohms, $V = 60$ volts, and $E = 30 \times 10^6$ psi,

(a) Determine the strain on each gage.
(b) Determine the maximum value of ΔE considering $n = 0$.
(c) Compute the nonlinearity term, n. Is it worth considering?

4.9. A steel beam is subjected to a bending moment of $M = 12\,500$ in-lb and a tensile force of $F = 18\,000$ lb, as shown in Fig. 4.20. Using $E = 30 \times 10^6$ psi, $G_F = 2.08$, $R_b = R_g = 120$ ohms, and $V = 25$ volts,

(a) Determine the strain on each gage.
(b) Determine the change in resistance of R_g and R_b.
(c) Determine the value of ΔE.
(d) Is this a constant-current circuit?
(e) If the load F is eliminated, will this be a constant-current circuit?

4.10. Using a T cross section, determine suitable dimensions for the beam in Example 4.3 if $F = 150$ lb, $a = 15$ in, and $L = 18$ in.

THE POTENTIOMETRIC CIRCUIT

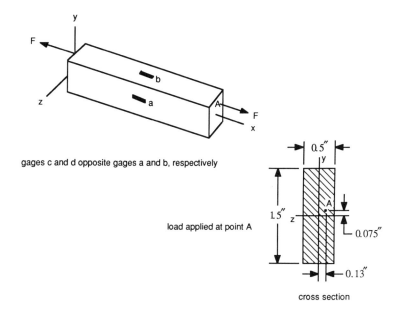

gages c and d opposite gages a and b, respectively

load applied at point A

cross section

FIG. 4.19.

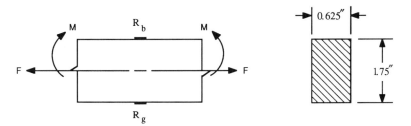

FIG. 4.20.

REFERENCES

1. Sanchez, J. C. and W. V. Wright, "Recent Developments in Flexible Silicon Strain Gages," in *Semiconductor and Conventional Strain Gages*, edited by Mills Dean III and Richard D. Douglas, New York, Academic Press, 1962, pp. 307–345.
2. Hines, Frank F. and Leon J. Weymouth, "Practical Aspects of Temperature Effects on Resistance Strain Gages," in *Semiconductor and Conventional Strain Gages*, edited by Mills Dean III and Richard D. Douglas, New York, Academic Press, 1962, pp. 143–168.
3. Wnuk, S. P. Jr., "Strain Gages for Cryogenics," *ISA Journal*, Vol. 11, No. 5, May 1964, pp. 67–71. *Reprinted by permission. Copyright © Instrument Society of America 1964. From Strain Gages of Cryogenics*, S. P. Wnuk, Jr.
4. Geldmacher, R. C., "Ballast Circuit Design," *SESA Proceedings*, Vol. XII, No. 1, 1954, pp. 27–38.

5

WHEATSTONE BRIDGE

5.1. Introduction

Although the potentiometric circuit, shown in Fig. 5.1 and discussed in Chapter 4, possesses many desirable characteristics for use with strain gages, nevertheless, it does present the difficulty that the strain signal, ΔE, must either be measured in combination with a very much larger voltage, E, or first isolated and then measured by itself.

When ΔE is determined by measuring the combined quantity $E + \Delta E$, and noting the change from a comparable indication of E, one runs into the problem that if ΔE is relatively small with respect to E, a small error in the observation of either E or $E + \Delta E$ may produce an excessively large and intolerable percentage error in the comparatively small change, ΔE. The importance of the reading error, of course, will depend upon the relative magnitudes of E and ΔE, and the instruments available for making the observations.

For large signals from semiconductor gages, there may be no problem in obtaining sufficiently precise values of ΔE from readings of $E + \Delta E$. In general, however, for strain gage work it will be preferable to isolate ΔE and measure this quantity entirely by itself. This approach is much more direct, since it involves making an observation immediately upon the quantity that is the real measure of the strain.

For dynamic strains, ΔE can be isolated by using a filter (condenser)

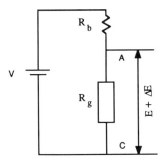

FIG. 5.1. Potentiometric circuit.

that will block the steady component E but still transmit the time-varying signal, ΔE. For static strains this system will not work because the filter will not transmit any constant value of ΔE. Therefore, another approach must be sought.

The real difficulty encountered in making static strain measurements with the potentiometric circuit is caused by the wide divergence in the relative magnitudes between E and ΔE, and so we now look into the possibility of overcoming this problem. We can achieve our objective either by increasing ΔE with respect to E (using semiconductor gages with large strains) or by reducing E relative to ΔE. The latter approach must be followed to develop a method for metallic gages, or for semiconductors when the strain level is low. If a scheme for reducing E relative to ΔE can be worked out, and if it can be carried to the ultimate so that E is finally reduced to zero, then we have achieved a means of isolating ΔE so that the strain signal can be measured directly by itself. The ideas just expressed are presented graphically in Fig. 5.2, which indicates qualitative relations between E and ΔE.

Figures 5.3 and 5.4 show various stages in the development of a method for reducing E to zero, and thereby facilitating the direct measurement of the strain signal, ΔE, by itself. The fundamental idea is to change the reference level from which $E + \Delta E$ is measured so that the numerical value of E will be reduced and, ultimately, brought to zero.

Instead of measuring $E + \Delta E$ as the potential drop across the gage, shown as points A and C in Fig. 5.1, we will establish a reference other than C with a steady potential level much nearer, or perhaps equal, to that prevailing at A. If an auxiliary battery with voltage V_1 (which is slightly less than the voltage drop across the gage) is introduced and connected as shown in Fig. 5.3a, then by measuring the voltage drop across terminals A and B

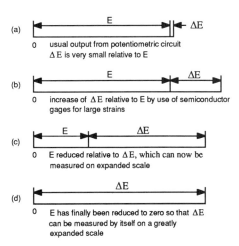

FIG. 5.2. Qualitative relations between E and ΔE.

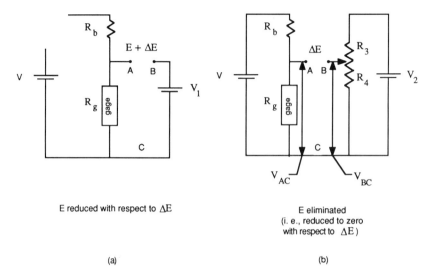

(a) E reduced with respect to ΔE

(b) E eliminated (i. e., reduced to zero with respect to ΔE)

FIG. 5.3. Methods of reducing E with respect to ΔE.

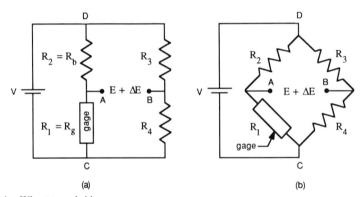

(a)   (b)

FIG. 5.4. Wheatstone bridge.

instead of across A and C, the steady-state component, E, will be reduced by an amount equal to V_1. Even though this is a move in the right direction, since ΔE may be exceedingly small, especially for metallic gages, the introduction of the auxiliary battery may not reduce E sufficiently with respect to ΔE. Consequently, we strive for something better.

An improved technique is shown in Fig. 5.3b. Here an auxiliary battery with voltage, V_2, that is greater than the potential drop across the gage, is connected to a potentiometer with which we can vary the voltage drop, V_{BC}, between points B and C. Thus, we can now control the voltage at terminal B. Furthermore, since, for zero strain conditions, the difference in potential between A and B represents the steady-state component, E, of the output, control of the voltage at B also provides control of E. Therefore, by adjusting the potentiometer until there is no potential difference between A and B, we can make E equal to zero and thereby eliminate it from the output. When this has been done, any change in the gage's resistance will produce a change in potential at terminal A. This change is equal to ΔE, which can be measured directly, and by itself, against the reference voltage at terminal B.

What has actually been accomplished by making the initial adjustment, which brings the potential difference across A–B to zero, is to make the voltage drop, V_{BC}, from B to C equal to the potential drop across the gage, V_{AC}. Then, when terminal B is used for reference we have, in effect, changed the level of reference voltage from the level at C to the original level at A. We can now read ΔE independently (because we are using E as the reference level of voltage).

Theoretically, this method provides us with a direct means of observing the strain signal, ΔE, for both static and dynamic strains. However, for practical reasons this procedure is not convenient to use (especially for long-time static readings) because it is subject to errors arising from differences in rate of decay (voltage drop) between the two batteries. Fortunately, this difficulty can be eliminated very easily.

Let us now see how the difficulty involved with the second battery can be overcome. The only requirement in regard to the voltage, V_2, of the auxiliary battery is that it should be larger than the potential drop across the gage. Since the presence of R_b requires that the voltage, V, must also be larger than the potential drop across the gage, it appears that a single battery can be used to power both circuits, which can be connected together as shown in Fig. 5.4a. This is the well-known Wheatstone bridge, which is shown in more conventional form in Fig. 5.4b. When the terminals A and B of the Wheatstone bridge are brought to the same potential, the bridge is said to be balanced ($E = 0$). However, since it is quite possible that the bridge might be initially unbalanced, the output indicated in Fig. 5.4 has been shown as $E + \Delta E$, where E represents the potential difference between A and B resulting from initial unbalance, and ΔE corresponds to the change in output due to the change in gage resistance.

5.2. Elementary bridge equations

As with the potentiometric circuit, the Wheatstone bridge circuit equations, and some discussion of them, will be presented first. Figure 5.5 shows an idealized Wheatstone bridge in which all four arms may contain strain gages. The bridge is supplied with a constant voltage, V, (from a source of zero internal resistance) at terminals D and C. The output voltage across A–B is measured with an instrument of infinite impedance which draws no current. Although this represents a theoretical situation, nevertheless, there are times when it can be very closely approximated.

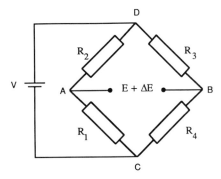

FIG. 5.5. Idealized Wheatstone bridge.

For an initially balanced bridge, $E = 0$. Thus,

$$R_1 R_3 = R_2 R_4 \tag{5.1}$$

From this

$$\frac{R_2}{R_1} = \frac{R_3}{R_4} = a \tag{5.2}$$

When the gages are strained, the incremental bridge output is given as

$$\Delta E_0 = V \frac{a}{(1+a)^2} \left[\frac{\Delta R_1}{R_1} - \frac{\Delta R_2}{R_2} + \frac{\Delta R_3}{R_3} - \frac{\Delta R_4}{R_4} \right] (1 - n) \tag{5.3}$$

where $n =$ the nonlinearity factor which, for this case, is very closely approximated by

$$n = \frac{1}{1 + \dfrac{a+1}{\dfrac{\Delta R_1}{R_1} + a \dfrac{\Delta R_2}{R_2} + a \dfrac{\Delta R_3}{R_3} + \dfrac{\Delta R_4}{R_4}}} \tag{5.4}$$

When the gages are all alike and of initial resistance R_g, then

$$R_1 = R_2 = R_3 = R_4 = R_g \tag{5.5}$$

For this case $a = 1$ and Eqs. (5.3) and (5.4) simplify to

$$\Delta E_0 = \frac{V}{4} \left[\frac{\Delta R_1 - \Delta R_2 + \Delta R_3 - \Delta R_4}{R_g} \right] (1 - n) \tag{5.6}$$

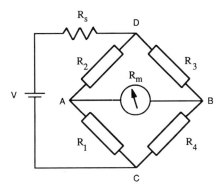

FIG. 5.6. Generalized bridge.

and

$$n = \frac{1}{1 + \dfrac{2R_g}{\Delta R_1 + \Delta R_2 + \Delta R_3 + \Delta R_4}} \quad (5.7)$$

Equations (5.1) through (5.7) are the elementary bridge equations. For a more general concept of the Wheatstone bridge, we examine Fig. 5.6. Here allowance is made for the internal resistance, R_s, of the power supply and the fact that the meter resistance, R_m, may not be infinite. Since R_s is treated merely as a resistance in series with the bridge, this might include resistance of leads, a voltage control, or any other resistance, including the actual internal resistance of the power supply itself. When the resistance in series with the power supply and the resistance of the meter (or galvanometer) are taken into account, the expression for the incremental output, ΔE_0, from an initial condition of balance is given by

$$\Delta E_0 = V \left[\frac{1}{1 + \dfrac{R_s}{R_{BI}}} \right] \left[\frac{1}{1 + \dfrac{R_{BO}}{R_m}} \right] \left[\frac{a}{(1+a)^2} \right]$$
$$\times \left[\frac{\Delta R_1}{R_1} - \frac{\Delta R_2}{R_2} + \frac{\Delta R_3}{R_3} - \frac{\Delta R_4}{R_4} \right] (1 - n) \quad (5.8)$$

where R_{BI} = bridge input resistance as seen between terminals D and C (not including R_s)

R_{BO} = bridge output resistance as seen by the meter across terminals A and B (this includes the series resistance R_s)

Analysis of the circuit shows that for the unbalanced bridge,

$$R_{BI} = \frac{R_m(A+B)(C+D) + AB(C+D) + CD(A+B)}{R_m(A+B+C+D) + (A+D)(B+C)} \quad (5.9)$$

and

$$R_{BO} = \frac{R_s(B+C)(A+D) + BC(A+D) + AD(B+C)}{R_s(A+B+C+D) + (A+B)(C+D)} \quad (5.10)$$

where
$$A = R_1 + \Delta R_1 \quad (5.11)$$
$$B = R_2 + \Delta R_2 \quad (5.12)$$
$$C = R_3 + \Delta R_3 \quad (5.13)$$
$$D = R_4 + \Delta R_4 \quad (5.14)$$

If each arm of the bridge now contains one of four identical strain gages whose initial resistance is R_g, as given by Eq. (5.5), and the bridge is initially balanced, then, under this special condition,

$$R_{BI} = R_{BO} = R_g \quad (5.15)$$

This means that when the changes in gage resistance are small, as usually occurs with metallic gages, we can write the expression for ΔE_0 to a very good approximation as

$$\Delta E_0 = V \left[\frac{1}{1 + \dfrac{R_s}{R_g}} \right] \left[\frac{1}{1 + \dfrac{R_g}{R_m}} \right] \left[\frac{\Delta R_1 - \Delta R_2 + \Delta R_3 - \Delta R_4}{4 R_g} \right] (1 - n) \quad (5.16)$$

From Eq. (5.16), we see that the maximum output, $(\Delta E)_{max}$, will occur when $R_s = 0$ and $R_m = \infty$.

Galvanometer current

For the unbalanced bridge, it can be shown that the current through the meter (galvanometer) can be expressed as

$$I_{galvo} = \frac{V(AC - BD)}{\begin{bmatrix} R_m R_s(A+B+C+D) + R_m(A+B)(C+D) \\ + R_s(A+D)(B+C) + AB(C+D) + CD(A+B) \end{bmatrix}} \quad (5.17)$$

When $R_s = 0$, Eq. (5.17) reduces to

$$I_{\text{galvo}} = \frac{V(AC - BD)}{[R_m(A + B)(C + D) + AB(C + D) + CD(A + B)]} \quad (5.18)$$

Ways of using the Wheatstone bridge

There are three different ways in which the Wheatstone bridge is usually employed to obtain indications from strain gages: the null balance system; the unbalance system; the reference system.

The null balance system. In this system there is provision for adjusting the resistance in one or more arms of the bridge to compensate for the effect of change in gage resistance. The bridge is brought to initial balance by manipulating the adjustable resistances. Then, after the gages have been subjected to strain, a further adjustment of the variable resistances is made to restore the condition of balance. The amount of the adjustment required to reestablish the balance is a measure of the change in gage resistance, or the strain. This method has the advantage of giving an indication independent of variations in bridge supply voltage and, under certain conditions, it will eliminate some nonlinearities.

On the other hand, its use is limited to static, or exceedingly low-frequency dynamic, observations. This is due to the fact that it takes appreciable time to rebalance the bridge and, in consequence, it is impossible to follow rapidly fluctuating changes. Furthermore, depending upon the manner in which the rebalancing of the bridge is accomplished, the readout may be a nonlinear quantity requiring a conversion chart for determining strain. In the event that all four bridge arms contain strain gages, it may be impossible to avoid an appreciable amount of desensitization (loss in effective gage factor) caused by the balancing network. If a direct calibration can be made, this should not present a serious difficulty.

The unbalance system. The bridge is directly connected to the readout device, which may be a galvanometer, a cathode-ray oscilloscope, or some type of recording oscillograph producing a record of the strain signal (usually, although not necessarily) as a function of time. This system has the advantage that it is suited for both static and dynamic observations. However, since its indication is directly proportional to the applied voltage, a stable power supply is required. For measurements conducted over long periods of time, this is particularly important.

The reference system. There are certain instruments combining the advantages of both the null balance and the unbalance systems, and at the same time eliminating some of the undesirable features of each procedure. These instruments incorporate an internal bridge that is separate from the strain gage bridge but powered from the same source. Provision is made for

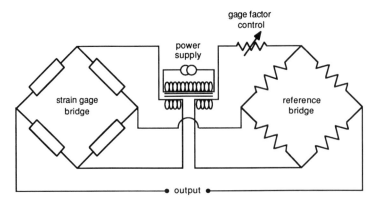

FIG. 5.7. Schematic diagram of reference bridge with gage factor control.

adjustment of the resistances in the internal bridge so that its output can be set at some fixed value or controlled to match the output of the external bridge. It is thus possible to employ the internal bridge as a reference from which to establish the strain indication. A schematic diagram of a circuit with a reference bridge is shown in Fig. 5.7.

When the output of the internal bridge is calibrated, then, by comparison, one is able to evaluate the indication from the external bridge. The comparison and evaluation can be carried out by one or the other of the following two arrangements.

1. *The null balance reference bridge.* With this system, the output of the reference bridge is initially adjusted to cancel the output from the strain gage bridge. Any subsequent change in output from the strain gage bridge will require a readjustment of the reference bridge in order to restore equality of outputs from the two bridges. The amount of the readjustment of the reference bridge (in order to restore equality of outputs from the two bridges) is a measure of the change in strain, or other indications, from the strain gage bridge (1).
2. *The unbalance reference bridge.* The reference bridge in this system is initially adjusted so that its output just cancels, or balances, the output from the strain gage bridge. Any subsequent change in the strain gage bridge will then be indicated by an unbalance or difference in output between the two bridges. This unbalance is a measure of the change which has taken place in the strain gage bridge. Calibration of this signal can be achieved by making a known change in the reference bridge, and then comparing the signal from the strain gage bridge with the signal produced by the change in the reference bridge (2).

With both the null balance and unbalance reference bridges, we are merely comparing the output of the strain gage bridge with a calibrated reference. From this, then, the indication from the strain gage is evaluated.

The null balance reference system is suited to static and low-frequency dynamic conditions. The unbalance reference system can be used for both static and dynamic observations.

Some of the advantages of the reference bridge methods are as follows:

1. The strain indication is independent of the power supply voltage that is connected to the two bridges. In the case of the unbalance reference method, it is necessary that the calibration indication should be made with the same applied voltage as that employed for the strain indication.
2. The system lends itself conveniently to the inclusion of a gage factor adjustment.
3. The reference bridge can be set up and calibrated, then left alone.
4. The strain gage bridge can be closed, and, since its output is compared with that from the reference bridge, it is not necessary to provide further adjustment by adding series or parallel resistance in any one of the arms. This is a great convenience when all four arms of the bridge contain strain gages, because it overcomes the necessity for including trimming resistances to achieve initial balance.

Summary of properties of the Wheatstone bridge

1. For strain gage applications, probably the most attractive characteristic of the Wheatstone bridge is its ability to provide the means for measuring both static and dynamic strains, or combinations thereof, conveniently.
2. In comparison with the potentiometric circuit, the Wheatstone bridge is more elaborate. This is to be expected since it actually contains two potentiometric circuits connected together. Furthermore, due to the nature of the Wheatstone bridge, a measuring system employing it cannot have all components connected to a common ground. If one side of the input is grounded, then the output must be floating, or vice versa. This requires complete isolation of one part of the system relative to the remainder.
3. Temperature compensation. Under suitable conditions the Wheatstone bridge will provide an electrical method for temperature compensation of strain gages as well as many other convenient properties of the potentiometric circuit. One will observe that for a single active strain gage, the equations representing the output from the Wheatstone bridge reduce to exactly the same form as the corresponding expressions for the potentiometric circuit.
4. Optimum bridge ratio (for a single gage). When the Wheatstone bridge is to be used with a single gage, we have the opportunity of making an arbitrary decision regarding the choice of the bridge ratio, which is the ratio of the resistances in the half bridge connected across the power supply and containing the strain gage. This ratio is represented by the symbol a in Eqs. (5.3) and (5.8). Examination of the relations expressed

in the equations for the bridge output indicates that the value of the bridge ratio, a, necessary for optimum output per unit change in resistance (or per unit change in strain) will depend upon the character of the power supply as follows:

Character of power supply	Value of ratio a for maximum output per unit strain
a. Fixed voltage (V = constant)	$a = 1$
b. Variable voltage (max. gage current = constant)	$a = \infty$ [1]

For the fixed-voltage power supply, it can be proven analytically that for optimum output, $a = 1$. However, for the variable-voltage power supply, since theory predicts the optimum bridge output for the largest possible value of a, we will have to proceed from practical considerations in order to establish a definite and convenient value for the bridge ratio.

We commence by selecting a strain gage of resistance R_g and deciding upon the maximum permissible gage current and the maximum voltage that can be safely employed. From the maximum permissible current and voltage, the total resistance in the half bridge, $R_1 + R_2$, can be computed. If R_g corresponds to R_1, the bridge ratio, $a = R_2/R_1$, can be calculated.

Since approximately 90 percent of the ultimate output can be achieved with a bridge ratio of 10, there is little incentive to make the bridge ratio, a, larger than 10 because the required increase in applied voltage goes up much faster than the gain in output. Many investigators prefer to use a value of about 5 for the bridge ratio, since this will yield an output of about 85 percent of the ultimate. Correspondingly, the voltage required is only about three times as great as that needed when $a = 1$.

If a carrier system is employed, the power requirements will usually necessitate keeping the value of the bridge ratio near one (3).

5. Computing characteristics. Equations (5.8) and (5.16) also indicate that, by appropriate control of the parameters, the Wheatstone bridge can be employed to perform certain additions, subtractions, multiplications, and divisions. The relationships can be summarized by the following statements:

Subject to the possibility of some nonlinearities, the bridge output, ΔE, will be:

a. Directly proportional to the applied voltage.
b. Directly proportional to the sums and differences of the unit changes in the resistances in the four arms.
c. Directly proportional to the product of the applied voltage and the net unit change in resistance of all four arms.

[1] Practical considerations will usually place an upper limit of about 10 on the maximum usable value of the bridge ratio.

d. Inversely proportional to functions involving resistance in series with the bridge and power supply, and the resistance of the instrument which is used to determine the output voltage or current. Stein (4) gives a detailed discussion.

5.3. Derivation of elementary bridge equations

Figure 5.8 shows an elementary and idealized Wheatstone bridge in which all four arms may contain strain gages. In the succeeding analysis, the following assumptions have been made:

1. The bridge is supplied with a constant voltage, V, from a source whose impedance is negligible.
2. The resistances of the leads from the power supply to the bridge, and of all the leads connecting the internal components of the bridge, can be neglected.
3. The output from the bridge is represented by the difference in voltage between terminals A and B. The instrument used to measure the output has infinite impedance and draws no current.

The bridge output is the difference in voltage between A and B, which is also the voltage drop from A to C minus the voltage drop from B to C. According to assumption (3), no current flows from A to B; thus, current I_1 flows through R_1 and R_2, while current I_2 flows through R_3 and R_4. Since the voltage around each loop must sum to zero, we can write

$$V - I_1(R_1 + R_2) = 0 \qquad (5.19)$$

$$V - I_2(R_3 + R_4) = 0 \qquad (5.20)$$

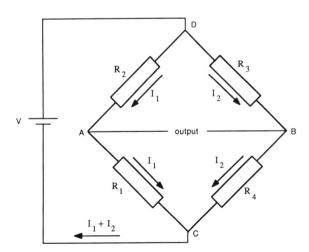

FIG. 5.8. Elementary Wheatstone bridge.

From these

$$I_1 = \frac{V}{R_1 + R_2} \tag{5.21}$$

$$I_2 = \frac{V}{R_3 + R_4} \tag{5.22}$$

Let us now consider the situation in which all four arms of the bridge contain strain gages whose initial resistances are R_1, R_2, R_3, and R_4, as shown in Fig. 5.8. The corresponding initial output, E, is then

$$E = E_{AC} - E_{BC} = I_1 R_1 - I_2 R_4 \tag{5.23}$$

Substituting the values of I_1 and I_2, given by Eqs. (5.21) and (5.22), respectively, into Eq. (5.23), we have

$$E = V \frac{R_1}{R_1 + R_2} - V \frac{R_4}{R_3 + R_4} \tag{5.24}$$

or

$$E = V \frac{R_1 R_3 - R_2 R_4}{(R_1 + R_2)(R_3 + R_4)} \tag{5.25}$$

If each gage undergoes a change in resistance such that $R_1 \to R_1 + \Delta R_1$, $R_2 \to R_2 + \Delta R_2$, $R_3 \to R_3 + \Delta R_3$, and $R_4 \to R_4 + \Delta R_4$, then the bridge output will change from E to $E + \Delta E$. Equation (5.25) can be written, using the new resistances and new output, as

$$E + \Delta E = V \frac{(R_1 + \Delta R_1)(R_3 + \Delta R_3) - (R_2 + \Delta R_2)(R_4 + \Delta R_4)}{(R_1 + \Delta R_1 + R_2 + \Delta R_2)(R_3 + \Delta R_3 + R_4 + \Delta R_4)} \tag{5.26}$$

With the full bridge, just as in the case of the half bridge (potentiometric circuit), we can show that the change in output, ΔE, is a function of the unit changes in gage resistance, or the strains in the material to which the gages are attached.

The value of ΔE can now be determined, in terms of resistances, by subtracting E, or its equivalent as expressed by Eq. (5.25), from both sides of Eq. (5.26). This results in

$$\Delta E = V \frac{(R_1 + \Delta R_1)(R_3 + \Delta R_3) - (R_2 + \Delta R_2)(R_4 + \Delta R_4)}{(R_1 + \Delta R_1 + R_2 + \Delta R_2)(R_3 + \Delta R_3 + R_4 + \Delta R_4)}$$

$$- V \frac{R_1 R_3 - R_2 R_4}{(R_1 + R_2)(R_3 + R_4)} \tag{5.27}$$

Equation (5.27) is a perfectly general expression for the change in bridge output from any initial condition. It specifies no particular relation between the initial resistances of the bridge arms, but unfortunately it is somewhat cumbersome to handle.

For the special situation in which the bridge is initially balanced, the initial output, E, will be zero and the expression for the change in output, ΔE_0, will be much simpler than the general relation given by Eq. (5.27).

When the bridge is initially balanced, the initial output is

$$E = 0 = V \frac{R_1 R_3 - R_2 R_4}{(R_1 + R_2)(R_3 + R_4)} \tag{5.28}$$

This means that

$$R_1 R_3 - R_2 R_4 = 0 \tag{5.29}$$

Equation (5.29) indicates that, for a balanced bridge (output = 0), a definite relation must exist among the resistances of the four arms. This relationship can be expressed in the three following ways:

1. From Eq. (5.29), we see that the cross products of the resistances in the arms must be equal. Thus

$$R_1 R_3 = R_2 R_4 \tag{5.30}$$

2. We also see that

$$\frac{R_2}{R_1} = \frac{R_3}{R_4} \tag{5.31}$$

Equation (5.31) indicates that the ratios of the resistances in the two halves of the bridge, which are in series with the power supply (*DAC* and *DBC* in Fig. 5.8), must be equal. This ratio, frequently called the bridge ratio, is equivalent to the ratio of ballast resistance to gage resistance in the potentiometric circuit. It is represented by the symbol a. Hence,

$$\text{Bridge ratio} = \frac{R_2}{R_1} = \frac{R_3}{R_4} = a \tag{5.32}$$

3. If we divide the bridge into two halves with respect to the two output terminals (*ADB* and *ACB* in Fig. 5.8), the ratio of the resistances in

these two halves must also be equal. Letting the symbol b represent this ratio, we have

$$\frac{R_1}{R_4} = \frac{R_2}{R_3} = b \tag{5.33}$$

When the values of the ratios a and b have been chosen, and also the resistance in one of the bridge arms (for example, R_1), the other three resistances can be computed.

For any values of a and b

$$R_2 = aR_1, \qquad R_3 = (a/b)R_1, \qquad R_4 = R_1/b$$

For any values when $a = b$

$$R_2 = aR_1, \qquad R_3 = R_1, \qquad R_4 = R_1/a$$

For any value of a when $b = 1$

$$R_2 = R_3 = aR_1, \qquad R_4 = R_1$$

For any value of b when $a = 1$

$$R_2 = R_1, \qquad R_3 = R_4 = R_1/b$$

When $a = b = 1$

$$R_1 = R_2 = R_3 = R_4 \qquad \text{(equal arm bridge)}$$

Choice of ratios a and b

$a = b = 1$ Since it is frequently desired to use strain gages in two and four arms of the bridge, the equal-arm arrangement is probably the most usual, in spite of the fact that, for a single gage, its efficiency is only 50 percent.

$a > 1$ For operation with a single gage, and under some conditions with two gages, the efficiency can be improved by increasing the bridge ratio. There is relatively little to be gained, however, by going beyond a ratio of about 10, which will yield approximately 90 percent of the ultimate. Many investigators prefer to use a maximum value of 5, which allows considerably lower voltage for the power supply with an efficiency that is above 80 percent.

$b \neq 1$ The choice of the value of b is not critical. It is often taken as unity for convenience. We should avoid making this ratio so large

Output of the initially balanced bridge

When the bridge is initially balanced, the expression for the change in output from the initial condition is simplified. By referring to Eq. (5.27), we see that the second term corresponding to the initial output drops out, because $E = 0$ for the condition of balance.

In order to be specific we will use the symbol ΔE_0 for the change in output from the initial condition of balance. This makes a distinction with respect to the symbol ΔE which has been used for the change in output from any initial condition. Therefore, ΔE_0 corresponds only to the special case of initial bridge balance. Since the output of the bridge is usually nonlinear, this distinction between the general case and a particular case is necessary. Furthermore, it becomes more important with larger resistance changes, such as those that may be encountered with semiconductor gages.

We now rewrite Eq. (5.27) in the simplified form corresponding to initial bridge balance. It becomes

$$\Delta E_0 = V \frac{(R_1 + \Delta R_1)(R_3 + \Delta R_3) - (R_2 + \Delta R_2)(R_4 + \Delta R_4)}{(R_1 + \Delta R_1 + R_2 + \Delta R_2)(R_3 + \Delta R_3 + R_4 + \Delta R_4)} \quad (5.34)$$

Since the strain gage indicates strain in terms of unit changes in resistance, we now proceed to convert Eq. (5.34) into terms of ratios and unit changes. If both numerator and denominator are divided by the product $R_1 R_3$, Eq. (5.34) can be rewritten as

$$\Delta E_0 = V \frac{\left[1 + \dfrac{\Delta R_1}{R_1}\right]\left[1 + \dfrac{\Delta R_3}{R_3}\right] - \dfrac{R_2 R_4}{R_1 R_3}\left[1 + \dfrac{\Delta R_2}{R_2}\right]\left[1 + \dfrac{\Delta R_4}{R_4}\right]}{\left[1 + \dfrac{\Delta R_1}{R_1} + \dfrac{R_2}{R_1} + \dfrac{\Delta R_2}{R_1}\right]\left[1 + \dfrac{\Delta R_3}{R_3} + \dfrac{R_4}{R_3} + \dfrac{\Delta R_4}{R_3}\right]} \quad (5.35)$$

As we are now dealing with conditions of initial bridge balance, we introduce the relations given by Eqs. (5.29) and (5.32). They are

$$\frac{R_2 R_4}{R_1 R_3} = 1, \quad \frac{1}{R_1} = \frac{a}{R_2}, \quad \frac{1}{R_3} = \frac{1}{aR_4}$$

These relations are now substituted into Eq. (5.35) to arrive at

$$\Delta E_0 = V \frac{\left[1 + \dfrac{\Delta R_1}{R_1}\right]\left[1 + \dfrac{\Delta R_3}{R_3}\right] - \left[1 + \dfrac{\Delta R_2}{R_2}\right]\left[1 + \dfrac{\Delta R_4}{R_4}\right]}{\left[1 + \dfrac{\Delta R_1}{R_1} + a + a\dfrac{\Delta R_2}{R_1}\right]\left[1 + \dfrac{\Delta R_3}{R_3} + \dfrac{1}{a} + \dfrac{1}{a}\dfrac{\Delta R_4}{R_3}\right]}$$

THE BONDED ELECTRICAL RESISTANCE STRAIN GAGE

Multiply the numerator and denominator by a to obtain

$$\Delta E_0 = aV \frac{\left[1 + \frac{\Delta R_1}{R_1}\right]\left[1 + \frac{\Delta R_3}{R_3}\right] - \left[1 + \frac{\Delta R_2}{R_2}\right]\left[1 + \frac{\Delta R_4}{R_4}\right]}{\left[1 + a + \frac{\Delta R_1}{R_1} + a\frac{\Delta R_2}{R_2}\right]\left[1 + a + a\frac{\Delta R_3}{R_3} + \frac{\Delta R_4}{R_4}\right]} \quad (5.36)$$

Next, Eq. (5.36) can further be rearranged.

$$\Delta E_0 = aV \frac{\left[\frac{\Delta R_1}{R_1} - \frac{\Delta R_2}{R_2} + \frac{\Delta R_3}{R_3} - \frac{\Delta R_4}{R_4}\right] + \left[\left(\frac{\Delta R_1}{R_1}\right)\left(\frac{\Delta R_3}{R_3}\right) - \left(\frac{\Delta R_2}{R_2}\right)\left(\frac{\Delta R_4}{R_4}\right)\right]}{\left[1 + a + \left(\frac{1+a}{1+a}\right)\left(\frac{\Delta R_1}{R_1} + a\frac{\Delta R_2}{R_2}\right)\right]\left[1 + a + \left(\frac{1+a}{1+a}\right)\left(a\frac{\Delta R_3}{R_3} + \frac{\Delta R_4}{R_4}\right)\right]}$$

This reduces to

$$\Delta E_0 = V \frac{a}{(1+a)^2}$$

$$\times \frac{\left[\frac{\Delta R_1}{R_1} - \frac{\Delta R_2}{R_2} + \frac{\Delta R_3}{R_3} - \frac{\Delta R_4}{R_4}\right] + \left[\left(\frac{\Delta R_1}{R_1}\right)\left(\frac{\Delta R_3}{R_3}\right) - \left(\frac{\Delta R_2}{R_2}\right)\left(\frac{\Delta R_4}{R_4}\right)\right]}{\left[1 + \left(\frac{1}{1+a}\right)\left(\frac{\Delta R_1}{R_1} + a\frac{\Delta R_2}{R_2}\right)\right]\left[1 + \left(\frac{1}{1+a}\right)\left(a\frac{\Delta R_3}{R_3} + \frac{\Delta R_4}{R_4}\right)\right]}$$

(5.37)

In order to put Eq. (5.37) into a more desirable form, let

$$A = \frac{\Delta R_1}{R_1} - \frac{\Delta R_2}{R_2} + \frac{\Delta R_3}{R_3} - \frac{\Delta R_4}{R_4}$$

$$B = \left(\frac{\Delta R_1}{R_1}\right)\left(\frac{\Delta R_3}{R_3}\right) - \left(\frac{\Delta R_2}{R_2}\right)\left(\frac{\Delta R_4}{R_4}\right)$$

$$C = 1 + \frac{1}{1+a}\left(\frac{\Delta R_1}{R_1} + a\frac{\Delta R_2}{R_2}\right)$$

$$D = 1 + \frac{1}{1+a}\left(a\frac{\Delta R_3}{R_3} + \frac{\Delta R_4}{R_4}\right)$$

Equation (5.37) becomes

$$\Delta E_0 = \frac{aV}{(1+a)^2}\frac{A+B}{CD} = \frac{aVA}{(1+a)^2}\left[\frac{1+\frac{B}{A}}{CD}\right] \quad (5.38)$$

WHEATSTONE BRIDGE

The bracketed term in Eq. (5.38) is the nonlinearity factor, and so ΔE_0 can now be written as

$$\Delta E_0 = \frac{aV}{(1+a)^2}\left[\frac{\Delta R_1}{R_1} - \frac{\Delta R_2}{R_2} + \frac{\Delta R_3}{R_3} - \frac{\Delta R_4}{R_4}\right](1-n) \quad (5.39)$$

where the nonlinearity factor, $(1-n)$, is

$$1-n = \frac{1 + \left[\left(\frac{\Delta R_1}{R_1}\right)\left(\frac{\Delta R_3}{R_3}\right) - \left(\frac{\Delta R_2}{R_2}\right)\left(\frac{\Delta R_4}{R_4}\right)\right]\Big/\left[\frac{\Delta R_1}{R_1} - \frac{\Delta R_2}{R_2} + \frac{\Delta R_3}{R_3} - \frac{\Delta R_4}{R_4}\right]}{\left[1 + \frac{1}{1+a}\left(\frac{\Delta R_1}{R_1} + a\frac{\Delta R_2}{R_2}\right)\right]\left[1 + \frac{1}{1+a}\left(a\frac{\Delta R_3}{R_3} + \frac{\Delta R_4}{R_4}\right)\right]} \quad (5.40)$$

Equation (5.40) is exact and will yield correct values of n, or $(1-n)$, for all values of the unit changes in resistance of the bridge arms. It is, however, somewhat inconvenient to handle.

When the unit changes in resistance are small relative to unity (let us say less than 10 percent), their products will be even smaller (less than 1 percent) and can be neglected. From Eq. (5.40) we can therefore develop a much simpler and very good approximate relationship if we disregard the second-order quantities in the numerator and in the expansion of the denominator. This procedure will give us

$$1 - n = \frac{1}{1 + \frac{1}{1+a}\left[\frac{\Delta R_1}{R_1} + a\frac{\Delta R_2}{R_2} + a\frac{\Delta R_3}{R_3} + \frac{\Delta R_4}{R_4}\right]} \quad (5.41)$$

Equation (5.41) can be solved for n by letting

$$D = \left[\frac{\Delta R_1}{R_1} + a\frac{\Delta R_2}{R_2} + a\frac{\Delta R_3}{R_3} + \frac{\Delta R_4}{R_4}\right]$$

Thus,

$$n = 1 - \frac{1}{1 + \frac{D}{1+a}} = \frac{1}{1 + \frac{1+a}{D}}$$

The relation for n becomes

$$n = \cfrac{1}{1 + \cfrac{1+a}{\left[\dfrac{\Delta R_1}{R_1} + a\dfrac{\Delta R_2}{R_2} + a\dfrac{\Delta R_3}{R_3} + \dfrac{\Delta R_4}{R_4}\right]}} \qquad (5.42)$$

If we compare Eq. (5.39) for the bridge output and Eq. (5.42) for the nonlinearity factor with the corresponding expressions for the potentiometric circuit, a marked similarity will be observed. Furthermore, if the bridge arms corresponding to R_3 and R_4 contain fixed resistors, ΔR_3 and ΔR_4 will both be zero. Equations (5.39) and (5.42), then, become identical with those of the potentiometric circuit. In additon, Eq. (5.42) loses its approximate nature and becomes exact.

Equations (5.39) and (5.42) can be written in terms of strain, since $\Delta R/R = G_F \varepsilon$. With like gages in all four bridge arms, $a = 1$, and Eq. (5.39) can be written as

$$\Delta E_0 = \frac{G_F V}{4}[\varepsilon_1 - \varepsilon_2 + \varepsilon_3 - \varepsilon_4](1 - n) \qquad (5.43)$$

Equation (5.42) becomes

$$n = \cfrac{1}{1 + \cfrac{2}{G_F(\varepsilon_1 + \varepsilon_2 + \varepsilon_3 + \varepsilon_4)}} \qquad (5.44)$$

When measuring elastic strains in metals, the error due to nonlinearity is generally small and is usually ignored. As a rule of thumb, the error, in percent, is approximately equal to the strain, in percent.

When nonlinearity must be taken into account, its influence for any bridge arrangement can be readily computed through the use of Eqs. (5.39) and (5.40). To illustrate this, a quarter-bridge circuit can be examined, where $\Delta R_1/R_1 = G_F \varepsilon$. Using this value of $\Delta R_1/R_1$ and $a = 1$, Eq. (5.39) produces

$$\Delta E_0 = V\left(\frac{G_F}{4}\right)\varepsilon(1 - n) \qquad (5.45)$$

The nonlinearity factor, $(1 - n)$, is obtained from Eq. (5.40). Thus,

$$1 - n = \frac{1}{1 + \frac{1}{2}(G_F \varepsilon)} = \frac{2}{2 + G_F \varepsilon} \qquad (5.46)$$

Substituting the value of $(1 - n)$ given by Eq. (5.46) into Eq. (5.45) yields

$$\Delta E_0 = V\left(\frac{G_F}{4}\right)(\varepsilon)\left(\frac{2}{2 + G_F\varepsilon}\right)$$

This expression can be rewritten as

$$\frac{\Delta E_0}{V} = \frac{G_F\varepsilon}{4 + 2G_F\varepsilon} \qquad (5.47)$$

The strain, ε, in these equations must be entered as $\varepsilon \times 10^{-6}$ in/in. The second term in the denominator of Eq. (5.47) produces the nonlinearity in $\Delta E_0/V$. Thus, a compressive strain will produce an indicated value of $\Delta E_0/V$ that is too large in magnitude, while a tensile strain will produce an indicated value that is too low in magnitude.

Reference 5 gives a tabulation of the effect of nonlinearity for various bridge arrangements. Furthermore, it also gives the ratio of the actual strain, ε, to the indicated strain, ε_i. In order to show this, we know that $(\Delta E_0/V)/\varepsilon_i$ is equal to the constant $G_F/4$, and so the following can be written:

$$\frac{\Delta E_0}{V} = \frac{G_F\varepsilon_i}{4} = \frac{G_F\varepsilon}{4 + 2G_F\varepsilon}$$

From this, the indicated strain is

$$\varepsilon_i = \frac{2\varepsilon}{2 + G_F\varepsilon} \qquad (5.48)$$

Solving Eq. (5.48) for ε produces

$$\varepsilon = \frac{2\varepsilon_i}{2 - G_F\varepsilon_i} \qquad (5.49)$$

The ratio of $\varepsilon/\varepsilon_i$ can be written as

$$\frac{\varepsilon}{\varepsilon_i} = \frac{2}{2 - G_F\varepsilon_i} \qquad (5.50)$$

or

$$\frac{\varepsilon}{\varepsilon_i} = 1 + \frac{G_F\varepsilon_i}{2 - G_F\varepsilon_i} \qquad (5.51)$$

THE BONDED ELECTRICAL RESISTANCE STRAIN GAGE

In the following two example problems, two bridge arrangements are developed, while others are left as problems at the end of the chapter.

Example 5.1. A cantilever beam with four gages arranged in a full bridge is shown in Fig. 5.9. Each page will read the same magnitude of strain, with gages 1 and 3 in tension, and gages 2 and 4 in compression. Using Eqs. (5.39) and (5.40), determine ΔE_0. Also determine $\varepsilon/\varepsilon_i$.

Solution

$$\varepsilon_1 = \varepsilon_3 = \varepsilon, \qquad \varepsilon_2 = \varepsilon_4 = -\varepsilon, \qquad \text{bridge ratio} = 1, \qquad \Delta R/R = G_F \varepsilon$$

From Eq. (5.39),

$$\Delta E_0 = \frac{aV}{(1+a)^2}\left[\frac{\Delta R_1}{R_1} - \frac{\Delta R_2}{R_2} + \frac{\Delta R_3}{R_3} - \frac{\Delta R_4}{R_4}\right](1-n)$$

$$= \frac{V}{4} G_F[\varepsilon - (-\varepsilon) + \varepsilon - (-\varepsilon)](1-n) = VG_F\varepsilon(1-n)$$

Equation (5.40) is

$$1 - n = \frac{1 + \left[\left(\dfrac{\Delta R_1}{R_1}\right)\left(\dfrac{\Delta R_3}{R_3}\right) - \left(\dfrac{\Delta R_2}{R_2}\right)\left(\dfrac{\Delta R_4}{R_4}\right)\right] \Big/ \left[\dfrac{\Delta R_1}{R_1} - \dfrac{\Delta R_2}{R_2} + \dfrac{\Delta R_3}{R_3} - \dfrac{\Delta R_4}{R_4}\right]}{\left[1 + \dfrac{1}{1+a}\left(\dfrac{\Delta R_1}{R_1} + a\dfrac{\Delta R_2}{R_2}\right)\right]\left[1 + \dfrac{1}{1+a}\left(a\dfrac{\Delta R_3}{R_3} + \dfrac{\Delta R_4}{R_4}\right)\right]}$$

Substituting the gage factor and appropriate strains for $\Delta R/R$, we have

$$1 - n = \frac{1 + \dfrac{(G_F\varepsilon)(G_F\varepsilon) - (-G_F\varepsilon)(-G_F\varepsilon)}{G_F[\varepsilon - (-\varepsilon) + \varepsilon - (-\varepsilon)]}}{\left[1 + \dfrac{G_F}{2}(\varepsilon - \varepsilon)\right]\left[1 + \dfrac{G_F}{2}(\varepsilon - \varepsilon)\right]} = 1$$

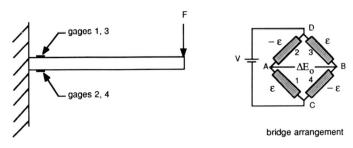

FIG. 5.9. Cantilever beam with strain gages aligned parallel to the longitudinal axis.

The circuit is linear, and so the output is

$$\Delta E_0 = V G_F \varepsilon$$

Since the circuit is linear and $(1 - n) = 1$, then

$$\frac{\varepsilon}{\varepsilon_i} = 1$$

Example 5.2. A round rod in tension has four gages mounted on it in order to form a full bridge. Gages 1 and 3 are mounted in the axial direction 180° apart. Gages 2 and 4 are mounted transverse to gages 1 and 2, respectively, as shown in Fig. 5.10. Determine ΔE_0, using Eqs. (5.39) and (5.40), as well as $\varepsilon/\varepsilon_i$.

Solution

$$\varepsilon_1 = \varepsilon_3 = \varepsilon, \qquad \varepsilon_2 = \varepsilon_4 = -v\varepsilon, \qquad \text{bridge ratio} = 1$$

From Eq. (5.39),

$$\Delta E_0 = \frac{aV}{(1+a)^2} \left[\frac{\Delta R_1}{R_1} - \frac{\Delta R_2}{R_2} + \frac{\Delta R_3}{R_3} - \frac{\Delta R_4}{R_4} \right] (1-n)$$

$$= \frac{V}{4} G_F [\varepsilon - (-v\varepsilon) + \varepsilon - (-v\varepsilon)](1-n)$$

$$= \frac{V}{2} G_F \varepsilon (1+v)(1-n)$$

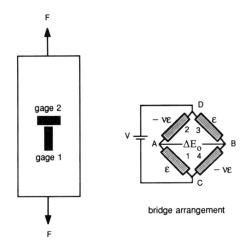

gages 3 and 4 diametrically opposite

FIG. 5.10. Tension member with strain gages.

THE BONDED ELECTRICAL RESISTANCE STRAIN GAGE

From Eq. (5.40),

$$1 - n = \frac{1 + \left[\left(\frac{\Delta R_1}{R_1}\right)\left(\frac{\Delta R_3}{R_3}\right) - \left(\frac{\Delta R_2}{R_2}\right)\left(\frac{\Delta R_4}{R_4}\right)\right] / \left[\frac{\Delta R_1}{R_1} - \frac{\Delta R_2}{R_2} + \frac{\Delta R_3}{R_3} - \frac{\Delta R_4}{R_4}\right]}{\left[1 + \frac{1}{1+a}\left(\frac{\Delta R_1}{R_1} + a\frac{\Delta R_2}{R_2}\right)\right]\left[1 + \frac{1}{1+a}\left(a\frac{\Delta R_3}{R_3} + \frac{\Delta R_4}{R_4}\right)\right]}$$

Substituting in the gage factor and appropriate strains, we have

$$1 - n = \frac{1 + \dfrac{(G_F\varepsilon)(G_F\varepsilon) - (-G_F v\varepsilon)(-G_F v\varepsilon)}{G_F[\varepsilon - (-v\varepsilon) + \varepsilon - (-v\varepsilon)]}}{\left[1 + \dfrac{G_F}{2}(\varepsilon - v\varepsilon)\right]\left[1 + \dfrac{G_F}{2}(\varepsilon - v\varepsilon)\right]} = \frac{2}{2 + G_F(1-v)\varepsilon}$$

Multiplying the expression for ΔE_0 by $(1 - n)$, we obtain

$$\Delta E_0 = \left[\frac{V}{2} G_F(1+v)\varepsilon\right]\left[\frac{2}{2 + G_F(1-v)\varepsilon}\right]$$

Thus,

$$\Delta E_0 = \frac{V G_F(1+v)\varepsilon}{2 + G_F(1-v)\varepsilon}$$

The value of the indicated strain can be written as

$$\Delta E_0 = V\left(\frac{G_F}{2}\right)(1+v)\varepsilon_i$$

Equating this to the value of ΔE_0 when nonlinearity is considered gives

$$\Delta E_0 = V\left(\frac{G_F}{2}\right)(1+v)\varepsilon_i = \frac{V(G_F)(1+v)\varepsilon}{2 + G_F(1-v)\varepsilon}$$

Thus,

$$\varepsilon_i = \frac{2\varepsilon}{2 + G_F(1-v)\varepsilon}$$

Solving for ε,

$$\varepsilon = \frac{2\varepsilon_i}{2 - G_F(1-v)\varepsilon_i}$$

In terms of the ratio of the actual strain, ε, to the indicated strain, ε_i, we have

$$\frac{\varepsilon}{\varepsilon_i} = \frac{2}{2 - G_F(1-v)\varepsilon_i} = 1 + \frac{G_F(1-v)\varepsilon_i}{2 - G_F(1-v)\varepsilon_i}$$

Other bridge arrangements can be handled in the same manner.

Alternate method for the derivation of elementary bridge equations

An alternate method for developing the expression for the output, ΔE_0, of an initially balanced Wheatstone bridge will be shown. Consider the possibility of connecting two potentiometric (half-bridge) circuits together in parallel, as shown in Fig. 5.11.

The initial resistances are R_1, R_2, R_3, and R_4. Since the two half-bridges are to be joined together, they will both be subjected to the same voltage, V. The potential drops across R_1 and R_4 are represented as E_{2-1} and E_{3-4}, respectively, and can be expressed as

$$E_{2-1} = V \frac{R_1}{R_1 + R_2} = V \frac{1}{1 + \frac{R_2}{R_1}} = V \frac{1}{1 + a_{2-1}} \qquad (5.52)$$

$$E_{3-4} = V \frac{R_4}{R_3 + R_4} = V \frac{1}{1 + \frac{R_3}{R_4}} = V \frac{1}{1 + a_{3-4}} \qquad (5.53)$$

where

$$a_{2-1} = \frac{R_2}{R_1} \quad \text{and} \quad a_{3-4} = \frac{R_3}{R_4}$$

When the two half-bridges are put together to form a Wheatstone bridge, as shown in Fig. 5.12, and then initially balanced, the voltage drops

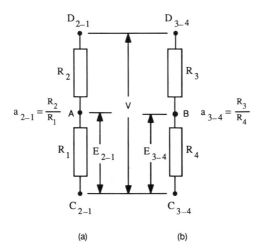

FIG. 5.11. Two potentiometric circuits (or two half bridges).

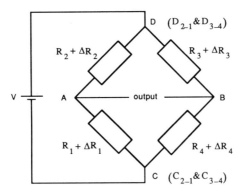

FIG. 5.12. Wheatstone bridge formed from two half bridges.

across resistance R_1 and R_4 must be equal. Thus, from Eqs. (5.52) and (5.53),

$$E_{2-1} = E_{3-4} = V\frac{1}{1+a_{2-1}} = \frac{1}{1+a_{3-4}}$$

From this, it is evident that $a_{2-1} = a_{3-4}$. This means that

$$\frac{R_2}{R_1} = \frac{R_3}{R_4} = a \quad (5.54)$$

where

$$a_{2-1} = a_{3-4} = a$$

For initial bridge balance, the ballast ratio must be the same for both sides, as expressed by Eq. (5.54). When changes take place in each arm by the appropriate ΔR, the potential drop across R_1 and R_4 will be

$$E_{A-C} = E_{2-1} + \Delta E_{2-1}$$
$$E_{B-C} = E_{3-4} + \Delta E_{3-4}$$

The bridge output, ΔE_0, will be equal to the difference in voltage between A and B. Therefore,

$$\Delta E_0 = E_{A-C} - E_{B-C} = (E_{2-1} + \Delta E_{2-1}) - (E_{3-4} + \Delta E_{3-4}) \quad (5.55)$$

For the condition of initial balance, however, $E_{2-1} = E_{3-4}$, so that

$$\Delta E_0 = \Delta E_{2-1} - \Delta E_{3-4} \quad (5.56)$$

WHEATSTONE BRIDGE

From the relations for the potentiometric (half-bridge) circuit, as given by Eq. (4.21),

$$\Delta E_{2-1} = V \frac{a}{(1+a)^2} \left[\frac{\Delta R_1}{R_1} - \frac{\Delta R_2}{R_2} \right] \left[\frac{1+a}{1+a+\dfrac{\Delta R_1}{R_1}+a\dfrac{\Delta R_2}{R_2}} \right]$$

Rearranging,

$$\Delta E_{2-1} = V \frac{a}{(1+a)^2} \frac{\left[\dfrac{\Delta R_1}{R_1} - \dfrac{\Delta R_2}{R_2}\right]}{\left[1+\dfrac{1}{1+a}\left(\dfrac{\Delta R_1}{R_1}+a\dfrac{\Delta R_2}{R_2}\right)\right]} \qquad (5.57)$$

In a like manner, ΔE_{3-4}, is written

$$\Delta E_{3-4} = V \frac{a}{(1+a)^2} \frac{\left[\dfrac{\Delta R_4}{R_4} - \dfrac{\Delta R_3}{R_3}\right]}{\left[1+\dfrac{1}{1+a}\left(\dfrac{\Delta R_4}{R_4}+a\dfrac{\Delta R_3}{R_3}\right)\right]} \qquad (5.58)$$

Note that, in Eqs. (5.57) and (5.58), R_2 and R_3 are the ballast resistances. If the values of ΔE_{2-1} and ΔE_{3-4} given by Eqs. (5.57) and (5.58), respectively, are substituted into Eq. (5.56), the output voltage will be

$$\Delta E_0 = V \frac{a}{(1+a)^2} \left\{ \frac{\left[\dfrac{\Delta R_1}{R_1} - \dfrac{\Delta R_2}{R_2}\right]}{\left[1+\dfrac{1}{1+a}\left(\dfrac{\Delta R_1}{R_1}+a\dfrac{\Delta R_2}{R_2}\right)\right]} - \frac{\left[\dfrac{\Delta R_4}{R_4} - \dfrac{\Delta R_3}{R_3}\right]}{\left[1+\dfrac{1}{1+a}\left(\dfrac{\Delta R_4}{R_4}+a\dfrac{\Delta R_3}{R_3}\right)\right]} \right\} \qquad (5.59)$$

If the bracketed term only is considered, it can be expressed as

$$\frac{\left[\dfrac{\Delta R_1}{R_1} - \dfrac{\Delta R_2}{R_2}\right]\left[1+\dfrac{1}{1+a}\left(\dfrac{\Delta R_4}{R_4}+a\dfrac{\Delta R_3}{R_3}\right)\right] - \left[\dfrac{\Delta R_4}{R_4}-\dfrac{\Delta R_3}{R_3}\right]\left[1+\dfrac{1}{1+a}\left(\dfrac{\Delta R_1}{R_1}+a\dfrac{\Delta R_2}{R_2}\right)\right]}{\left[1+\dfrac{1}{1+a}\left(\dfrac{\Delta R_1}{R_1}+a\dfrac{\Delta R_2}{R_2}\right)\right]\left[1+\dfrac{1}{1+a}\left(\dfrac{\Delta R_4}{R_4}+a\dfrac{\Delta R_3}{R_3}\right)\right]}$$

If the numerator is expanded, it becomes

$$\left[\frac{\Delta R_1}{R_1} - \frac{\Delta R_2}{R_2} + \frac{\Delta R_3}{R_3} - \frac{\Delta R_4}{R_4}\right]$$

$$+ \frac{1}{1+a}\left[(1+a)\left(\frac{\Delta R_1}{R_1}\right)\left(\frac{\Delta R_3}{R_3}\right) - (1+a)\left(\frac{\Delta R_2}{R_2}\right)\left(\frac{\Delta R_4}{R_4}\right)\right]$$

$$= \left[\frac{\Delta R_1}{R_1} - \frac{\Delta R_2}{R_2} + \frac{\Delta R_3}{R_3} - \frac{\Delta R_4}{R_4}\right] + \left[\left(\frac{\Delta R_1}{R_1}\right)\left(\frac{\Delta R_3}{R_3}\right) - \left(\frac{\Delta R_2}{R_2}\right)\left(\frac{\Delta R_4}{R_4}\right)\right]$$

Combining all terms, the output voltage is

$$\Delta E_0 = V \frac{a}{(1+a)^2}$$

$$\times \frac{\left[\dfrac{\Delta R_1}{R_1} - \dfrac{\Delta R_2}{R_2} + \dfrac{\Delta R_3}{R_3} - \dfrac{\Delta R_4}{R_4}\right] + \left[\left(\dfrac{\Delta R_1}{R_1}\right)\left(\dfrac{\Delta R_3}{R_3}\right) - \left(\dfrac{\Delta R_2}{R_2}\right)\left(\dfrac{\Delta R_4}{R_4}\right)\right]}{\left[1 + \dfrac{1}{1+a}\left(\dfrac{\Delta R_1}{R_1} + a\dfrac{\Delta R_2}{R_2}\right)\right]\left[1 + \dfrac{1}{1+a}\left(a\dfrac{\Delta R_3}{R_3} + \dfrac{\Delta R_4}{R_4}\right)\right]}$$

(5.60)

Equation (5.60) is exactly the same as Eq. (5.37).

5.4. General bridge equations

We will now consider a somewhat more elaborate arrangement of the Wheatstone bridge. This will include the following items that were omitted in the previous section:

1. The effect of resistance in series with the bridge. This will include the internal resistance of the power supply as well as the resistance of the leads connecting the bridge to the energy source. In the analysis, both of these resistances will be lumped together and considered as though they presented a single combined resistance in series with the bridge.
2. The influence of the meter (or galvanometer) resistance on the bridge output voltage. In the previous section, the analysis of the bridge output was made on the assumption that the meter presented an infinite impedance and, in consequence, would draw no current from the bridge. We will now examine the situation in which the meter has a finite impedance and draws some current from the bridge.

Fortunately, the results of the analysis of the idealized, or simplified, bridge circuit can be used in building up the general case, which includes the preceding considerations.

Effect of resistance in series with the bridge

The bridge input resistance, R_{BI}, and the bridge output resistance, R_{BO}, are given by Eqs. (5.9) and (5.10), respectively. Here we will outline the method of computing them. Since Thévenin's theorem will be used, it is stated as follows (6):

> Any two-terminal netork of fixed resistances and sources of e.m.f. may be replaced by a single source of e.m.f. having an equivalent e.m.f. equal to the open-circuit e.m.f. at the terminals of the original network and having an internal resistance equal to the resistance looking back into the network from the two terminals, and with all sources of e.m.f. replaced by their internal resistance.

The resistance in series with the bridge will include the internal resistance of the power supply as well as the resistance of the leads connecting the bridge to the source of energy. In the analysis, both of these resistances are lumped together and considered as a single combined resistance in series with the bridge. The effect of the series resistance is to reduce the voltage actually received at the bridge compared with that available at the power supply, since the total voltage must be apportioned across the series and bridge resistances rather than being applied entirely to the bridge. The circuit is again shown in Fig. 5.13.

In order to compute R_{BI}, the circuit is opened at points D and C. The resistance, R_s, is no longer in the circuit being considered, and neither are there energy sources. Looking into the bridge from points D and C, we see a circuit with resistances R_1, R_2, R_3, R_4 and R_m. Since the circuit is not a combination of series and parallel resistances, it must be changed into such a combination. Figure 5.14a shows the original circuit being considered, while Fig. 5.14b shows the converted circuit.

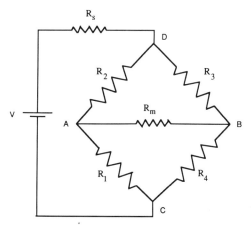

FIG. 5.13. Wheatstone bridge with supply resistance and meter resistance.

174   THE BONDED ELECTRICAL RESISTANCE STRAIN GAGE

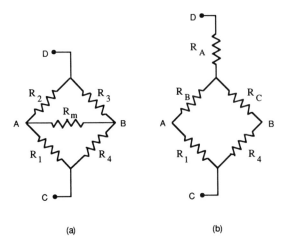

FIG. 5.14. Original circuit (a) and equivalent circuit (b).

The means of obtaining the circuit of Fig. 5.14b will be outlined. The resistances R_2, R_3, and R_m form a Delta network that must be converted to a Wye network consisting of resistances R_A, R_B, and R_C. The resistances in the Wye network (6) are given as

$$R_A = \frac{R_2 R_3}{R_2 + R_3 + R_m} \tag{5.61}$$

$$R_B = \frac{R_m R_2}{R_2 + R_3 + R_m} \tag{5.62}$$

$$R_C = \frac{R_m R_3}{R_2 + R_3 + R_m} \tag{5.63}$$

Referring to Fig. 5.14b, the resistances $R_B + R_1$ and $R_C + R_4$ are in parallel, and their equivalent resistance is then in series with R_A. The bridge input resistance is then

$$R_{CD} = \frac{(R_B + R_1)(R_C + R_4)}{R_B + R_C + R_1 + R_4} + R_A \tag{5.64}$$

Equation (5.64) can be expressed in terms of the original resistances shown in Fig. 5.14. Although considerable algebra is involved, the final result is

$$R_{CD} = \frac{R_m(R_1 + R_2)(R_3 + R_4) + R_1 R_2(R_3 + R_4) + R_3 R_4(R_1 + R_2)}{R_m(R_1 + R_2 + R_3 + R_4) + (R_1 + R_4)(R_2 + R_3)} \tag{5.65}$$

If the resistances R_1, R_2, R_3, and R_4 are increased by their individual ΔR values, then Eq. (5.65) becomes Eq. (5.9), the expression for R_{BI}. Furthermore, if $R_1 = R_2 = R_3 = R_4 = R_g$, then Eq. (5.65) reduces to $R_{CD} = R_g$, regardless of the value of R_m.

Since R_s is in series with the bridge, the bridge voltage, V_{DC}, is

$$V_{DC} = V \frac{R_{BI}}{R_{BI} + R_s} = V \left[\frac{1}{1 + \dfrac{R_s}{R_{BI}}} \right] \quad (5.66)$$

Equation (5.66) shows that when a resistance is in series with the bridge, the voltage must be multiplied by the desensitization factor, $1/(1 + R_s/R_{BI})$, in order to determine the actual bridge voltage.

Since the bridge output is directly proportional to the applied voltage, the voltage, V_{DC}, can be substituted for the voltage, V, in Eq. (5.39). The value of ΔE_0 then becomes

$$\Delta E_0 = V \left[\frac{1}{1 + \dfrac{R_s}{R_{BI}}} \right] \left[\frac{1}{(1+a)^2} \right] \left[\frac{\Delta R_1}{R_1} - \frac{\Delta R_2}{R_2} + \frac{\Delta R_3}{R_3} - \frac{\Delta R_4}{R_4} \right] (1-n) \quad (5.67)$$

It should be noted in Eq. (5.67) that R_{BI} is not a constant, since it varies with the ΔR quantities. If the unit changes in resistance are large, then, depending on the relative magnitude of R_s, some allowance for the variation in R_{BI} may be required.

Influence of meter resistance

So far we have examined the bridge output voltage when the meter, or indicating device, was considered as having infinite input impedance. We now look at what happens when the meter (or galvanometer) has a finite resistance and draws current from the bridge. To do this, the circuit is opened between the meter and one of the output terminals of the bridge, as shown in Fig. 5.15. Thevenin's theorem will then be applied in order to get an equivalent circuit.

According to Thevenin's theorem, we first find the open-circuit potential between points A and B. In order to do this, the loop, or mesh, equations can be written by referring to Fig. 5.15. As we see, there will be two equations. They are

$$I_1 R_s + (I_1 - I_2)(R_1 + R_2) = V$$
$$I_2(R_3 + R_4) + (I_2 - I_1)(R_1 + R_2) = 0$$

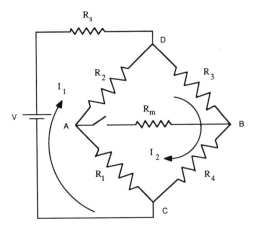

FIG. 5.15. Wheatstone bridge with supply resistance and output meter disconnected.

Rearranging, we have

$$I_1(R_1 + R_2 + R_s) - I_2(R_1 + R_2) = V \quad (5.68)$$

$$-I_1(R_1 + R_2) + I_2(R_1 + R_2 + R_3 + R_4) = 0 \quad (5.69)$$

Solving Eqs. (5.68) and (5.69) simultaneously for I_1 and I_2 results in

$$I_1 = V \frac{R_1 + R_2 + R_3 + R_4}{R_s(R_1 + R_2 + R_3 + R_4) + (R_1 + R_2)(R_3 + R_4)} \quad (5.70)$$

$$I_2 = \frac{R_1 + R_2}{R_s(R_1 + R_2 + R_3 + R_4) + (R_1 + R_2)(R_3 + R_4)} \quad (5.71)$$

The potential, E, across AB is

$$E = E_{DB} - E_{DA} = I_2 R_3 - (I_1 - I_2)R_2 \quad (5.72)$$

Substituting the values of I_1 and I_2 given by Eqs. (5.70) and (5.71), respectively, into Eq. (5.72), we have

$$E = V \frac{R_1 R_3 - R_2 R_4}{R_s(R_1 + R_2 + R_3 + R_4) + (R_1 + R_2)(R_3 + R_4)} \quad (5.73)$$

Thus, Eq. (5.73) is the voltage source applied to the equivalent circuit.

The internal resistance of the equivalent circuit must be determined. This is accomplished by looking back into the network from terminals A and B with the potential, V, shorted. The internal resistance of V is added

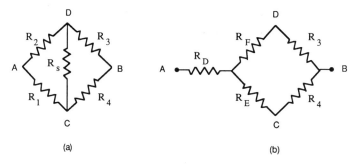

FIG. 5.16. Original circuit (a) and equivalent circuit (b).

to the resistance R_s. The original network and the equivalent network are shown in Fig. 5.16. Figure 5.16a shows R_s across terminals D and C so that resistances R_1, R_2, and R_s form a Delta network that is to be converted to the Wye network, shown by resistances R_D, R_E, and R_F in Fig. 5.16b. We see that R_D is now in series with the parallel resistance formed by $R_F + R_3$ and $R_E + R_4$. The Wye resistances are

$$R_D = \frac{R_1 R_2}{R_1 + R_2 + R_s} \tag{5.74}$$

$$R_E = \frac{R_s R_1}{R_1 + R_2 + R_s} \tag{5.75}$$

$$R_F = \frac{R_s R_2}{R_1 + R_2 + R_s} \tag{5.76}$$

The equivalent resistance, R_{AB}, is

$$R_{AB} = \frac{(R_F + R_3)(R_E + R_4)}{R_E + R_F + R_3 + R_4} + R_D \tag{5.77}$$

The resistance R_{AB} can be expressed in terms of the original resistances shown in Fig. 5.16a. Carrying out the necessary algebra, the final result is

$$R_{AB} = \frac{R_s(R_2 + R_3)(R_1 + R_4) + R_2 R_3(R_1 + R_4) + R_1 R_4(R_2 + R_3)}{R_s(R_1 + R_2 + R_3 + R_4) + (R_1 + R_2)(R_3 + R_4)} \tag{5.78}$$

Again, if resistances R_1, R_2, R_3, and R_4 are increased by their individual ΔR values, then Eq. (5.78) becomes Eq. (5.10), the expression for R_{BO}. Also, if $R_1 = R_2 = R_3 = R_4 = R_g$, then Eq. (5.78) reduces to $R_{AB} = R_g$, regardless of the value of R_s.

The circuit can now be drawn as shown in Fig. 5.17. The voltage source, E, is given by Eq. (5.73). The current flowing through the circuit is the meter current, I_{galvo}. Thus, we can write

$$E = I_{galvo}(R_{BO} + R_m) = I_{galvo} R_m \left(1 + \frac{R_{BO}}{R_m}\right) \quad (5.79)$$

Equation (5.79) can be rewritten as

$$E = E_m \left(1 + \frac{R_{BO}}{R_m}\right) \quad (5.80)$$

where E_m is the voltage drop across the meter.

If we consider the special case in which the bridge has been initially balanced, then Eq. (5.80) can be expressed as

$$\Delta E_0 = \Delta E_{m0} \left(1 + \frac{R_{BO}}{R_m}\right) \quad (5.81)$$

Rearranging Eq. (5.81) to obtain ΔE_{m0}, the change in voltage drop across the meter from a condition of initial balance, we have

$$\Delta E_{m0} = \Delta E_0 \left[\frac{1}{1 + \dfrac{R_{BO}}{R_m}}\right] \quad (5.82)$$

Equation (5.82) shows tht the output is further desensitized by the factor $1/(1 + R_{BO}/R_m)$. Also note that R_{BO} is not a constant, since it varies with the ΔR quantities. If the unit changes in resistance are large, depending upon the relative magnitudes of R_m, some allowance for variation in R_{BO} may be required.

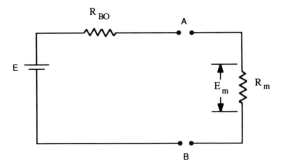

FIG. 5.17. Equivalent circuit for the Wheatstone bridge.

There are two desensitization factors involved, one concerning the resistance in the power supply, R_s, and the other concerning the meter resistance, R_m. Multiplying the right side of Eq. (5.67) by the desensitization factor containing R_m, the change in voltage drop across the meter, ΔE_{m0}, from a condition of initial balance, becomes

$$\Delta E_{m0} = V \left[\frac{1}{1 + \frac{R_s}{R_{BI}}} \right] \left[\frac{1}{1 + \frac{R_{BO}}{R_m}} \right] \left[\frac{1}{(1+a)^2} \right]$$

$$\times \left[\frac{\Delta R_1}{R_1} - \frac{\Delta R_2}{R_2} + \frac{\Delta R_3}{R_3} - \frac{\Delta R_4}{R_4} \right] (1 - n) \quad (5.83)$$

If $R_1 = R_2 = R_3 = R_4 = R_g$, then $R_{BI} = R_{BO} = R_g$, and the bridge ratio is $a = 1$. Using $\Delta R/R = G_F \varepsilon$, Eq. (5.83) becomes

$$\Delta E_{m0} = V \left[\frac{1}{1 + \frac{R_s}{R_g}} \right] \left[\frac{1}{1 + \frac{R_g}{R_m}} \right] \left[\frac{G_F}{4} \right] [\varepsilon_1 - \varepsilon_2 + \varepsilon_3 - \varepsilon_4](1 - n) \quad (5.84)$$

Meter current

The current drawn by the meter, or galvanometer, can be computed by referring to Fig. 5.17. The voltage, E, is given by Eq. (5.73). If the resistances in Eq. (5.73) are increased by the ΔR quantities, as per Eqs. (5.11) through (5.14), to make it compatible with R_{BO}, then the galvanometer current for the unbalanced bridge is

$$I_{galvo} = \frac{E}{R_{BO} + R_m} \quad (5.85)$$

When expanded, Eq. (5.85) becomes Eq. (5.17). For the balanced bridge,

$$I_{galvo} = \frac{\Delta E_{m0}}{R_m} \quad (5.86)$$

Example 5.3. A full bridge is made up of four 120-ohm gages, each with a gage factor of $G_F = 2.05$. The gages are mounted on a cantilever beam, with gages 1 and 3 on the top surface and gages 2 and 4 on the bottom surface directly underneath. Thus, $\varepsilon_1 = \varepsilon_3 = \varepsilon$ and $\varepsilon_2 = \varepsilon_4 = -\varepsilon$. Assume that n may be neglected.

(a) Using an instrument such that $R_m \to \infty$ and $R_s = 0$, determine ΔE_{m0}.
(b) Using an instrument such that $R_m = 350$ ohms and $R_s = 0$, determine ΔE_{m0}.

180   THE BONDED ELECTRICAL RESISTANCE STRAIN GAGE

Solution. (a) Equation (5.84) reduces to Eq. (5.43). Thus,

$$\Delta E_{m0} = \Delta E_0 = \frac{G_F V}{4}[\varepsilon_1 - \varepsilon_2 + \varepsilon_3 - \varepsilon_4](1-n) = \frac{2.05V}{4}[4\varepsilon] = 2.05V\varepsilon$$

(b) Since $R_s = 0$ and $R_{BI} = R_{BO} = R_g$, Eq. (5.84) is

$$\Delta E_{m0} = V\left[\frac{1}{1+\dfrac{R_g}{R_m}}\right]\left[\frac{G_F}{4}\right][\varepsilon_1 - \varepsilon_2 + \varepsilon_3 - \varepsilon_4](1-n)$$

$$= V\left[\frac{1}{1+(120/350)}\right]\left[\frac{2.05}{4}\right][4\varepsilon] = 1.53V\varepsilon$$

The output signal is reduced by approximately 25 percent when a meter with $R_m = 350$ ohms is used.

5.5. Effect of lead-line resistance

When strain gages are located at a test area remote from the instrumentation, lead-line resistance densensitizes the system and produces strain readings lower than those actually occurring. These resistances will not only desensitize the circuit, but they will affect calibration and may also introduce a temperature-compensation problem. The objective now is to examine several common circuit arrangements and determine to what extent each is desensitized by lead-line resistance.

Full bridge

As pointed out in Section 5.4, the internal resistance of the power supply, R_s, could also have been included in the lead-line resistance that is in series with the power supply. Reserving now the symbol R_s for the power supply internal resistance, there is in series with it the lead-line resistance, $2R_{sL}$, as shown in Fig. 5.18. While not stated explicitly in Section 5.4, the meter resistance, R_m, could also have included the lead-line resistance on the output side of the circuit. Again, this is evident in Fig. 5.18. The resistances, R_{sL} on the power side and R_{mL} on the output side, could also contain switch, and other, resistances.

Lead-line resistance can be accounted for without a new analysis by replacing R_s with $R_s + 2R_{sL}$ and R_m with $R_m + 2R_{mL}$ in Eq. (5.83). Thus,

$$\Delta E_{m0} = V\left[\frac{1}{1+\dfrac{R_s+2R_{sL}}{R_{BI}}}\right]\left[\frac{1}{1+\dfrac{R_{BO}}{R_m+2R_{mL}}}\right]\left[\frac{a}{(1+a)^2}\right]$$

$$\times \left[\frac{\Delta R_1}{R_1} - \frac{\Delta R_2}{R_2} + \frac{\Delta R_3}{R_3} - \frac{\Delta R_4}{R_4}\right](1-n) \qquad (5.87)$$

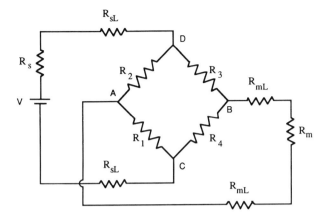

FIG. 5.18. Wheatstone bridge with lead-line resistance.

If $R_1 = R_2 = R_3 = R_4 = R_g$, then $R_{BI} = R_{BO} = R_g$, the bridge ratio is $a = 1$, and using $\Delta R/R = G_F \varepsilon$, Eq. (5.87) can be rewritten as

$$\Delta E_{m0} = V \left[\frac{1}{1 + \dfrac{R_s + 2R_{sL}}{R_g}} \right] \left[\frac{1}{1 + \dfrac{R_g}{R_m + 2R_{mL}}} \right]$$

$$\times \left[\frac{G_F}{4} \right] [\varepsilon_1 - \varepsilon_2 + \varepsilon_3 - \varepsilon_4](1 - n) \qquad (5.88)$$

For the case in which R_m is very large, there is no correction for lead-line resistance on the output side. Thus, for a system where $R_m \to \infty$ (open circuit) and R_s is negligible, Eq. (5.88) reduces to

$$\Delta E_0 = V \left[\frac{R_g}{R_g + 2R_{sL}} \right] \left[\frac{G_F}{4} \right] [\varepsilon_1 - \varepsilon_2 + \varepsilon_3 - \varepsilon_4](1 - n) \qquad (5.89)$$

Therefore, for the remote full bridge the output signal is desensitized (attenuated) by the factor $R_g/(R_g + 2R_{sL})$.

In the circuits that follow, the internal resistance in the power supply, R_s, will be considered negligible and the meter resistance, R_m, will be large enough so that the output side is taken as open.

Half bridge—four wire

In this arrangement, R_1 and R_2 are the active gages and are located at a distance from the instrument. Each lead has a resistance of R_L. The circuit is shown in Fig. 5.19.

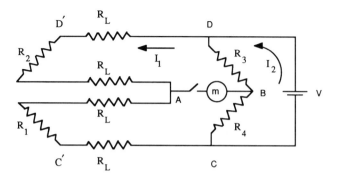

FIG. 5.19. Half bridge with four lead wires.

If the loop equations are written and then solved for the currents, I_1 and I_2, the result is

$$I_1 = \frac{V}{R_1 + R_2 + 4R_L} \tag{5.90}$$

$$I_2 = \frac{V}{R_3 + R_4} \tag{5.91}$$

The potential difference between points A and B is

$$E = I_1(R_1 + 2R_L) - I_2 R_4 \tag{5.92}$$

Substituting the values of I_1 and I_2 given by Eqs. (5.90) and (5.91), respectively, into Eq. (5.92) produces

$$E = V \frac{R_3(R_1 + 2R_L) - R_4(R_2 + 2R_L)}{(R_3 + R_4)(R_1 + R_2 + 2R_L)} \tag{5.93}$$

Equation (5.93) gives the initial output, E, for the unbalanced bridge.

If gages R_1 and R_2 undergo a change in resistance such that R_1 changes to $R_1 + \Delta R_1$ and R_2 changes to $R_2 + \Delta R_2$, then the bridge output will change from E to $E + \Delta E$, and so Eq. (5.93) becomes

$$E + \Delta E = V \frac{R_3(R_1 + \Delta R_1 + 2R_L) - R_4(R_2 + \Delta R_2 + 2R_L)}{(R_3 + R_4)(R_1 + \Delta R_1 + R_2 + \Delta R_2 + 4R_L)} \tag{5.94}$$

If we start with an initially balanced bridge, the initial output, E, is

$$E = 0 = V\frac{R_3(R_1 + 2R_L) - R_4(R_2 + 2R_L)}{(R_3 + R_4)(R_1 + R_2 + 2R_L)} \quad (5.95)$$

From this,

$$R_3(R_1 + 2R_L) - R_4(R_2 + 2R_L) = 0 \quad (5.96)$$

Thus, Eq. (5.94) can be rewritten, for an initially balanced bridge, as

$$\Delta E_0 = V\frac{R_3(R_1 + \Delta R_1 + 2R_L) - R_4(R_2 + \Delta R_2 + 2R_L)}{(R_3 + R_4)(R_1 + \Delta R_1 + R_2 + \Delta R_2 + 4R_L)} \quad (5.97)$$

Equation (5.97) can be written in terms of unit changes in resistance by multiplying and dividing ΔR_1 by R_1 and ΔR_2 by R_2. Doing this, and using Eq. (5.96), the end result is

$$\Delta E_0 = V\frac{R_1 R_3 \dfrac{\Delta R_1}{R_1} - R_2 R_4 \dfrac{\Delta R_2}{R_2}}{(R_3 + R_4)\left(R_1 + R_1 \dfrac{\Delta R_1}{R_1} + R_2 + R_2 \dfrac{\Delta R_2}{R_2} + 4R_L\right)} \quad (5.98)$$

The resistance R_4 can be eliminated from Eq. (5.98) by again using Eq. (5.96). Making this substitution and carrying out the intervening algebra, Eq. (5.98) can finally be rewritten as

$$\Delta E_0 = V\frac{R_1(R_2 + 2R_L)\dfrac{\Delta R_1}{R_1} - R_2(R_1 + 2R_L)\dfrac{\Delta R_2}{R_2}}{(R_1 + R_2 + 4R_L)^2 + (R_1 + R_2 + 4R_L)\left(R_1 \dfrac{\Delta R_1}{R_1} + R_2 \dfrac{\Delta R_2}{R_2}\right)}$$

(5.99)

Equation (5.99) can be put into a more desirable form if we let

$$A = R_1(R_2 + 2R_L)\frac{\Delta R_1}{R_1} - R_2(R_1 + 2R_L)\frac{\Delta R_2}{R_2} \quad (a)$$

$$B = R_1 + R_2 + 4R_L \quad (b)$$

$$C = R_1\frac{\Delta R_1}{R_1} + R_2\frac{\Delta R_2}{R_2} \quad (c)$$

Using Eqs. (a), (b), and (c), Eq. (5.99) becomes

$$\Delta E_0 = V \frac{A}{B^2 + BC} = V \frac{A}{B^2}\left(1 - \frac{C}{B + C}\right) \quad (5.100)$$

The bracketed term in Eq. (5.100) is the nonlinearity factor, $(1 - n)$, and so by substituting the values of A, B, and C given by Eqs. (a), (b), and (c), respectively, back into Eq. (5.100), the output, ΔE_0, becomes

$$\Delta E_0 = V \frac{R_1(R_2 + 2R_L)\dfrac{\Delta R_1}{R_1} - R_2(R_1 + 2R_L)\dfrac{\Delta R_2}{R_2}}{(R_1 + R_2 + 4R_L)^2}(1 - n) \quad (5.101)$$

where

$$n = \frac{1}{1 + \dfrac{R_1 + R_2 + 4R_L}{R_1 \dfrac{\Delta R_1}{R_1} + R_2 \dfrac{\Delta R_2}{R_2}}} \quad (5.102)$$

Letting $R_1 = R_2 = R_g$ and knowing $\Delta R/R = G_F \varepsilon$, the output, ΔE_0, from Eq. (5.101) and n, from Eq. (5.102), can be written in terms of strains. These two equations then become

$$\Delta E_0 = V\left[\frac{R_g}{R_g + 2R_L}\right]\left[\frac{G_F}{4}\right][\varepsilon_1 - \varepsilon_2](1 - n) \quad (5.103)$$

$$n = \frac{1}{1 + \dfrac{2(R_g + 2R_L)}{G_F R_g(\varepsilon_1 + \varepsilon_2)}} \quad (5.104)$$

For this half-bridge arrangement, the output is desensitized (attenuated) by the factor $R_g/(R_g + 2R_L)$. Although R_1 and R_2 were considered active gages, one could be active and the other used as a compensating (dummy) gage for temperature compensation. The dummy gage is mounted on an unstrained piece of material similar to that on which the active gage is mounted, with both gages subjected to the same temperature.

Half bridge—three wire

In this circuit, R_1 and R_2 are located some distance from the instrument, but R_1 and R_2 are joined at A' so that only one lead is brought from this juncture to the instrument. Each lead has a resistance of R_L. The circuit is shown in Fig. 5.20.

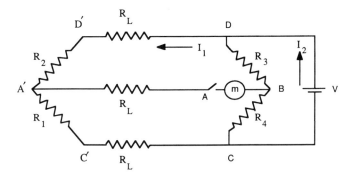

FIG. 5.20. Half bridge with three lead wires.

If the loop equations are written and then solved for the currents, I_1 and I_2, we obtain

$$I_1 = \frac{V}{R_1 + R_2 + 2R_L} \quad (5.105)$$

$$I_2 = \frac{V}{R_3 + R_4} \quad (5.106)$$

The potential difference between points A and B is

$$E = I_1(R_1 + R_L) - I_2 R_4 \quad (5.107)$$

Substituting the values of I_1 and I_2 given by Eqs. (5.105) and (5.106), respectively, into Eq. (5.107) gives the output, E, for the unbalanced bridge. Thus,

$$E = V \frac{R_3(R_1 + R_L) - R_4(R_2 + R_L)}{(R_3 + R_4)(R_1 + R_2 + 2R_L)} \quad (5.108)$$

If gages R_1 and R_2 undergo a change in resistance such that R_1 changes from R_1 to $R_1 + \Delta R_1$ and R_2 changes from R_2 to $R_2 + \Delta R_2$, then the bridge output will change from E to $E + \Delta E$. Equation (5.108) then becomes

$$E + \Delta E = V \frac{R_3(R_1 + \Delta R_1 + R_L) - R_4(R_2 + \Delta R_2 + R_L)}{(R_3 + R_4)(R_1 + \Delta R_1 + R_2 + \Delta R_2 + 2R_L)} \quad (5.109)$$

If we start with an initially balanced bridge and write the output, ΔE_0, in terms of the unit changes in resistance, the final result is

$$\Delta E_0 = V \frac{R_1(R_2 + R_L)\dfrac{\Delta R_1}{R_1} - R_2(R_1 + R_L)\dfrac{\Delta R_2}{R_2}}{(R_1 + R_2 + 2R_L)^2 + (R_1 + R_2 + 2R_L)\left(R_1 \dfrac{\Delta R_1}{R_1} + R_2 \dfrac{\Delta R_2}{R_2}\right)}$$

$$(5.110)$$

Equation (5.110) can be put into a more desirable form, and so it can be rewritten as

$$\Delta E_0 = V \frac{R_1(R_2 + R_L)\dfrac{\Delta R_1}{R_1} - R_2(R_1 + R_L)\dfrac{\Delta R_2}{R_2}}{(R_1 + R_2 + 2R_L)^2}(1 - n) \quad (5.111)$$

where

$$n = \frac{1}{1 + \dfrac{R_1 + R_2 + 2R_L}{R_1 \dfrac{\Delta R_1}{R_1} + R_2 \dfrac{\Delta R_2}{R_2}}} \quad (5.112)$$

Letting $R_1 = R_2 = R_g$ and knowing $\Delta R/R = G_F \varepsilon$, ΔE_0 and n from Eqs. (5.111) and (5.112), respectively, become

$$\Delta E_0 = V \left[\frac{R_g}{R_g + R_L}\right]\left[\frac{G_F}{4}\right](\varepsilon_1 - \varepsilon_2)(1 - n) \quad (5.113)$$

$$n = \frac{1}{1 + \dfrac{2(R_g + R_L)}{G_F R_g (\varepsilon_1 + \varepsilon_2)}} \quad (5.114)$$

The output, ΔE_0, of this circuit is desensitized by the factor $R_g/(R_g + R_L)$; thus, we see that the desensitization of the three-wire half bridge differs from the four-wire half bridge. This circuit can be used in the same manner as the circuit with four wires. Table 5.1 compares the desensitization factors of the two circuits.

Table 5.1. Comparison of desensitization factors for three-wire and four-wire half bridges

Wire resistance, R_L	Three-wire, $R_g/(R_g + R_L)$	Four-wire, $R_g/(R_g + 2R_L)$
0	0	0
5	0.960	0.923
10	0.923	0.857
15	0.889	0.800
20	0.857	0.750
25	0.828	0.706
30	0.800	0.667

Quarter bridge—three wire

In this circuit R_1 is the only active gage and it is located at a distance from the instrument. Three leads of resistance R_L are used in this circuit, with the third lead being brought from the gage to the center point connection, A, at the instrument, as shown in Fig. 5.21. The two lead wires in adjacent arms should be of the same length and maintained at the same temperature. This three-wire circuit is the standard method for a single active temperature-compensated strain gage in this arrangement (7).

If the loop equations are written and then solved for the currents, I_1 and I_2, we have

$$I_1 = \frac{V}{R_1 + R_2 + 2R_L} \tag{5.115}$$

$$I_2 = \frac{V}{R_3 + R_4} \tag{5.116}$$

The potential difference between points A and B is

$$E = I_1(R_1 + R_L) - I_2 R_4 \tag{5.117}$$

Substituting the values of I_1 and I_2 given by Eqs. (5.115) and (5.116), respectively, into Eq. (5.117) gives the output, E, for the unbalanced bridge. Thus,

$$E = V \frac{R_3(R_1 + R_L) - R_4(R_2 + R_L)}{(R_3 + R_4)(R_1 + R_2 + 2R_L)} \tag{5.118}$$

If gage R_1 undergoes a change in resistance from R_1 to $R_1 + \Delta R_1$, then

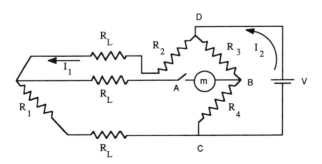

FIG. 5.21. Quarter bridge with three lead wires.

the bridge output will change from E to $E + \Delta E$. Equation (5.118) then becomes

$$E + \Delta E = V \frac{R_3(R_1 + \Delta R_1 + R_L) - R_4(R_2 + R_L)}{(R_3 + R_4)(R_1 + \Delta R_1 + R_2 + 2R_L)} \quad (5.119)$$

If we start from an initially balanced bridge and write the output, ΔE_0, in terms of unit changes in resistances, we have the final result as

$$\Delta E_0 = V \frac{R_1(R_2 + R_L)\dfrac{\Delta R_1}{R_1}}{(R_1 + R_2 + 2R_L)^2}(1 - n) \quad (5.120)$$

where

$$n = \frac{1}{1 + \dfrac{R_1 + R_2 + 2R_L}{R_1 \dfrac{\Delta R_1}{R_1}}} \quad (5.121)$$

Letting $R_1 = R_2 = R_g$ and using $\Delta R/R = G_F \varepsilon$, ΔE_0 and n from Eqs. (5.120) and (5.121), respectively, become

$$\Delta E_0 = V \left[\frac{R_g}{R_g + R_L} \right] \left[\frac{G_F}{4} \right] (\varepsilon_1)(1 - n) \quad (5.122)$$

$$n = \frac{1}{1 + \dfrac{2(R_g + R_L)}{R_g G_F \varepsilon_1}} \quad (5.123)$$

In this circuit the resistor, R_2, is equal to R_g and is located at the instrument. The equations are identical to those for the three-wire half bridge if R_2 in that circuit is a dummy gage. In that case $\Delta R_2/R_2$ and ε_2 are zero; thus, Eqs. (5.111), (5.112), (5.113), and (5.114) reduce to Eqs. (5.120), (5.121), (5.122), and (5.123), respectively.

Quarter bridge—two wire

As in the three-wire quarter bridge, R_1 is the only active gage and it is located some distance from the instrument by two lead wires, each having a resistance of R_L. In this circuit, temperature compensation is lost, and for R_L on the order of 0.5 ohms the bridge will not balance, and so the initial reading will be that for an unbalanced bridge. A value of R_L on the order

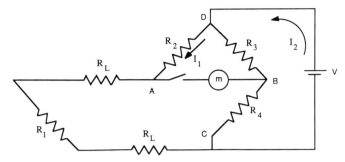

FIG. 5.22. Quarter bridge with two lead wires.

of several ohms will generally be out of the instrument's range and readings cannot be obtained. The circuit is shown in Fig. 5.22.

As before, the potential difference between points A and B is found. It is

$$E = V \frac{R_3(R_1 + 2R_L) - R_2 R_4}{(R_3 + R_4)(R_1 + R_2 + 2R_L)} \quad (5.124)$$

If R_1 undergoes a change in resistance from R_1 to $R_1 + \Delta R_1$, the bridge output will change from E to $E + \Delta E$. Thus,

$$E + \Delta E = V \frac{R_3(R_1 + \Delta R_1 + 2R_L) - R_2 R_4}{(R_3 + R_4)(R_1 + \Delta R_1 + R_2 + 2R_L)} \quad (5.125)$$

If we start with a balanced bridge, the output, ΔE_0, can be written in terms of the unit change in resistance, and so the final result is

$$\Delta E_0 = V \frac{R_1 R_2 \dfrac{\Delta R_1}{R_1}}{(R_1 + R_2 + 2R_L)^2} (1 - n) \quad (5.126)$$

where

$$n = \frac{1}{1 + \dfrac{R_1 + R_2 + 2R_L}{R_1 \dfrac{\Delta R_1}{R_1}}} \quad (5.127)$$

If $R_1 = R_2 = R_g$, then ΔE_0 and n, from Eqs. (5.126) and (5.127), respectively, can be written as

$$\Delta E_0 = V\left[\frac{R_g^2}{(R_g + R_L)^2}\right]\left[\frac{G_F}{4}\right](\varepsilon_1)(1 - n) \tag{5.128}$$

$$n = \frac{1}{1 + \dfrac{2(R_g + R_L)}{G_F R_g \varepsilon_1}} \tag{5.129}$$

If the lead-line resistance in a particular circuit is known, the output voltage, ΔE_0, can be corrected by multiplying it by the reciprocal of the desensitization factor for that circuit. Corrections for the circuits discussed are listed, where ΔE_{0c} is the corrected output voltage.

Full bridge

$$\Delta E_{0c} = \frac{R_g + 2R_L}{R_g} \Delta E_0 \tag{5.130}$$

Half bridge—four wire

$$\Delta E_{0c} = \frac{R_g + 2R_L}{R_g} \Delta E_0 \tag{5.131}$$

Half bridge—three wire

$$\Delta E_{0c} = \frac{R_g + R_L}{R_g} \Delta E_0 \tag{5.132}$$

Quarter bridge—three wire

$$\Delta E_{0c} = \frac{R_g + R_L}{R_g} \Delta E_0 \tag{5.133}$$

Quarter bridge—two wire

$$\Delta E_{0c} = \left(\frac{R_g + R_L}{R_g}\right)^2 \Delta E_0 \tag{5.134}$$

Figure 5.23 shows the influence of lead-line resistance on a half-bridge four-wire circuit. The information plotted is from a cantilever beam test, with one gage on top of the beam and the second gage on the bottom directly underneath.

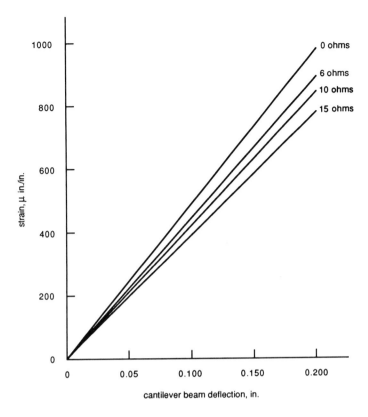

FIG. 5.23. Influence of lead-line resistance, R_L, on a half-bridge, four-wire circuit.

Figure 5.24 shows the influence of lead-line resistance on a half-bridge three-wire circuit. The same cantilever beam was used, but it is apparent that the attenuation of this circuit is less than that of the four wires. A comparison of Eqs. (5.131) and (5.132) shows the reason for this.

Example 5.4. The linear drive tube of a machine has four 120-ohm gages, forming a full bridge, mounted on it in order to determine the longitudinal force acting on the tube. Gages 1 and 3 are aligned parallel to the longitudinal axis and are 180° apart, while gages 2 and 4 are mounted transverse to the longitudinal axis. The bridge is connected to the instrumentation, located in a control booth, with 100 ft of No. 26 copper wire having a resistance of 4.081 ohms/100 ft. Figure 5.25 shows the drive tube and bridge arrangement. Determine the output voltage.

Solution. With gages 1 and 3 in opposite arms, as well as gages 2 and 4, bending strains will be canceled and only longitudinal compressive strains will be recorded. Furthermore, the nonlinearity factor will be small and can be

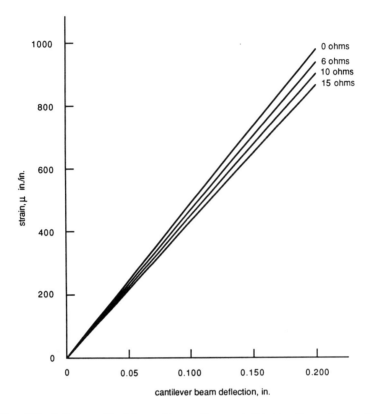

FIG. 5.24. Influence of lead-line resistance, R_L, on a half-bridge, three-wire circuit.

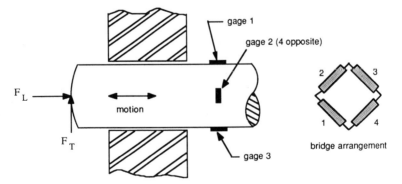

FIG. 5.25. Drive tube with bridge arrangement for measuring axial force.

disregarded. Thus,

$$\varepsilon_1 = \varepsilon_3 = -\varepsilon \quad \text{and} \quad \varepsilon_2 = \varepsilon_4 = v\varepsilon$$

Ignoring the lead-line resistance for the moment, Eq. (5.43) can be used to compute ΔE_0:

$$\Delta E_0 = \frac{G_F V}{4}(\varepsilon_1 - \varepsilon_2 + \varepsilon_3 - \varepsilon_4)$$

$$= \frac{G_F V}{4}[(-\varepsilon) - (v\varepsilon) + (-\varepsilon) - (v\varepsilon)] = -\frac{G_F V}{2}(1+v)\varepsilon$$

The voltage, ΔE_0, can be corrected by using Eq. (5.130).

$$\Delta E_{0c} = \frac{R_g + 2R_L}{R_g}\Delta E_0 = \left[\frac{120 + 2(4.081)}{120}\right]\Delta E_0$$

$$\Delta E_{0c} = -1.068\left[\frac{G_F V}{2}\right](1+v)\varepsilon$$

This result shows that the signal was reduced by approximately 6.8 percent. In passing, note that if 350-ohm gages were used, the correction factor would be

$$\frac{R_g + 2R_L}{R_g} = \frac{350 + 2(4.081)}{350} = 1.023$$

Thus, the signal would be reduced by approximately 2.3 percent, and so, if long lead lines are used, it would be better to use higher-resistance gages.

5.6. Circuit calibration

The two basic methods of calibrating a strain gage circuit are mechanical and electrical (8, 9). The mechanical calibration method, while good for establishing the validity of the measuring system, is inconvenient and costly for regular use. In this section, electrical calibration only will be considered, where a calibration resistor, R_c, is shunted across one of the gages. Furthermore, it will be assumed that the permissible error will be such that the nonlinearity of the Wheatstone bridge can be neglected. As a further restriction, only arm R_1 will be shunted, as shown in Fig. 5.26. For a detailed analysis of shunt calibration, for both small and large strains, Reference 9 is recommended.

When the resistor R_c is shunted across R_1, where $R_1 = R_g$, the total resistance in that arm is reduced. The equivalent resistance is

$$R = \frac{R_c R_g}{R_g + R_c} \quad (5.135)$$

THE BONDED ELECTRICAL RESISTANCE STRAIN GAGE

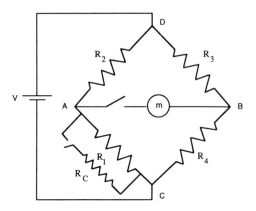

FIG. 5.26. Wheatstone bridge with calibration resistor.

The change in resistance in the bridge arm is

$$\Delta R_g = R - R_g = \frac{R_c R_g}{R_g + R_c} - R_g = -\frac{R_g^2}{R_g + R_c} \quad (5.136)$$

Dividing both sides of Eq. (5.136) by R_g gives

$$\frac{\Delta R_g}{R_g} = -\frac{R_g}{R_g + R_c} \qquad \frac{350}{87150+350} \quad (5.137)$$

-0.004   -1.4 / 350

Since $\Delta R/R = G_F \varepsilon$, the equivalent strain produced by shunting R_c across R_1 is

$$\varepsilon_{eq} = -\frac{R_g}{G_F(R_g + R_c)} \qquad -\frac{350}{2.15(350+87150)} \quad (5.138)$$

= -0.0018604651

The negative sign tells us that this calibration method produces an equivalent strain that is compressive in sense. Precision calibration resistors can be purchased, using $G_F = 2.0$, that will give microstrains of even values, such as 500, 1000, etc. This method can be employed whether or not a quarter-, half-, or full-bridge circuit is being used. Knowing the bridge arrangement, the surface strain at the primary gage can be found by calculation. It should be noted that the shunt is applied at the gage and not at the instrument.

Example 5.5. Determine the value of R_c that will produce an equivalent strain of -500 μin/in when $G_F = 2.0$ and $R_g = 120$ ohms.

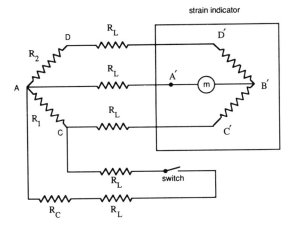

FIG. 5.27. Circuit with calibration resistor, R_C, shunted across resistor R_1.

Solution. Solving Eq. (5.138) for R_c produces

$$R_c = -\frac{R_g}{G_F \varepsilon_{eq}} - R_g = -\frac{120}{2(-500 \times 10^{-6})} - 120 = 119\,880 \text{ ohms}$$

The calibration of a circuit with gages mounted remote from the instrument and that have equal resistance, R_L, in each lead line will be considered. Equations (5.130) through (5.134) show the factors by which the indicated output voltage (or indicated strain) will have to be multiplied in order to obtain the true output value.

Figure 5.27 shows a half-bridge arrangement with R_1 an active gage and R_2 being either an active or a dummy gage. The calibration resistor, R_c, can be located at either R_1 or back near the instrument, but in either case its leads also have the same resistance, R_L. In general, for high values of R_c, its lead resistances will have little effect on the calibration strain. When R_L is now shunted across R_1, the gain (gage factor setting) of the instrument can be adjusted so that the indicated strain reads the calibration strain. For subsequent loading, the instrument will now read the strains directly. Although a half bridge has been shown, the method also applies to a quarter, half, or full bridge.

5.7. Comments

In the development of the bridge equations, the output of the bridge has been in terms of voltage, specified either as ΔE_0 or ΔE_{m0}. In the strain instrumentation generally used, the instrument is calibrated to read directly in strain. Furthermore, if a full bridge is considered, as shown in Fig. 5.28, we have learned, starting with arm 1, that the arms alternate in sign. Thus,

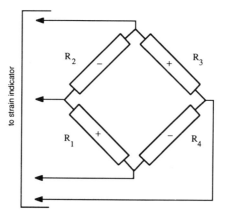

FIG. 5.28. Wheatstone bridge showing the signs of the respective arms.

if a gage connected in arm 2 is subjected to a compressive strain, the sign will be changed and the indicator will give a positive value. Because of this property of the Wheatstone bridge, bridge circuits can be arranged in such a manner that we can isolate, for instance, the effect of an axial force and null out the effects of bending. The bridge arrangement in Fig. 5.10, as an example, will do just that by canceling bending strains and producing the strains of the direct axial force.

Many times, a number of strain gages, used in quarter-bridge circuits, may be bonded at various locations on a structure. Because it would be time-consuming and awkward to connect each strain gage, in turn, to a strain indicator and then load the structure, a switching and balancing unit is used in conjunction with the strain indicator. A typical multichannel application is shown in Fig. 5.29 with a strain indicator and its companion switching and balancing unit.

In this application, a number of gages are connected to the switching and balancing unit which, in turn, is connected to the strain indicator. Here, six of a total of ten channels are used. The switch is turned to each channel and the strain indicator is balanced by using the balancing potentiometer of the individual circuit. Then, at every load level, the switch is turned to each channel and that strain recorded.

Since the strain gages may not all be alike (single gages and rosettes may be mixed), there will be several different gage factors. In this case, set one value of G_F on the strain indicator and correct the indicated strain by calculation. Since $\Delta R/R$ will be the same regardless of the value of G_F used, we can write

$$\frac{\Delta R}{R} = G'_F \varepsilon' = G_F \varepsilon \tag{5.139}$$

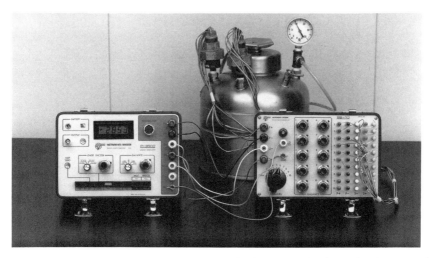

FIG. 5.29. Multichannel arrangement using a switching and balancing unit. (Courtesy of Measurements Group, Inc.)

This is rearranged to

$$\varepsilon = \frac{G'_F}{G_F} \varepsilon' \qquad (5.140)$$

where G'_F = gage factor set on the strain indicator

G_F = gage factor of the strain gage

ε' = indicated strain

ε = corrected (actual) strain

Therefore, once all of the indicated strains are recorded, Eq. (5.140) can be used to determine the actual strains.

Problems

5.1. A full bridge, made up of 120-ohm gages, has a constant-voltage power supply of 10 volts. The following resistors are shunted, in turn, across arm R_1: 119 880, 11 880, 1080, 360, 120, 40, and 10 ohms. Using Eqs. (5.39) and (5.40), plot ΔE_0 vs. $\Delta R_1/R_1$.

In Probs. 5.2 through 5.7, use Eqs. (5.39) and (5.40) to determine an expression for $\Delta E_0/V$ and $\varepsilon/\varepsilon_i$.

5.2. In Fig. 5.9, gage 1 is the only active gage, so that $\varepsilon_1 = \varepsilon$.
5.3. In Fig. 5.10, gages 1 and 2 are the active gages, thus $\varepsilon_1 = \varepsilon$ and $\varepsilon_2 = -\nu\varepsilon$.
5.4. In Fig. 5.9, gages 1 and 2 are the active gages. In this case $\varepsilon_1 = \varepsilon$ and $\varepsilon_2 = -\varepsilon$.
5.5. In Fig. 5.9, gages 1 and 3 are active gages, so that $\varepsilon_1 = \varepsilon_3 = \varepsilon$.

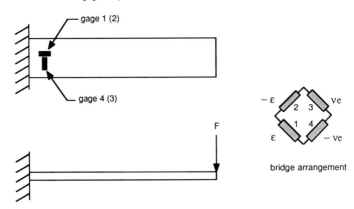

Fig. 5.30.

5.6. A cantilever beam has gages arranged as shown in Fig. 5.30. Gages 1 and 2 are longitudinal gages, mounted top and bottom, respectively. Gage 3 is mounted on the bottom transverse to gage 2, while gage 4 is mounted on the top transverse to gage 1.

5.7. The gages of the cantilever beam in Fig. 5.30 are rewired into the bridge arrangement shown in Fig. 5.31.

5.8. A small assembly machine has the dimensions shown in Fig. 5.32. Gages 1 and 2 are bonded at the inner and outer radius, respectively, in a longitudinal direction. Each gage is read individually, with $\varepsilon_1 = 1083\ \mu\text{in/in}$ and $\varepsilon_2 = -652\ \mu\text{in/in}$. Determine the stresses at each gage location as well as the load acting on the machine. The material is steel.

5.9. Two steel sleeves are shrunk together, as shown in Fig. 5.33. The nominal radii are $a = 2.00$ in, $b = 2.75$ in, and $c = 3.25$ in. After assembly, a strain gage is bonded to the outer cylinder in the hoop (tangential) direction, the strain indicator is balanced, and then the inner cylinder is pushed out. After disassembly, the strain gage gives a reading of $-840\ \mu\text{in/in}$. Determine the shrink-fit pressure and the amount of interference.

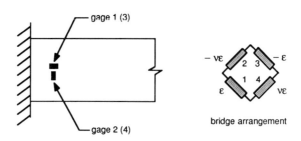

Fig. 5.31.

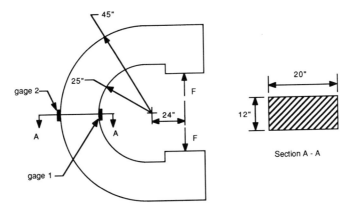

FIG. 5.32.

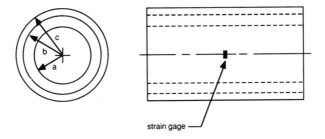

FIG. 5.33.

5.10. A cantilever beam, shown in Fig. 5.34, has a width of 2 in and a thickness of 0.250 in. A weight of 25 lb can be positioned at any point between 10 in and 18 in from the support. Strain gages are to be placed at 1 in and 8 in from the support.

(a) Show that the difference in the moments at the strain gage locations will be the same for any position of the load within its range; that is, $\Delta M = M_1 - M_8$.

(b) Determine a suitable full-bridge arrangement that will give the strain associated with ΔM and determine its magnitude.

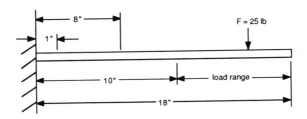

FIG. 5.34.

200   THE BONDED ELECTRICAL RESISTANCE STRAIN GAGE

5.11. A thick-walled cylinder of steel with capped ends is subjected to an internal pressure. The inner radius is 2 in and the outer radius is 3.125 in. On the outside surface at mid-length, two strain gages are bonded. Gage 1 is in the circumferential (hoop) direction and gage 2 is in the longitudinal direction. After pressurization the following readings are obtained:

$$\varepsilon_1 = 590 \text{ μin/in}, \qquad \varepsilon_2 = 139 \text{ μin/in}$$

Determine the stress state and the internal pressure.

5.12. Figure 5.35 shows a cantilever beam with offset loading. Four longitudinal strain gages are bonded to the beam at section A–A and then arranged into the bridge circuits illustrated in A, B, C, and D. Beneath each bridge circuit is the strain indicator reading. Determine the loads, F_x, F_y, and F_z, as well as the total strain at each gage.

5.13. When a shaft is in pure torsion, the principal stresses, and therefore the principal strains, lie at $\pm 45°$ to the longitudinal axis. If a pair of strain gages are bonded to the shaft in these directions and another pair are bonded diametrically opposite, then, if they are arranged into a proper full bridge, only the torsional effect will be measured by the bridge. Furthermore, if the shaft is subjected to bending moments or axial forces, their effect will be canceled. Figure 5.36 shows a section of the shaft.

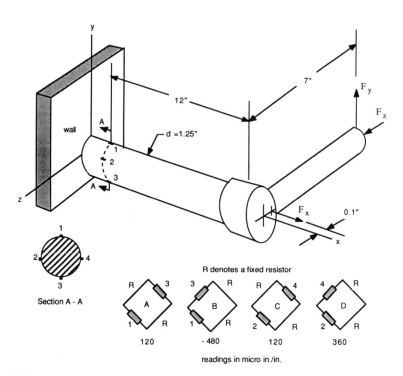

FIG. 5.35.

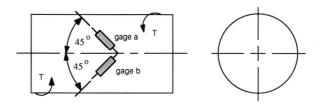

gages c and d diametrically opposite gages a and b

FIG. 5.36.

(a) Sketch a Mohr's circle and verify the strain directions.
(b) Show how the gages should be arranged into a full bridge.
(c) Explain why strains due to bending moments or axial forces will cancel.

5.14. The dies on a two-post casting machine are to be set so that each post has an equal axial force. Two gages are bonded, 180° apart, to each post as shown in Fig. 5.37. The gages are arranged in turn to form the bridge circuits shown, along with their respective readings after loading. If the posts are 3.0 in in diameter, determine the following:
  (a) The axial force in each post.
  (b) The bending moment in each post in the plane containing gages.

5.15. A round tension link made of steel carries a maximum load of 50 000 lb.
  (a) Arrange four strain gages into a full bridge so that temperature compensation is achieved and only tensile loading is measured.

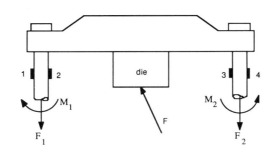

R denotes a fixed resistor

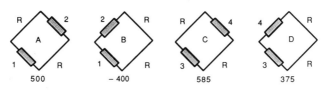

readings in micro in./in.

FIG. 5.37.

(b) Determine the link diameter if no individual strain gage is subjected to more than 1500 μin/in.

(c) Using a gage factor of 2.0, determine $\Delta E_0/V$ at the maximum load.

5.16. A circuit has the following resistances:

$$R_1 = R_2 = 120 \text{ ohms}, \qquad R_3 = R_4 = 500 \text{ ohms}, \qquad R_m = 750 \text{ ohms},$$

$$R_s = 0$$

If R_1 and R_2 change by 15 percent, what is the percentage change in R_{BI}?

5.17. Compute the bridge resistance for arm resistances of R_1, R_2, R_3, and R_4 when $R_s = 0$ and $R_m = \infty$. Use Thevenin's theorem.

5.18. A Wheatstone bridge has the following resistances:

$$R_1 = R_3 = 120 \text{ ohms}, \qquad R_2 = R_4 = 600 \text{ ohms}, \qquad R_m = 500 \text{ ohms},$$

$$R_s = 7 \text{ ohms}$$

If the bridge is initially balanced and $V = 10$ volts, determine ΔE_{m0} for the following conditions:

(a) Resistances R_1 and R_3 increase by 1 percent.

(b) Resistances R_1 and R_3 increase by 15 percent.

5.19. If R_1, R_2, R_3, and R_4 in Problem 5.18 each *increase* by 15 percent, determine ΔE_{m0}.

5.20. Using the values given in Problem 5.18 for R_1, R_2, R_3, R_4, and V, let R_1 have the following percentage changes: 0.5, 1.0, 2.0, 5.0, 10.0, and 15.0.

(a) For $R_s = 0$ and $R_m = \infty$, plot ΔE_{m0} vs. the percentage change in R_1.

(b) For $R_s = 0$ and $R_m = 750$ ohms, plot ΔE_{m0} vs. the percentage change in R_1.

5.21. An aluminum cantilever beam, shown in Fig. 5.38, has four strain gages bonded to it. Gages a and b are on the top of the beam, with gage a being a longitudinal gage and gage b being a transverse gage. Gage c (longitudinal) and gage d (transverse) are directly underneath. The following data are given:

$$R_g = 120 \text{ ohms}, \qquad G_F = 2.08, \qquad E = 10 \times 10^6 \text{ psi}, \qquad \nu = 0.33,$$

$$R_s = 0, \qquad V = 10 \text{ volts}$$

(a) Arrange the gages into a full bridge in order to get the maximum reading.

(b) When the end of the beam is deflected 0.225 in, determine ΔE_{m0} if $R_m = \infty$; if $R_m = 3000$ ohms; if $R_m = 450$ ohms.

(c) Is it worthwhile considering the change in R_{BO} or to compute n?

5.22. A round, hollow shaft of steel has four 120-ohm gages bonded to it. The gages are arranged in a full bridge in order to function as a torque meter (see Fig. 5.36). The shaft has an outer diameter of 1.500 in, an inner diameter of 1.125 in, and is subjected to 7000 in-lb of torque. If $V = 10$ volts, $G_F = 2.07$, and the lead-line resistance is 2.0 ohms, determine ΔE_0.

5.23. A weight, W, is at rest as shown in Fig. 5.39. When the weight is released it falls onto the stop, where a latch is engaged that keeps it from rebounding.

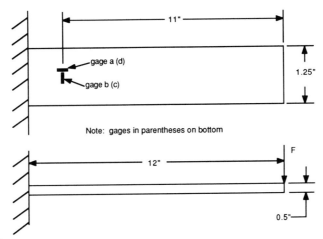

FIG. 5.38.

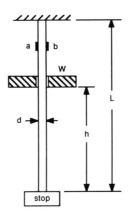

FIG. 5.39.

Strain gages a and b are bonded longitudinally to the vertical bar and wired into opposite arms of a full bridge, with the gages in adjacent arms being dummy gages. Each lead of the bridge has a resistance of $R_L = 1.5$ ohms. Assuming the stress is uniformly distributed throughout the length of the vertical bar, determine W if the maximum strain recorded by the bridge is 1520 μin/in, $d = 0.505$ in, $v = 0.3$, $E = 30 \times 10^6$ psi, $R_g = 120$ ohms, $h = 18$ in, and $L = 30$ in.

5.24. The dummy gages in Problem 5.23 are replaced with gages bonded transverse to gages a and b. The lead wires are also extended so that each lead has a resistance of $R_L = 2.5$ ohms. Using the value of W from Problem 5.23, determine the maximum indicated strain that the meter would record for a repeated test.

5.25. Four 120-ohm gages are bonded to a machine element and individually connected to a strain indicator through a switching and balancing unit, using the three-wire quarter-bridge circuit shown in Fig. 5.21. The following data are given:

Gage No.	G_F	R_L	ε, µin/in
1	1.95	0	1950
2	2.075	0	1245
3	2.00	0	−500
4	2.15	4	975

A gage factor of 2.0 is set on the strain indicator. Determine the actual strain at each gage.

REFERENCES

1. "Portable Digital Strain Indicator P-350A," Bulletin 130-A, Measurements Group, Inc., P.O. Box 27777, Raleigh, NC 27611, 1980. (Now out of print.)
2. "Portable Strain Indicator P-3500," Bulletin 245, Measurements Group, Inc., P.O. Box 27777, Raleigh, NC 27611, 1983.
3. *Handbook of Experimental Stress Analysis*, edited by M. Hetenyi, New York, Wiley, 1950, pp. 191–193.
4. Stein, Peter K., "Strain-Gage-Based Computers," *Strain Gage Readings*, Vol. IV, No. 4, Oct.–Nov. 1961, pp. 17–50. Also, Chap. 26 in *The Strain Gage Encyclopaedia*, Vol. II of *Measurement Engineering*, by Peter K. Stein, 1962, 2d edition, Stein Engineering Services, Inc., Phoenix, AZ 85018-4646. (Now out of print.) (1960, 1st edition.)
5. "Errors Due to Wheatstone Bridge Nonlinearity," TN-507, Measurements Group, Inc., P.O. Box 27777, Raleigh, NC 27611, 1982.
6. Herbert W. Jackson and Preston A. White, III, *Introduction to Electric Circuits*, 7e, © 1989, pp. 213, 236. Adapted by permission of Prentice-Hall, Englewood Cliffs, New Jersey.
7. "Student Manual for Strain Gage Technology," Bulletin 309B, Measurements Group, Inc., P.O. Box 27777, Raleigh, NC 27611, 1983, p. 24.
8. *Handbook on Experimental Mechanics*, edited by A. S. Kobayashi, Englewood Cliffs, Prentice-Hall, 1987, pp. 102–104.
9. "Shunt Calibration of Strain Gage Instrumentation," TN-514, Measurements Group, Inc., P.O. Box 27777, Raleigh, NC 27611, 1988.

6
SENSITIVITY VARIATION

6.1. Introduction

Reasons for varying strain sensitivity

Why should one desire to vary the sensitivity of strain gages, or the circuits of which they form a part? In general, this requirement stems from a need to put the indications from two or more strain gages on a common basis, or in the correct relative proportions. There are numerous special situations which may show up. However, a few of the more common cases requiring sensitivity variation are listed as follows (1):

1. To allow for differences in gage factor among individual gages when the readout for all gages is to be made directly in terms of strain on a single scale. For example, the gage factor dial adjustment on strain indicators.
2. To combine the indications from several strain gages in different relative proportions. For example, the direct and automatic computation of some quantity whose indication depends upon a combination of two, or more, strain indications in specified relative proportions.
3. To facilitate the use of an instrument which has a limited input range with a strain gage that develops an output which is larger than the maximum that can be accepted by the instrument. For example, the use of a standard strain indicator designed for metallic gages with a semiconductor gage that is subjected to a reasonably large strain. The same sort of situation may also prevail when a metal gage is used to measure post-yield strains of several percent.
4. To adjust the calibration factor of a transducer to some convenient round number. For example, to adjust to a readout of 1000 on the indicator scale for 1000 units of the quantity being measured, as contrasted with an indicator reading of 981 per 1000 units being measured.
5. For automatically correcting an indication for some uncontrolled variable which may change by unknown amounts. For example, the compensation of a load cell or torque meter indication for the influence of temperature changes on the modulus of elasticity of the load-carrying member.

6. For producing a direct readout of some quantity which is indicated by the product of two independent quantities. For example, the measurement of the instantaneous value of power being transmitted by a circular shaft. This can be accomplished by using a strain gage bridge to sense the torque and energizing it with a variable applied voltage (variable sensitivity) that is proportional to the speed of rotation.

Indicated strain vs. actual strain

One will recall that strain is sensed through a change in gage resistance according to the following relationship:

$$\varepsilon = \frac{1}{G_F}\left(\frac{\Delta R_g}{R_g}\right) \tag{6.1}$$

Provided there are no inactive resistances in series (or parallel) with the gage, the readout instrument will be able to indicate the correct value of strain in accordance with Eq. (6.1). However, if there are inactive resistances (relative to strain) in series and (or) parallel with the gage, these will, to some extent, mask the observation the instrument is making so that the indicated strain being read out is only a fraction of that actually prevailing at the gage. The corresponding relation for the indicated strain is given by

$$\varepsilon_i = \frac{1}{G_F}\left(\frac{\Delta R_t}{R_t}\right) = \frac{Q_t}{G_F}\left(\frac{\Delta R_g}{R_g}\right) \tag{6.2}$$

where ε_i = the indicated strain

Q_t = the desensitization factor, whose numerical value is less than 1

From Eq. (6.2),

$$\frac{\Delta R_t}{R_t} = Q_t \frac{\Delta R_g}{R_g} \tag{6.3}$$

The reason for the desensitization, or reduction in indicated strain, when series and parallel resistances are connected to the gage, is that these additional resistances contribute nothing to the change in resistance in spite of the fact that they have an influence on the total overall value as seen by the indicating device. This desensitization became apparent when lead-line resistance was considered in Section 5.5.

SENSITIVITY VARIATION

Kinds of desensitization

Strain gage desensitization due to the effects of resistances in series and in parallel with the gage can be considered from two points of view, depending upon whether the effect represents an inconvenience that must be overcome or an advantage that can be employed for some specific purpose. One may therefore look upon desensitization as falling into one or the other of the two categories that follow.

1. *Parasitic desensitization.* This is caused by such things as lead-wire resistance and parallel resistances which are brought into the circuit for trimming and balancing purposes. This is something that must be accepted. Usually (although not always) the parasitic desensitization produces a small deviation from the theoretical calibration factor. The important thing is to appreciate that this condition prevails and to be able to make a reasonably good estimate of the magnitude of its effect.
2. *Planned desensitization.* This involves the understanding of the factors which contribute to desensitization and the deliberate manipulation of them in order to produce certain desired results, such as those indicated in the introduction in the reasons for varying strain sensitivity.

Other approaches to sensitivity variation

Since the resistance change of a strain gage is actually determined by the corresponding effect on voltage or current, we may also approach the problem of sensitivity variation by control of the applied voltage, or the gage current.

One may consider the use of a resistance network connected to the gage as a primary means of achieving sensitivity control since this produces a direct effect upon the indicated relation between strain and unit change in gage resistance, independently of gage current or applied voltage.

On the other hand, variation of sensitivity through control of applied voltage, or gage current, means that we have to express the indication of sensitivity in terms of voltage or current changes per unit of strain. Furthermore, for those systems, such as null balance and some of the reference bridge arrangements, which produce an indication that is independent of variations in applied voltage or gage current, this method of varying the sensitivity is inapplicable.

6.2. Analysis of single gage desensitization (1, 2)

Resistance in series

Figure 6.1 shows a strain gage, R_g, desensitized by placing a resistor, R_s, in series with it. The initial total resistance is

$$R_t = R_g + R_s \tag{6.4}$$

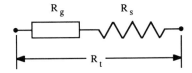

FIG. 6.1. Resistance in series with gage.

After a change in gage resistance, ΔR_g, we have

$$R_t + \Delta R_t = R_g + \Delta R_g + R_s$$

Dividing all terms by R_t and rearranging, the result is

$$1 + \frac{\Delta R_t}{R_t} = \frac{R_g + R_s}{R_t} + \frac{\Delta R_g}{R_t} \tag{6.5}$$

or

$$\frac{\Delta R_t}{R_t} = \frac{\Delta R_g}{R_t} + \frac{R_g + R_s}{R_t} - 1$$

Since $R_t = R_g + R_s$, this reduces to

$$\frac{\Delta R_t}{R_t} = \frac{\Delta R_g}{R_g + R_s} \tag{6.6}$$

If the numerator and denominator of the right-hand side are divided by R_g, then

$$\frac{\Delta R_t}{R_t} = \left[\frac{1}{1 + (R_s/R_g)}\right] \frac{\Delta R_g}{R_g} \tag{6.7}$$

Letting $R_s/R_g = s$, then

$$\frac{\Delta R_t}{R_t} = \left[\frac{1}{1 + s}\right] \frac{\Delta R_g}{R_g} \tag{6.8}$$

In this case, the nonlinearity factor, n, is zero, and the desensitization factor is

$$Q_t = Q = \frac{1}{1 + s} \tag{6.9}$$

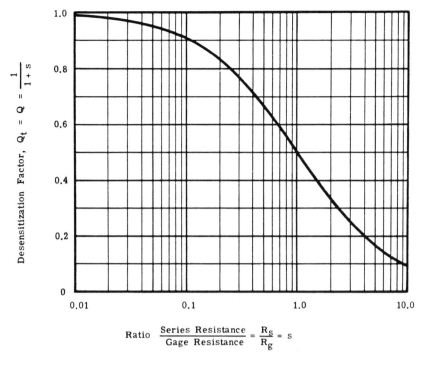

FIG. 6.2. Desensitization of a single gage with series resistance.

Figure 6.2 shows the value of the desensitization factor, Q, as a function of the ratio, s, of series resistance to gage resistance.

Resistance in parallel

Figure 6.3 shows a resistor, R_p, in parallel with the strain gage, R_g, in order to desensitize the strain gage. Initially, the total resistance is

$$R_t = \frac{R_g R_p}{R_p + R_g} \tag{6.10}$$

After straining,

$$R_t + \Delta R_t = \frac{(R_g + \Delta R_g) R_p}{R_p + R_g + \Delta R_g} \tag{6.11}$$

Dividing both sides by R_t results in

$$1 + \frac{\Delta R_t}{R_t} = \left[\frac{(R_g + \Delta R_g) R_p}{R_p + R_g + \Delta R_g} \right] \left[\frac{R_p + R_g}{R_g R_p} \right] \tag{6.12}$$

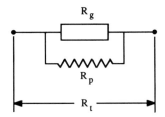

FIG. 6.3. Resistance in parallel with gage.

From this,

$$\frac{\Delta R_t}{R_t} = \frac{(R_g + \Delta R_g)(R_p + R_g)}{R_g(R_p + R_g + \Delta R_g)} - 1 \tag{6.13}$$

Expanding the right-hand side of Eq. (6.13) results in

$$\frac{\Delta R_t}{R_t} = \frac{R_p}{R_p + R_g + \Delta R_g}\left(\frac{\Delta R_g}{R_g}\right)$$

or

$$\frac{\Delta R_t}{R_t} = \frac{R_p}{R_g\left[1 + \dfrac{R_p}{R_g} + \dfrac{\Delta R_g}{R_g}\right]}\left(\frac{\Delta R_g}{R_g}\right) \tag{6.14}$$

For simplicity in writing, let $p = R_p/R_g$. Using this, Eq. (6.14) is rewritten as

$$\frac{\Delta R_t}{R_t} = \left[\frac{p}{1 + p + \dfrac{\Delta R_g}{R_g}}\right]\left(\frac{\Delta R_g}{R_g}\right)$$

Dividing the numerator and denominator of the right-hand side by $(1 + p)$ results in

$$\frac{\Delta R_t}{R_t} = \left[\frac{p}{1 + p}\right]\left[\frac{1}{1 + \dfrac{1}{1+p}\dfrac{\Delta R_g}{R_g}}\right]\left(\frac{\Delta R_g}{R_g}\right) \tag{6.15}$$

From Eq. (6.15),

$$Q_t = \left[\frac{p}{1 + p}\right]\left[\frac{1}{1 + \dfrac{1}{1+p}\dfrac{\Delta R_g}{R_g}}\right] \tag{6.16}$$

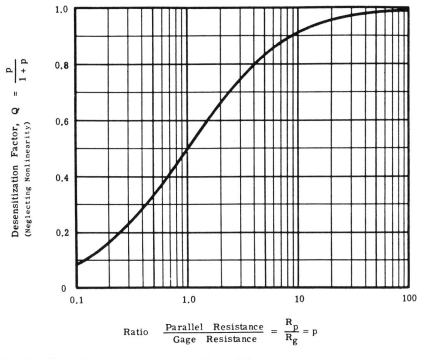

FIG. 6.4. Desensitization of a single gage with parallel resistance.

and

$$Q = \frac{p}{1+p} = \frac{1}{\frac{1}{p}+1} \quad (6.17)$$

Also,

$$n = \left[\frac{1}{1+p}\right]\frac{\Delta R_g}{R_g} \quad (6.18)$$

The value of Q, the desensitization factor exclusive of nonlinearities, is shown as a function of p in Fig. 6.4. Here it is seen that for values of p greater than 100, the desensitization will be less than 1 percent.

Equation (6.18) indicates that, as long as the parallel resistance, R_p, is greater than the gage resistance, the nonlinearity factor will be less than $(0.5)(\Delta R_g/R_g)$.

Combination of series and parallel resistances

An examination of the work covering resistances in series and resistances in parallel reveals that R_t is greater than R_g when series resistance is employed, and R_t is less than R_g when parallel resistance is used.

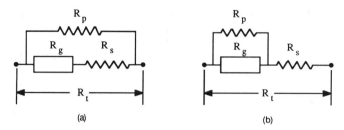

FIG. 6.5. Two arrangements of connecting series and parallel resistances to a gage.

In the event that it is desired to desensitize a strain gage in one arm of a bridge when the resistance in each of the other arms corresponds to R_g, the two previous methods of desensitization are unsuitable because the bridge cannot be initially balanced. We will now investigate how series and parallel resistances may be combined so that $R_t = R_g$, which condition will permit initial balance of the bridge. Figure 6.5 illustrates two alternative methods for connecting series and parallel resistances to a strain gage. In the following analysis, the arrangement in Fig. 6.5a will be analyzed.

The initial resistance, R_t, as seen by the readout instrument, is

$$R_t = \frac{(R_g + R_s)R_p}{R_g + R_s + R_p} \tag{6.19}$$

After straining, the gage resistance changes to $R_g + \Delta R_g$, and so Eq. (6.19) becomes

$$R_t + \Delta R_t = \frac{(R_g + \Delta R_g + R_s)R_p}{R_g + \Delta R_g + R_s + R_p} \tag{6.20}$$

If both sides of Eq. (6.20) are divided by R_t, the result is

$$1 + \frac{\Delta R_t}{R_t} = \left[\frac{R_g + \Delta R_g + R_s}{R_g + \Delta R_g + R_s + R_p}\right]\left[\frac{R_g + R_s + R_p}{R_g + R_s}\right]$$

Divide the numerator and denominator of each bracketed term on the right-hand side by R_g, then

$$1 + \frac{\Delta R_t}{R_t} = \left[\frac{1 + \dfrac{\Delta R_g}{R_g} + \dfrac{R_s}{R_g}}{1 + \dfrac{\Delta R_g}{R_g} + \dfrac{R_s}{R_g} + \dfrac{R_p}{R_g}}\right]\left[\frac{1 + \dfrac{R_s}{R_g} + \dfrac{R_p}{R_g}}{1 + \dfrac{R_s}{R_g}}\right] \tag{6.21}$$

SENSITIVITY VARIATION

Since $s = R_s/R_g$ and $p = R_p/R_g$, Eq. (6.21) can be rewritten as

$$1 + \frac{\Delta R_t}{R_t} = \left[\frac{1 + \frac{\Delta R_g}{R_g} + s}{1 + \frac{\Delta R_g}{R_g} + s + p}\right]\left[\frac{1 + s + p}{1 + s}\right] \quad (6.22)$$

For initial bridge balance, however, $R_t = R_g$, so Eq. (6.19) becomes

$$R_g = \frac{(R_g + R_s)R_p}{R_g + R_s + R_p} \quad (6.23)$$

In terms of ratios,

$$R_g = \frac{(1 + s)R_p}{1 + s + p} \quad (6.24)$$

If both sides of Eq. (6.24) are divided by R_p, then

$$\frac{R_g}{R_p} = \frac{1}{p} = \frac{1 + s}{1 + s + p} \quad (6.25)$$

Equation (6.25) shows that the last bracketed term on the right-hand side of Eq. (6.22) is equal to p, and so Eq. (6.22) is rewritten as

$$1 + \frac{\Delta R_t}{R_t} = \frac{\left[1 + s + \frac{\Delta R_g}{R_g}\right]p}{\left[1 + s + p + \frac{\Delta R_g}{R_g}\right]} \quad (6.26)$$

The expression for the unit change in resistance as seen by the readout instrument can now be written as

$$\frac{\Delta R_t}{R_t} = \frac{\left[1 + s + \frac{\Delta R_g}{R_g}\right]p}{\left[1 + s + p + \frac{\Delta R_g}{R_g}\right]} - 1 \quad (6.27)$$

Expanding Eq. (6.27) produces

$$\frac{\Delta R_t}{R_t} = \frac{(1 + s)p + p\left(\frac{\Delta R_g}{R_g}\right) - (1 + s + p) - \frac{\Delta R_g}{R_g}}{1 + s + p + \frac{\Delta R_g}{R_g}}$$

From Eq. (6.25), $(1 + s)p = 1 + s + p$, and so

$$\frac{\Delta R_t}{R_t} = \frac{(p - 1)\dfrac{\Delta R_g}{R_g}}{1 + s + p + \dfrac{\Delta R_g}{R_g}} \tag{6.28}$$

Equation (6.25) also shows that

$$s = \frac{1}{p - 1} \tag{6.29}$$

and

$$1 + s + p = \frac{p^2}{p - 1} \tag{6.30}$$

Substituting the value of $1 + s + p$ given by Eq. (6.30) into Eq. (6.28),

$$\frac{\Delta R_t}{R_t} = \frac{(p - 1)\dfrac{\Delta R_g}{R_g}}{\dfrac{p^2}{p - 1} + \dfrac{\Delta R_g}{R_g}}$$

This expression can also be written as

$$\frac{\Delta R_t}{R_t} = \frac{(p - 1)^2 \dfrac{\Delta R_g}{R_g}}{p^2 + (p - 1)\dfrac{\Delta R_g}{R_g}}$$

Further rearrangement gives

$$\frac{\Delta R_t}{R_t} = \left[\frac{p - 1}{p}\right]^2 \left[\frac{1}{1 + \left(\dfrac{p - 1}{p^2}\right)\dfrac{\Delta R_g}{R_g}}\right]\left(\frac{\Delta R_g}{R_g}\right) \tag{6.31}$$

Equation (6.31) tells us that the desensitization factor is

$$Q_t = \left[\frac{p - 1}{p}\right]^2 \left[\frac{1}{1 + \left(\dfrac{p - 1}{p^2}\right)\dfrac{\Delta R_g}{R_g}}\right] \tag{6.32}$$

Neglecting nonlinearities,

$$Q = \left[\frac{p-1}{p}\right]^2 \tag{6.33}$$

The nonlinearity factor, n, is

$$n = \left(\frac{p-1}{p^2}\right)\frac{\Delta R_g}{R_g} \tag{6.34}$$

Together with the knowledge that s and p must always be positive, Eqs. (6.29), (6.33), and (6.34) provide us with some interesting facts.

1. From Eq. (6.29) one sees that p must always be larger than 1, because s becomes larger as p becomes smaller and would have to be infinite if p became unity. Also, if p were less than unity, s would be negative, which is impossible.
2. Equations (6.32) and (6.33) indicate that the desensitization factor approaches zero as p approaches unity. This is to be expected, of course, because s is approaching infinity as p approaches 1, and consequently any changes in gage resistance have less overall influence.
3. The nonlineartiy factor, n, approaches zero as p approaches 1, and also as p becomes very large. By differentiation we find that the maximum value occurs for $p = 2$, so that

$$n_{max} = \left(\frac{1}{4}\right)\frac{\Delta R_g}{R_g}$$

So far the desensitization factor has been determined for a given value of p, or s. There are, however, other situations in which it will be necessary to determine the values of p and s that will be required to produce a given desensitization. For this purpose we will need to find p and s in terms of Q. From Eq. (6.33),

$$\sqrt{Q} = \frac{p-1}{p} = 1 - \frac{1}{p}$$

From this

$$p = \frac{1}{1 - \sqrt{Q}} \tag{6.35}$$

From Eq. (6.29),

$$S = \frac{1}{p-1} = \frac{1}{\dfrac{1}{1-\sqrt{Q}} - 1} = \frac{1-\sqrt{Q}}{\sqrt{Q}}$$

or

$$S = \frac{1}{\sqrt{Q}} - 1 \tag{6.36}$$

Substituting the value of p given by Eq. (6.35) into Eq. (6.34), the nonlinearity factor, n, can be written as

$$n = \sqrt{Q}(1 - \sqrt{Q})\frac{\Delta R_g}{R_g} \tag{6.37}$$

From Eq. (6.37) we see that $n = 0$ when Q equals 0 or 1, and by differentiation we find that the maximum value occurs when $Q = 0.25$ ($p = 2$), which, as previously, produces the maximum nonlinearity represented as

$$n_{max} = \left(\frac{1}{4}\right)\frac{\Delta R_g}{R_g}$$

For rapid evaluation of the ratios p and s that are required to produce a given desensitization, Fig. 6.6, in which the values are plotted in terms of Q, will be found helpful. Since Fig. 6.6 neglects the effect of nonlinearity, the value of $\sqrt{Q}(1 - \sqrt{Q})$ has been plotted as a function of Q, in Fig. 6.7. In most cases, however, it will be sufficient if we know that the maximum value is 0.25.

A note on temperature effects

The derivations of this section all assume that the temperature remains constant. However, if there is a temperature change, a false indication of strain will be produced unless *all* of the following are independent of temperature changes: (1) the gage resistance, (2) the resistance of the leads, (3) the auxiliary series and parallel resistances. Theoretically, these conditions can be fulfilled by using a self-temperature-compensated strain gage with leads and auxiliary resistances having a zero temperature coefficient of resistance. Obtaining a suitable strain gage should present no problem, but acquiring lead wire (including soldered joints), and auxiliary resistances, with zero response to temperature may present a difficult problem. On this account it will be preferable to take another approach using the half-bridge arrangement as discussed in the following section.

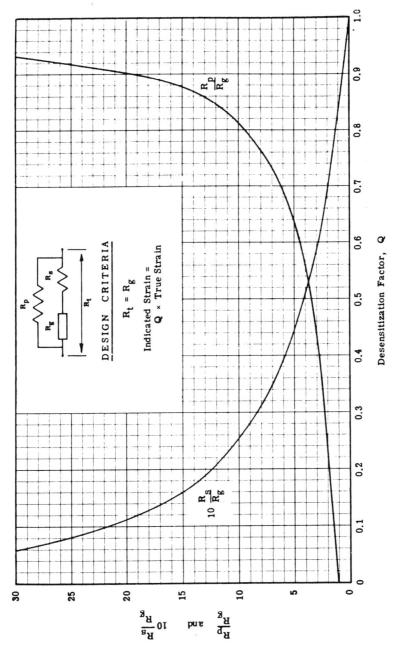

FIG. 6.6. Series and parallel resistances for single gage desensitization. (From ref. 2.)

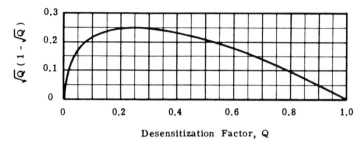

FIG. 6.7. $\sqrt{Q}(1 - \sqrt{Q})$ as a function of Q.

6.3. Analysis of half-bridge desensitization

In this section, methods of eliminating the effects of temperature changes by employing active and dummy gages in adjacent arms of a half bridge will be discussed.

Duplicating the system for a single gage

The most direct approach is to set up duplicate arrangements in the two adjacent arms of the half bridge and to make sure that corresponding components are subjected to exactly the same temperature conditions.

When the temperature variation at the gage is greater than that at the readout instrument, it will be best to locate the series and parallel resistances near the instrument and to run the leads out to the gages, making sure that the leads from the parallel resistances around the gages are equal in length and tied into the system at equivalent locations in both arms of the half bridge.

In the event that the half bridge is to be connected across the power supply, it will not be necessary to use both parallel and series resistance because the ratio of the total resistance in each arm can be maintained at unity for either series or parallel resistance connected to the gage. If the half bridge is connected across the bridge output, depending upon the resistances in the other two arms, the ratio of the resistances of the arms in series with the power supply may, or may not, be unity.

When the half bridge is connected across the power supply, although the adjustment may be a little more difficult, it will be preferable to desensitize with series resistance alone because the output will be linear and the complication of the extra leads from the parallel resistances can be eliminated. This means that the standard four-lead active-dummy system can be employed with a pair of equal series resistors in each arm adjacent to the readout instrument, as long as the total resistance in each arm does not exceed the capability of the instrument.

The concept of desensitization using series resistance alone in a half bridge becomes even more attractive when one wishes to use a single

SENSITIVITY VARIATION

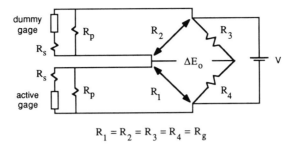

FIG. 6.8. Desensitization with temperature compensation.

self-temperature-compensated strain gage in the active arm, because one can then use the three-wire system with all the resistance (exclusive of leads) in the inactive arm in, or at, the readout instrument, and still maintain freedom from the influence of temperature.

Figure 6.8 shows a schematic layout for one arrangement of half-bridge desensitization with temperature compensation.

An alternate and superior method of desensitization

An alternate method of desensitization, which uses the half bridge to provide temperature compensation, is shown schematically in Fig. 6.9. This arrangement employs a common parallel resistor in both arms.

Some of the advantages of this system are as follows:

1. The total effective resistance in each of the two desensitized arms can be made equal to the gage resistance, R_g, if desired. This is merely a convenience. The only requirement is that, initially, the effective resistance should be the same in both arms.

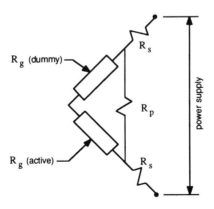

FIG. 6.9. Alternate method of half-bridge desensitization.

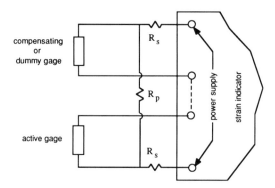

FIG. 6.10. Physical connections of gages to indicator. Note: On some indicators the relative positions of the terminals for the active and compensating gages are reversed with respect to this diagram.

2. One less resistor is required than for the previous method.
3. The equations for computing the series and parallel resistances, R_s and R_p, are simpler.
4. The network of resistances required is simple and easy to install at the strain indicator, as shown in Fig. 6.10.
5. All the advantages and simplicity of the standard active-dummy system are retained. No additional lead wires are required.

Limitations

In the analysis that follows, two limitations will become evident; however, these should cause no difficulty if one is cognizant of them. For emphasis the limitations will be summarized here.

1. The input impedance of the instrument connected to the bridge output must be extremely high.
2. The pair of gages (half bridge) containing the desensitizing network must be connected across the bridge power supply.

Analysis

Let us refer to Fig. 6.9, which shows a half-bridge diagram with a pair of like gages and a desensitizing network consisting of two series resistances and a single parallel resistance common to both gages.

The first step in analyzing the network will be to determine the equivalent of the combined gage and parallel resistances which should be considered in each of bridge arms 1 and 2. This can be done by means of a Delta–Wye transformation, as shown in Chapter 5.

Figure 6.11 shows the Delta network formed by the strain gages, R_{g1}

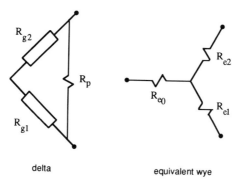

FIG. 6.11. Delta-wye transformation.

and R_{g2}, and the parallel resistor, R_p. Also in the same figure is the equivalent Wye network.

From this transformation,

$$R_{e1} = \frac{R_{g1} R_p}{R_{g1} + R_{g2} + R_p} \qquad (6.38)$$

$$R_{e2} = \frac{R_{g2} R_p}{R_{g1} + R_{g2} + R_p} \qquad (6.39)$$

$$R_{e0} = \frac{R_{g1} R_{g2}}{R_{g1} + R_{g2} + R_p} \qquad (6.40)$$

where R_{e1} = equivalent resistance in arm 1

R_{e2} = equivalent resistance in arm 2

R_{e0} = equivalent resistance in the output circuit

The equivalent total resistances, R_1 and R_2, in arms 1 and 2 of the bridge may now be expressed as

$$R_1 = R_s + R_{e1} = R_s + \frac{R_{g1} R_p}{R_{g1} + R_{g2} + R_p} \qquad (6.41)$$

$$R_2 = R_s + R_{e2} = R_s + \frac{R_{g2} R_p}{R_{g1} + R_{g2} + R_p} \qquad (6.42)$$

Examination of Fig. 6.12 indicates that the transformation has facilitated setting up relatively simple expressions for the equivalent resistances in arms 1 and 2 of the bridge. However, it also indicates that there is resistive effect,

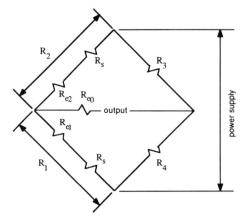

FIG. 6.12. Electrical equivalent of Fig. 6.9.

represented by R_{e0}, in the output circuit. This latter influence, R_{e0}, must be allowed for in some manner.

One way of allowing for R_{e0} is to make it ineffective by using a very high-impedance readout device so that essentially no current flows across the output from the bridge. This is the reason for statement (1) under Limitations. Item (2) of the limitations can be explained by considering what would happen if the half bridge containing the gages were not connected across the power supply. In this case, with the half bridge containing the gages across the output, the effect of R_{e0} will be the same as that of any other resistance in series with the bridge across the power supply. That is, a desensitization of the entire bridge will take place. Thus, to avoid this overall desensitizing effect, the half bridge with the gages must be connected across the power supply.

The derivation of the expression for the desensitization factor will now be considered. Note that even though two identical strain gages (both of resistance R_g) are used in the half bridge, their resistances have been designated separately by the symbols, R_{g1} and R_{g2}, to indicate their respective locations in the bridge. This is necessary because the two gages will have somewhat different functions if one is to do the strain measuring and the other to provide temperature compensation.

Since the series and parallel resistances, R_s and R_p, can be so chosen as to provide for a wide range of resistances in the bridge arms, let us consider that their values will be so chosen that, *numerically*,

$$R_1 = R_2 = R_g = R_{g1} = R_{g2} \qquad (6.43)$$

Thus, from Eqs. (6.41), (6.42), and (6.43), Eqs. (6.44) and (6.45) will result when initial conditions are used:

$$R_1 = R_s + \frac{R_{g1}R_p}{R_{g1} + R_{g2} + R_p} = R_g \tag{6.44}$$

and

$$R_2 = R_s + \frac{R_{g2}R_p}{R_{g1} + R_{g2} + R_p} = R_g \tag{6.45}$$

Let us now see what happens when the gage in arm 1 is strained and changes its resistance to $R_{g1} + \Delta R_{g1}$. Since R_{g1} appears in the expressions for both R_1 and R_2, this change will influence both arms 1 and 2 of the bridge and, consequently, we will have to consider changes in both of them simultaneously. Hence, after the change, Eqs. (6.44) and (6.45) will become

$$R_1 + \Delta R_1 = R_s + \frac{(R_{g1} + \Delta R_{g1})R_p}{R_{g1} + \Delta R_{g1} + R_{g2} + R_p} \tag{6.46}$$

and

$$R_2 + \Delta R_2 = R_s + \frac{R_{g2}R_p}{R_{g1} + \Delta R_{g1} + R_{g2} + R_p} \tag{6.47}$$

The unit changes in resistance in the arms of the bridge can now be found by dividing Eqs. (6.46) and (6.47) by R_1 and R_2, respectively, so that

$$1 + \frac{\Delta R_1}{R_1} = \frac{R_s}{R_1} + \left[\frac{(R_{g1} + \Delta R_{g1})R_p}{R_{g1} + \Delta R_{g1} + R_{g2} + R_p}\right]\left(\frac{1}{R_1}\right) \tag{6.48}$$

and

$$1 + \frac{\Delta R_2}{R_2} = \frac{R_s}{R_2} + \left[\frac{R_{g2}R_p}{R_{g1} + \Delta R_{g1} + R_{g2} + R_p}\right]\left(\frac{1}{R_2}\right) \tag{6.49}$$

From the relations expressed in Eq. (6.43), Eqs. (6.48) and (6.49) can be simplified to

$$1 + \frac{\Delta R_1}{R_1} = \frac{R_s}{R_g} + \frac{\left[1 + \frac{\Delta R_g}{R_g}\right]R_p}{2R_g + \Delta R_g + R_p} \tag{6.50}$$

and

$$1 + \frac{\Delta R_2}{R_2} = \frac{R_s}{R_g} + \frac{R_p}{2R_g + \Delta R_g + R_p} \tag{6.51}$$

Since the bridge output is proportional to the algebraic difference between the unit changes in resistance of adjacent arms, one can obtain a measure of this by subtracting Eq. (6.51) from Eq. (6.50). This means that what the instrument indicates is

$$\frac{\Delta R_1}{R_1} - \frac{\Delta R_2}{R_2} = \frac{[1 + (\Delta R_g/R_g)]R_p}{2R_g + \Delta R_g + R_p} - \frac{R_p}{2R_g + \Delta R_g + R_p} \quad (6.52)$$

Equation (6.52) simplifies to

$$\frac{\Delta R_1}{R_1} - \frac{\Delta R_2}{R_2} = \frac{R_p}{2R_g + \Delta R_g + R_p}\left(\frac{\Delta R_g}{R_g}\right) \quad (6.53)$$

From Eq. (6.53) it can be seen that the desensitization factor, Q_t, is given by

$$Q_t = \frac{R_p}{2R_g + \Delta R_g + R_p} \quad (6.54)$$

If the numerator and denominator on the right-hand side of Eq. (6.54) are divided by R_g, and the ratio R_p/R_g is expressed by the single symbol p, then

$$Q_t = \frac{p}{2 + p + \dfrac{\Delta R_g}{R_g}}$$

This expression can be rewritten as

$$Q_t = \left[\frac{p}{2+p}\right]\left[\frac{1}{1 + \left(\dfrac{1}{2+p}\right)\dfrac{\Delta R_g}{R_g}}\right] \quad (6.55)$$

If either Eq. (6.41) or Eq. (6.42) is used with the values given in Eq. (6.43), then we obtain

$$p = \frac{2(1-s)}{s} \quad \text{and} \quad s = \frac{2}{2+p}$$

From Eq. (6.55),

$$Q = \frac{p}{2+p} = 1 - s \quad (6.56)$$

SENSITIVITY VARIATION

and the nonlinearity factor is given by

$$n = \left[\frac{1}{2+p}\right]\frac{\Delta R_g}{R_g} \tag{6.57}$$

which will always be less than $(0.5)(\Delta R_g/R_g)$.

Sometimes it will be necessary to determine the sizes of the series and parallel resistances which will be required to produce a given desensitization. This can be done by solving Eq. (6.56) for p and s in terms of Q. This results in

$$s = \frac{R_s}{R_g} = 1 - Q \tag{6.58}$$

and

$$p = \frac{R_p}{R_g} = \frac{2Q}{1-Q} \tag{6.59}$$

For convenience, the values of p and s have been plotted against Q in Fig. 6.13. From Eq. (6.57) the value of the nonlinearity factor, n, has been determined in terms of Q as

$$n = \frac{1}{2}(1-Q)\frac{\Delta R_g}{R_g} \tag{6.60}$$

Example 6.1. A cantilever beam has four longitudinal strain gages ($R_g = 120$ ohms) bonded to it that are arranged into a full bridge. When a 5-lb weight is placed on the beam, the strain indicator reads 2140 μin/in.

(a) Use a series resistance in arms R_1 and R_2, as shown in Fig. 6.1, to desensitize the system so that the reading is reduced to 1500 μin/in.
(b) Use a parallel resistance in arms R_1 and R_2, as shown in Fig. 6.3, to desensitize the system so that the reading is reduced to 1500 μin/in.

Solution. (a) For each gage,

$$\varepsilon/\text{gage} = \frac{2140}{4} = 535 \text{ μin/in}$$

The two arms not desensitized will read a total of 1070 μin/in. The other two arms must read $1500 - 1070 = 430$ μin/in, or 215 μin/in per arm. Equation (6.8) is now used, but if both sides are divided by G_F, then $\Delta R_t/(G_F R_t) = \varepsilon_i$, the desired indicated strain, and $\Delta R_g/(G_F R_g) = \varepsilon$, the actual strain. In terms of strain, then, Eq. (6.8) is

$$\varepsilon_i = \frac{1}{1+s}\varepsilon$$

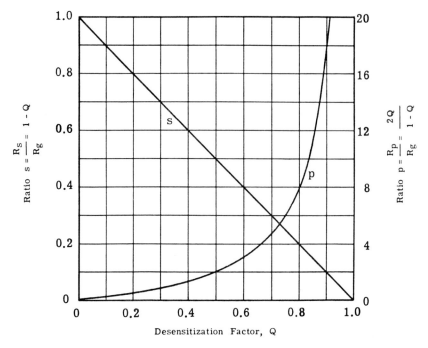

FIG. 6.13. Ratios p and s as functions of Q.

or

$$215 = \frac{1}{1+s}(535)$$

From this, $s = 1.488$, and so

$$R_s = 1.488 R_g = 1.488(120) = 178.56 \text{ ohms}$$

Therefore, use a series resistor of 178.56 ohms in arms R_1 and R_2.

(b) Refer to Fig. 6.3 for the parallel arrangement. Again, two arms will read 535 μin/in, while the two desensitized arms will each read 215 μin/in. Considering the nonlinearity portion to be unity, Eq. (6.15) gives

$$\varepsilon_i = \frac{p}{1+p}\varepsilon$$

or

$$215 = \frac{p}{1+p}(535)$$

From this, $p = 0.672$, and so

$$R_p = 0.672 R_g = 0.672(120) = 80.64 \text{ ohms}$$

Therefore, use a parallel resistor of 80.64 ohms in arms R_1 and R_2.

Example 6.2. A torque meter (four active arms with $R_g = 120$ ohms each) reads 1420 μin/in when subjected to a torsional moment of 1200 in-lb. Desensitize one arm, using parallel–series resistances in order to have the strain indicator read 1200 μin/in.

Solution. The meter reading must be reduced by $1420 - 1200 = 220$ μin/in. Since each arm reads $1420/4 = 355$ μin/in, then the arm that is desensitized must read $355 - 220 = 135$ μin/in. Thus, $\Delta R_t/(G_F R_t) = 135$ μin/in and $\Delta R_g/(G_F R_g) = 355$ μin/in. Again, considering the nonlinearity portion to be unity, Eq. (6.31) gives

$$\varepsilon_i = \left[\frac{p-1}{p}\right]^2 \varepsilon$$

or

$$135 = \left[\frac{p-1}{p}\right]^2 (355)$$

From this, $p = 2.609$, and so

$$R_p = 2.609 R_g = 2.609(120) = 313.08 \text{ ohms}$$

From Eq. (6.29),

$$s = \frac{1}{p-1} = \frac{1}{2.609 - 1} = 0.622$$

Thus,

$$R_s = 0.622 R_g = 0.622(120) = 74.64 \text{ ohms}$$

Use a series resistor of 74.64 ohms and a parallel resistor of 313.08 ohms.

6.4. Analysis of full-bridge sensitivity variation

There are certain situations in which it is desirable to vary the sensitivity of an entire bridge. For example:

1. To compensate the output from load cells for changes in modulus of elasticity of the load-carrying element due to variations in temperature.
2. To permit a standard strain indicator, which has been designed for use with metallic gages, to be employed with a semiconductor bridge whose output is in excess of the range of the instrument.

228   THE BONDED ELECTRICAL RESISTANCE STRAIN GAGE

3. To perform some computation automatically in order to obtain a direct readout of some desired quantity, as in the case of the torque meter that is made to indicate power transmitted by making the excitation voltage proportional to the speed of rotation.

Method of approach

Since the bridge output is directly proportional to the applied voltage, the sensitivity can be varied by means of voltage control. Frequently this is accomplished by using a power supply with a fixed voltage that is greater than that needed to energize the bridge and then reducing this to the necessary level by including a fixed, or variable, resistance in series with the bridge, according to the particular requirements at hand. The arrangement is shown in Fig. 6.14.

Limitation

Since this method of sensitivity variation depends upon controlling the voltage actually applied to the bridge, it is unsuited for use with a null balance system where indication is independent of the magnitude of the applied voltage. Likewise, it will not work for certain types of reference bridge instruments which have also been designed to produce readings that are independent of supply voltage, or fluctuations therein.

Derivation of equations

The voltage across the bridge can be expressed as

$$V = V_P \left[\frac{R_{BI}}{R_{BI} + R_s} \right] = V_P \left[\frac{1}{1 + \dfrac{R_s}{R_{BI}}} \right] \tag{6.61}$$

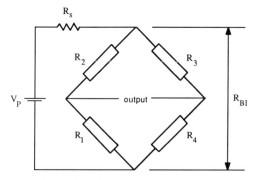

FIG. 6.14. Bridge with resistance in series.

SENSITIVITY VARIATION

where V_P = power supply voltage

R_s = resistance in series with the bridge

R_{BI} = input resistance of the bridge, excluding R_s

The bridge output (assuming initial balance and neglecting nonlinearity) is expressed as

$$\Delta E = V \frac{a}{(1+a)^2} \left[\frac{\Delta R_1}{R_1} - \frac{\Delta R_2}{R_2} + \frac{\Delta R_3}{R_3} - \frac{\Delta R_4}{R_4} \right] \qquad (6.62)$$

Note that, in Eq. (6.62), infinite impedance is assumed at the bridge output. Therefore, with respect to the voltage of the power supply,

$$\Delta E = V_P \left[\frac{1}{1 + \dfrac{R_s}{R_{BI}}} \right] \left[\frac{a}{(1+a)^2} \right] \left[\frac{\Delta R_1}{R_1} - \frac{\Delta R_2}{R_2} + \frac{\Delta R_3}{R_3} - \frac{\Delta R_4}{R_4} \right] \qquad (6.63)$$

For a constant-voltage power supply, this means that

$$\Delta E = \left[\frac{1}{1 + \dfrac{R_s}{R_{BI}}} \right] [\text{maximum output}] \qquad (6.64)$$

In other words, the desensitization factor for the entire bridge is given by

$$Q = \frac{1}{1 + \dfrac{R_s}{R_{BI}}} \qquad (6.65)$$

This value of Q assumes the bridge resistance remains constant.

When the bridge resistance remains constant, as in the case of certain transducers, such as torque meters, in which the resistances in adjacent bridge arms change by equal amounts but of opposite sign, or when changes are proportionately very small, as is usually the case with metallic strain gages, Eq. (6.65) is directly applicable. Also observe that Eq. (6.65) is of the same form as Eq. (6.9) when the symbol s is used to represent the ratio R_s/R_{BI}. On this account, Fig. 6.2 may be used to determine not only the desensitization factor for a single gage with resistance in series, but also the corresponding effect for an entire bridge with a resistance in series.

Effect of changes in bridge resistance

When there is an appreciable change in the bridge resistance, Eq. (6.65) will have to be modified by considering the actual bridge resistance, $R_{BI} + \Delta R_{BI}$, at any particular instant. In this case, Eq. (6.65) can be written in the modified form

$$Q_t = \frac{1}{1 + \dfrac{R_s}{R_{BI} + \Delta R_{BI}}} = \frac{1}{1 + \dfrac{R_s}{R_{BI}}\left[\dfrac{1}{1 + \dfrac{\Delta R_{BI}}{R_{BI}}}\right]} \qquad (6.66)$$

Except for the factor, $1/[1 + \Delta R_{BI}/R_{BI}]$ in the denominator of Eq. (6.66), Eqs. (6.65) and (6.66) are alike.

The error in Q can now be examined if the change in bridge resistance is neglected. Examination of Eq. (6.66) indicates that the error produced by neglecting the change in bridge resistance will be small for small ratios of both R_s/R_{BI} and $\Delta R_{BI}/R_{BI}$.

To get some idea of the numerical value of the error, we can investigate a particular situation for approximate values. The following values are given:

$$\frac{R_s}{R_{BI}} = 0.25$$

$\Delta R = 40$ percent for a single active arm

Neglecting the change in bridge resistance,

$$Q = \frac{1}{1 + \dfrac{R_s}{R_{BI}}} = \frac{1}{1 + 0.25} = 0.800$$

The change in bridge resistance can now be included. If the single active arm changes by 40 percent, then, for four equal arms, ΔR_{BI} will be about 10 percent. This can be verified by assuming 120-ohm gages and computing the bridge resistance with $R_m = \infty$. In this case,

$$Q_t = \frac{1}{1 + 0.25\left[\dfrac{1}{1 + 0.1}\right]} = \frac{1}{1 + 0.227} = 0.815$$

Thus, for the conditions given, the error in Q caused by neglecting the change in bridge resistance will be less than 2 percent.

For smaller ratios of R_s/R_{BI} and $\Delta R_{BI}/R_{BI}$, the variations will be even

SENSITIVITY VARIATION

less and, consequently, for a great many cases, we are justified in neglecting the effect of changes in total bridge resistance. Nevertheless, it is always desirable to check to be sure that the probable error from this source will fall within tolerable limits.

Discussion

Use of a bridge with unequal arms. The preceding example suggests that in a bridge containing a single strain gage (if one has the choice), there may be some advantages to be gained by having two of the arms of somewhat higher resistance than the strain gage. In addition to improving the linearity and increasing the output per unit strain, this procedure will enable us to reduce the ratio $\Delta R_{BI}/R_{BI}$, even for large values of resistance change in the one active arm, and thereby cut down on the variation Q with change in bridge resistance.

Temperature effects. A note of caution, especially in respect to transducers involving four active arms containing semiconductor gages, will be mentioned with regard to the total change in bridge resistance. Even though the gages all change by exactly the same amount and no bridge output results from this, nevertheless, as far as the total bridge resistance is concerned, this effect will be additive and will have some influence on the value of Q. If the ratio R_s/R_{BI} is small, the effect may not be noticeable, but for larger ratios of series to bridge resistance, the influence on Q should be checked.

Increasing and decreasing the sensitivity. Equations (6.63) and (6.64) show that the maximum output will occur when $R_s = 0$. For those applications in which one may wish to be able to increase, or decrease, the sensitivity from some usual value (such as modulus compensation of load cells), it will be necessary to design the system to provide for normal operation at somewhat less than the maximum output so that it will be possible to decrease R_s by the necessary amount in order to achieve the desired increase in sensitivity. When R_s has been reduced to zero, the maximum possible sensitivity will have been achieved.

Problems

6.1. A single strain gage records 1256 μin/in. If $G_F = 2.0$ and $R_g = 120$ ohms, determine the value of the series resistor, R_s, that is required in order to have $\Delta R_t/R_t = 0.002$. If this gage is used in a quarter-bridge circuit and $R_2 = R_3 = R_4 = 120$ ohms, can the bridge be initially balanced? If one is free to choose resistors R_2, R_3, and R_4, can the bridge be initially balanced?

6.2. Repeat Problem 6.1 using a parallel resistor, R_p.

6.3. Repeat Problem 6.1 using series and parallel resistors so that $R_t = R_g$.

6.4. Develop equations for the series–parallel arrangement shown in Fig. 6.5b. Follow the method used for Fig. 6.5a.

232   THE BONDED ELECTRICAL RESISTANCE STRAIN GAGE

6.5. In Fig. 6.8, a half-bridge circuit is shown with four lead wires. The bridge is desensitized with series resistors alone. If the active gage of 120 ohms is subjected to a strain of ε, determine the value of R_s needed to make the indicated strain, ε_i, equal to 0.75ε.

6.6. In Problem 6.5, the dummy gage becomes an active gage. If $\varepsilon_1 = \varepsilon$ and $\varepsilon_2 = -\nu\varepsilon$, will the value of R_s change if the indicated strain, ε_i, is to be 75 percent of the total strain?

6.7. In Example 6.1, desensitize arms R_1 and R_2 by using a combination of series and parallel resistances.

6.8. Figure 6.15 shows the small assembly machine used in Problem 5.8. In order to measure the load on the machine, add two gages on the centerline A–A so that gage 3 is transverse to gage 1 and gage 4 is transverse to gage 2. To get the strains in Problem 5.8, gages were used with $G_F = 2.08$ and $R_g = 120$ ohms. Using $F = 340\,800$ lb for the strains obtained in Problem 5.8, perform the following tasks:

(a) Arrange the gages into a full bridge in order to get the maximum output. Sketch the bridge arrangement.
(b) Since the sensitivity of the circuit can be altered by adjusting the gage factor, set the gage factor so that an indicated strain of 1 μin/in represents a force of 100 lb. The gage factor setting ranges from 1.15 to 3.50.

6.9. Using the data given in Problem 6.8, desensitize the circuit by adding series resistors in the two arms with gages 1 and 3. A force of 200 lb is to be represented by an indicated strain of 1 μin/in.

6.10. Rework Problem 6.9 but use parallel resistors.

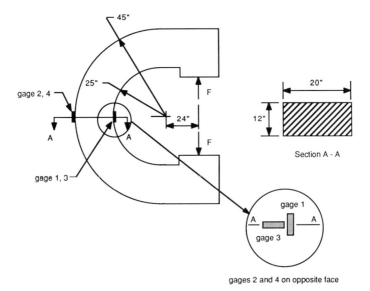

FIG. 6.15.

SENSITIVITY VARIATION

6.11. Two 120-ohm gages are arranged in a half-bridge circuit for temperature compensation. The gage in arm 1 is active while the gage in arm 2 is a dummy. Upon loading, the active gage reads 2695 μin/in. It is desired to desensitize the system using the arrangement shown in Fig. 6.9 so that the indicated strain is 2000 μin/in. Determine R_s and R_p.

REFERENCES

1. Murray, William M. and Peter K. Stein, *Strain Gage Techniques*, Lectures and laboratory exercises presented at MIT, Cambridge, MA: July 8–19, 1963, pp. 249–286.
2. Stein, Peter K., "Individual Strain Gage Desensitization," Letter to the Editor, *SESA Proceedings*, Vol. XIV, No. 2, 1957, pp. 33–36.

7

LATERAL EFFECTS IN STRAIN GAGES

7.1. Significance of strain sensitivity and gage factor

Strain sensitivity is a general term relating unit change in resistance and strain in an electrical conductor according to the following expression:

$$\text{Strain sensitivity} = \frac{\text{unit change in resistance}}{\text{strain}}$$

In symbols, this is

$$S = \frac{\Delta R/R}{\varepsilon} \tag{7.1}$$

where S = strain sensitivity

R = initial resistance

ΔR = change in resistance

ε = strain

The numerical value of the strain sensitivity will depend upon the conditions under which it has been determined.

For a straight conductor of uniform cross section that is subjected to simple tension, or compression, in the direction of its axis, and unstrained laterally, the strain sensitivity is a physical property of the material. The numerical value will be represented by S_t, which is determined by the relation

$$S_t = \frac{\Delta R/R}{\Delta L/L} = \frac{\Delta R/R}{\varepsilon} \tag{7.2}$$

where L is the initial length.

The transverse effect in strain gages (1–11). When a conductor is formed into a grid for a strain gage, the relationship between unit change in resistance of the conductor and the strain becomes much more complicated, and the numerical value of the strain sensitivity is influenced by a variety of conditions. The most important are the following:

(a) The strain sensitivity of the material of the sensing element.
(b) The geometry of the grid.
(c) The strain field in which the gage is used.
(d) The direction of the strain used in making the computation of the numerical value of the strain sensitivity.

In addition, there are also a number of other smaller effects.

Special cases of strain sensitivity. Since the strain sensitivity is influenced by so many factors, in stating a numerical value, the conditions under which this has been determined should also be known. Figure 7.1 shows a strain gage mounted on a surface which has reference axes, OA and ON, scribed on it. The reference axes are parallel and normal, respectively, to the gage axis. The corresponding strains in the axial and transverse (normal) directions will be represented by ε_a and ε_n, respectively.

Lateral effects in strain gages

Although there is an infinite variety of conditions under which the strain sensitivity of a strain gage might be determined, for practical purposes there are only three specific situations with which one must be concerned, as all other conditions can be represented in terms of these three special cases, which are as follows:

$$F_a = \text{axial strain sensitivity}$$

$$F_n = \text{normal strain sensitivity}$$

$$G_F = \text{the manufacturer's gage factor}$$

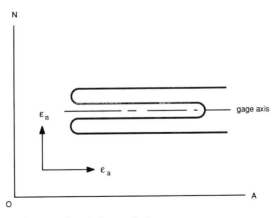

FIG. 7.1. Single strain gage aligned along axis OA.

These three specific values of strain sensitivity are defined in the following ways:

Axial strain sensitivity

$$\text{Axial strain sensitivity} = \frac{\text{unity change in resistance}}{\text{axial strain}}$$

when the normal strain is zero. This can be written as

$$F_a = \left(\frac{\Delta R/R}{\varepsilon_a}\right) \tag{7.3}$$

when $\varepsilon_n = 0$.

Normal strain sensitivity

$$\text{Normal strain sensitivity} = \frac{\text{unit change in resistance}}{\text{normal strain}}$$

when the axial strain is zero. This can be written as

$$F_n = \left(\frac{\Delta R/R}{\varepsilon_n}\right) \tag{7.4}$$

when $\varepsilon_a = 0$.

The manufacturer's gage factor. The manufacturer's gage factor, as determined in accordance with ASTM Standard E251-86(12), means the strain sensitivity, with reference to the axial strain on the gage when the gage is mounted in a uniaxial stress field, with the gage axis in the direction of the stress axis, and on a piece of material of known Poisson ratio ($v_0 = 0.285$). This procedure corresponds to calibrating the gage in a biaxial strain field in which the lateral strain, ε_n, is equal to $-v_0\varepsilon_a$. In symbols, the manufacturer's gage factor can be expressed as

$$G_F = \left(\frac{\Delta R/R}{\varepsilon_a}\right) \tag{7.5}$$

when $\varepsilon_n = -0.285\varepsilon_a$.

7.2. Basic equations for unit change in resistance

Since strain gages, in general, change their resistances for both axial and normal strains, let us proceed towards establishing a general relation for unit

LATERAL EFFECTS IN STRAIN GAGES

change in resistance by considering each of these effects alone, and then adding the individual influence to determine the result of both axial and transverse strains acting simultaneously.

Derivation. To develop the required expressions, we commence by writing Eq. (7.1) in terms of change in resistance as follows:

$$\Delta R = S\varepsilon R \qquad (7.6)$$

Equation (7.6) is general but needs further specification when applied to any particular condition to which the gage may be subjected. Equation (7.6) will be applied to the determination of the change in resistance produced under the following two conditions:

(a) When $\varepsilon_a \neq 0$ and $\varepsilon_n = 0$
(b) When $\varepsilon_a = 0$ and $\varepsilon_n \neq 0$

The two changes are then added together to determine the overall change in resistance resulting from the combined effect of the strains parallel and normal to the gage axis.

For the first condition, where there is strain *only* in the direction of the gage axis, the symbols of Eq. (7.6) will take on the following particular values:

$\Delta R = \Delta R_a$ Change of resistance
$S = F_a$ Strain sensitivity (by definition)
$\varepsilon = \varepsilon_a$ Strain

Substituting these values into Eq. (7.6) gives

$$\Delta R_a = F_a \varepsilon_a R \qquad (7.7)$$

For the second condition, when there is strain *only* in the direction normal to the direction of the gage axis, the symbols of Eq. (7.6) will take on the following particular values:

$\Delta R = \Delta R_n$ Change in resistance
$S = F_n$ Strain sensitivity (by definition)
$\varepsilon = \varepsilon_n$ Strain

Again, substituting these values into Eq. (7.6) yields

$$\Delta R_n = F_n \varepsilon_n R \qquad (7.8)$$

When the gage is subjected, simultaneously, to strains in the axial and normal directions, the expression for the total change in resistance can be written by adding Eqs. (7.7) and (7.8) together to give

$$\Delta R = \Delta R_a + \Delta R_n \tag{7.9}$$

Thus,

$$\Delta R = F_a \varepsilon_a R + F_n \varepsilon_n R \tag{7.10}$$

From this, the overall *unit* change in resistance can be found by dividing both sides of Eq. (7.10) by R, the resistance of the gage, so that

$$\frac{\Delta R}{R} = F_a \varepsilon_a + F_n \varepsilon_n \tag{7.11}$$

The transverse sensitivity factor, K. Although Eq. (7.11) presents the fundamental relation between unit changes in resistance and the axial and lateral strains, it is not in a convenient form for the user, since the manufacturers do not give the values of F_a and F_n directly. Instead, they provide the users with the equivalent information in terms of gage factor (determined under uniaxial stress) and the transverse sensitivity factor. Transverse sensitivity factor is a poorly chosen name, since it can easily be mistaken for the normal strain sensitivity represented by the symbol F_n. The meaning of the transverse sensitivity factor, which will be represented by the symbol K, can now be examined. In order to do this, Eq. (7.11) can be rewritten as

$$\frac{\Delta R}{R} = F_a \left(\varepsilon_a + \frac{F_n}{F_a} \varepsilon_n \right) \tag{7.12}$$

If the ratio F_n/F_a is represented by the single symbol K, then the unit change in resistance is expressed as

$$\frac{\Delta R}{R} = F_a (\varepsilon_a + K \varepsilon_n) \tag{7.13}$$

This means that the transverse sensitivity factor for a strain gage is defined as the ratio of the normal sensitivity to the axial sensitivity. It can be expressed as

$$K = \frac{F_n}{F_a} = \frac{\left(\frac{\Delta R/R}{\varepsilon_n} \right)_{\varepsilon_a = 0}}{\left(\frac{\Delta R/R}{\varepsilon_a} \right)_{\varepsilon_n = 0}} \tag{7.14}$$

The significance of the numerical value of the transverse sensitivity factor is that it indicates the proportion (or percentage) by which the transverse strain contributes to the total indicated strain from the gage. Table 7.1 lists values taken from the literature for the gage factor (approximate) and the transverse sensitivity factor for SR-4 wire gages (1). Tables 7.2 and 7.3 are gage and transverse sensitivity factors for foil gages from two manufacturers (13, 14).

Table 7.1. Typical values of gage factor and transverse sensitivity factor for SR-4 gages

Gage type	G_F(approx.)	$K(\%)$
A-1	2.0	2.0
A-5	2.0	3.5
A-6	2.0	1.75
A-8	1.8	−2.0
A-11	2.1	0.5
A-12	2.0	1.0
A-14	2.0	−0.75
A-18	1.9	−2.0
C-1	3.5	1.75
C-5	3.3	4.0
C-8	3.1	−2.0
C-10	3.2	−0.75

Source: reference 1.

Table 7.2. Typical values for gage factor and transverse sensitivity coefficients[a]

Gage type	G_F	$K(\%)$
FAE-03-12	1.90	1.3
FAE-03-35	1.88	−0.3
FAE-06-35	2.02	0.7
FAE-12-12	1.98	−0.8
FAE-12-100	2.04	−0.6
FAE-25-12	2.07	0.0
FAE-50-35	2.02	−1.7
FAB-12-12	2.02	−1.2
FAB-12-35	2.03	0.5
FAP-03-12	1.87	0.0
FAP-06-12	1.96	−0.7
FSM-03-12	1.94	0.4
FSM-12-12	2.00	−2.7
FSE-06-35	1.99	−1.4
FSE-25-35	2.03	−1.7

Source: reference 13.
[a] The listed values are typical only. Actual G_F and K values to be used depend on foil lot, and are provided on the engineering data form provided with each package of gages.

Table 7.3. Typical values for gage factor and transverse sensitivity coefficient[a]

Gage type	G_F	$K(\%)$
EA-06-0625AK-120	2.025 ± 0.5%	0.8
EA-06-125BT-120	2.085 ± 0.5%	0.7
WA-06-250BG-120	2.040 ± 0.5%	−1.1
CEA-06-250UW-120	2.045 ± 0.5%	0.6
CEA-06-250UW-350	2.085 ± 0.5%	0.4
EA-06-031CF-120	2.000 ± 1.0%	1.4
ED-DY-031CF-350	3.25 ± 3.0%	N/A
CEA-06-125UN-120	2.060 ± 0.5%	1.0
CEA-06-125UN-350	2.090 ± 0.5%	0.5
WA-06-500AE-350	2.065 ± 0.5%	−1.4
WK-06-500AE-10C	2.04 ± 1.0%	−5.9
EA-06-500BH-120	2.060 ± 0.5%	0.1
SA-06-250BK-10C	2.065 ± 0.5%	−0.5
SK-06-250BK-30C	2.06 ± 1.0%	−1.9
SK-06-031EC-350	1.99 ± 1.0%	0.5

Source: reference 14.
[a] The listed values are typical only. Actual G_F and K values to be used depend on foil lot, and are provided on the engineering data form provided with each package of gages.

For the standard types of gages, the numerical values of K will, in general, be less than about 4 percent, and for many gages the K factor is less than 2 percent. For comparable gage size, foil gages usually exhibit smaller values of K than wire gages, and some even indicate K equals zero. Flat-grid wire gages will always have a positive value of K. Wrap-around construction for wire gages produces negative values of K, due to the Poisson effect within the gage. Foil gages, depending upon the material of the foil, can exhibit either positive or small negative values of K.

Relations between gage factor and the axial and normal strain sensitivities. The relation between the axial strain sensitivity, F_a, of a strain gage, and the manufacturer's gage factor, G_F, can now be investigated. Since the gage factor is determined under uniaxial stress conditions, with the gage axis in the direction of the stress axis, we will consider the general aspects of this condition first, and then take up the special situation which prevails when the gages are calibrated. Figure 7.2 shows a gage in a uniaxial stress field.

For uniaxial stress in the direction of the gage axis,

$$\varepsilon_n = -v\varepsilon_a \qquad (7.15)$$

where v = Poisson's ratio for the material upon which the gage is mounted. For this situation, the expression for unit change in resistance, given by Eq.

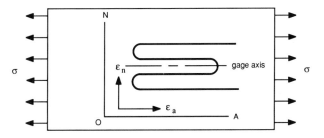

FIG. 7.2. Strain gage in a uniaxial stress field.

(7.13), can be written as

$$\frac{\Delta R}{R} = F_a[\varepsilon_a + K(-\nu\varepsilon_a)] = F_a\varepsilon_a(1 - \nu K) \tag{7.16}$$

When the gage factor is being determined, the Poisson ratio corresponds to $\nu_0 = 0.285$, which is the value for the bar on which the manufacturer makes the calibration. Therefore, for conditions of calibration,

$$\frac{\Delta R}{R} = F_a\varepsilon_a(1 - \nu_0 K) \tag{7.17}$$

For the same conditions, however, the manufacturer tells us that

$$\varepsilon_a = \frac{\Delta R/R}{\text{gage factor}} = \frac{\Delta R/R}{G_F} \tag{7.18}$$

From this we can write

$$\frac{\Delta R}{R} = G_F\varepsilon_a \tag{7.19}$$

Since the unit change in gage resistance, $\Delta R/R$, is independent of the mathematical relations which are used to express it, Eqs. (7.17) and (7.19) represent the same thing, so that

$$F_a\varepsilon_a(1 - \nu_0 K) = G_F\varepsilon_a$$

From this,

$$F_a(1 - \nu_0 K) = G_F \tag{7.20}$$

or

$$F_a = \frac{G_F}{1 - v_0 K} \qquad (7.21)$$

Since $F_n = KF_a$, then

$$F_n = \left[\frac{K}{1 - v_0 K}\right] G_F \qquad (7.22)$$

7.3. Determination of gage factor and transverse sensitivity factor (12)

Several methods of determining the gage factor for bonded resistance strain gages will be outlined. The two methods considered will be a beam in pure bending and a constant-stress cantilever beam.

Beam in pure bending

Figure 7.3 shows a typical system. The test beam is loaded by dead weights in such a manner that the beam is subjected to pure bending. The test beam, of a suitable material, has minimum dimensions of 0.75 in by 1.0 in by 30 in, and the minimum distance between the pivot points on the supports is 96 in. The assembly is symmetrical about a vertical line at its midpoint.

The pivots and weights are adjusted to give a strain on the beam surface of 1000 ± 50 μin/in. The strain over the usable portion of the test beam may not vary by more than 1 percent of the strain at the reference point. The need for measuring the strain directly can be eliminated by maintaining a calibration of the system with a Class A extensometer (15). However, the strain at the reference point may also be measured with a permanently

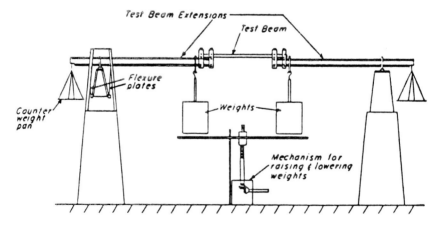

FIG. 7.3. Constant-bending-moment method for gage-factor determination. (From ref. 12 with permission. © ASTM.)

mounted strain gage that has been calibrated by spanning it with a Class A extensometer.

The usable portion of the beam is to be at least one-half of its exposed length. Measurements over each test station are made with the extensometer in order to verify the strain distribution over the beam width. Gages are installed on the unstrained test section and then the beam is loaded three times to the required strain level to $1000 \pm 50\ \mu\text{in/in}$. The gage factor of the individual gage is determined by dividing the unit change in gage resistance by the strain value determined from the beam calibration.

Constant-stress cantilever beam

A typical system using a constant-stress cantilever beam is shown in Fig. 7.4, while the beam details are given in Fig. 7.5. The size and arrangement of the equipment must be such that the beam can be deflected in either direction to produce a strain of $1200\ \mu\text{in/in}$. Two or more reference strain gages may be permanently bonded to the beam and calibrated by spanning them with a Class A extensometer. The constant-stress area is also explored with the Class A extensometer in order to determine the area where the strain is the same as that of the reference gages. Only areas where the differences in strain between the extensometer and the reference gage do not exceed $10\ \mu\text{in/in}$ are to be used.

Test gages are installed in the satisfactory areas, with the active axis of the gage parallel to the center line of the beam. The beam is deflected so that the surface strain is $1000 \pm 50\ \mu\text{in/in}$, and the unit resistance change recorded. Three such readings are taken, with the gage factor computed for each loading cycle.

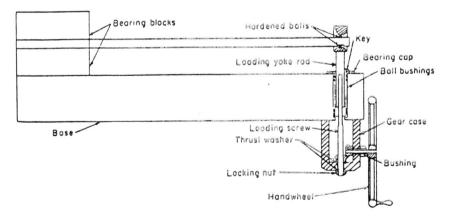

FIG. 7.4. Constant-stress cantilever beam method for gage-factor determination. (From ref. 12 with permission. © ASTM.).

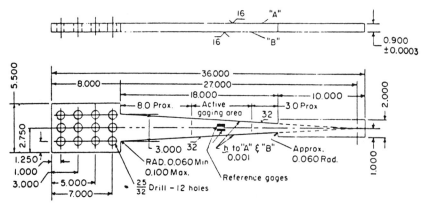

NOTE 1—All dimensions are in inches (1 in. = 25.4 mm).
NOTE 2—Surfaces "A" and "B" to be parallel to 0.0005 TIR and flat to 0.0002 TIR.
NOTE 3—Sides of beam must form triangle at apex as shown. Maximum allowable deviation of beam sides from correct line is ±0.001 in. in active area, 0.003 in. elsewhere.

FIG. 7.5. Constant-stress cantilever beam. All dimensions are in inches (1 in = 25.4 mm). Surfaces A and B to be parallel to 0.0005 TIR and flat to 0.0002 TIR. Sides of beam must form triangle at apex as shown. Maximum allowable deviation of beam sides from correct line is ±0.001 in in active area, 0.003 in elsewhere. (From ref. 12 with permission. © ASTM.)

Transverse sensitivity

Strain gage transverse sensitivity results in an undesired signal induced by strains in directions other than the one being measured. The errors induced in the plane of the gages depend on the stress distribution in the gaged areas. Figure 7.6 shows a typical test rig for determining transverse sensitivity, while Fig. 7.7 gives the test beam details and gage arrangements. The control gage may be either a Class A extensometer or a permanently installed and waterproofed resistance strain gage temperature compensated for the beam material and calibrated by a Class A extensometer.

The side plates fastened to the beam are loaded at their lower edge through the use of the crank mechanism, as shown in Fig. 7.6. This places the beam in compression as well as in bending. The transverse direction is in the long direction of the beam, and so, on the top surface, the transverse strain due to the compressive load is a tensile strain, while the transverse strain due to bending is a compressive strain. The dimensions of the apparatus are chosen so that these two strains cancel each other, thus leaving a plane strain condition across the beam.

The test beam has 16 defined stations. The difference between the strain measurements by the control gage and the actual strain at each station, both parallel and perpendicular to the principal strain direction, must be determined. The strain perpendicular to the principal strain (the transverse strain) must be less than 4 μin/in or 0.5 percent of the principal strain, with a maximum principal strain of 1000 ± 50 μin/in.

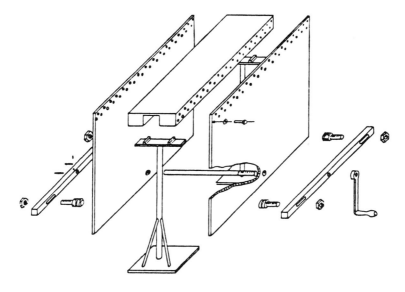

FIG. 7.6. Transverse-sensitivity test rig. (From ref. 12 with permission. © ASTM.)

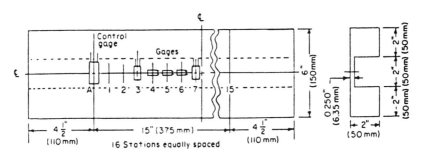

FIG. 7.7. Testing stations and gage arrangement for transverse-sensitivity test. (From ref. 12 with permission. © ASTM.)

A test requires a minimum of five identical gages of one type. At least three gages are mounted perpendicular to the principal strain direction and a minimum of two gages are mounted parallel to the principal strain direction. After gage installation, the beam is loaded to about 1000 μin/in at least three times before readings are taken. After these three load cycles, readings from the control gage and the test gages are taken in the unloaded condition, then the beam is loaded so that the surface strain is 1000 μin/in and readings taken again. This is repeated for three loading cycles. The transverse sensitivity is computed as

$$K = \left[\frac{\Delta R_t / R_{t0}}{\Delta R_L / R_{L0}}\right] \times 100 \qquad (7.23)$$

where $\Delta R_t/R_{t0}$ = unit resistance change in transverse gage

$\Delta R_L/R_{L0}$ = unit resistance change in gage parallel to the principal strain direction

The range of all values obtained is to be reported, while the transverse sensitivity of a gage type is taken as the average of all values recorded.

7.4. Use of strain gages under conditions differing from those corresponding to calibration

If a strain gage is used under biaxial conditions which differ from those prevailing during calibration, theoretically, there will be an error in the indicated value of the axial strain. Fortunately, this error is usually rather small and can be neglected. It can be shown that for gages whose K factor is less than 3 percent, the maximum error will not exceed about 4 percent as long as the numerical value of the normal strain does not exceed that of the axial strain.

The exact value of this error can now be examined, along with a simple means of correcting for it under any condition of biaxial strain. For this purpose, it will be convenient to represent the ratio of normal to axial strain by a single symbol. Thus,

$$\frac{\text{Normal strain}}{\text{Axial strain}} = \frac{\varepsilon_n}{\varepsilon_a} = \alpha \qquad (7.24)$$

From Eqs. (7.13) and (7.21), an expression for the unit change in resistance can be written as

$$\frac{\Delta R}{R} = \frac{G_F}{1 - v_0 K}(\varepsilon_a + K\varepsilon_n) \qquad (7.25)$$

From Eq. (7.24), $\varepsilon_n = \alpha\varepsilon_a$. Substituting this value of ε_n into Eq. (7.25) yields

$$\frac{\Delta R}{R} = \left[\frac{G_F}{1 - v_0 K}\right]\varepsilon_a(1 + K\alpha) \qquad (7.26)$$

Solving Eq. (7.26) for ε_a gives

$$\varepsilon_a = \frac{\Delta R/R}{G_F}\left[\frac{1 - v_0 K}{1 + K\alpha}\right] \qquad (7.27)$$

The significance of the result given in Eq. (7.27) is represented by the following observations:

1. The quantity $(\Delta R/R)/G_F$ corresponds to the indication of strain as determined by the manufacturer. That is,

$$\text{Strain} = \frac{\text{unit change in resistance}}{\text{gage factor}}$$

2. The term $(1 - v_0 K)/(1 + \alpha K)$ represents a modifying factor whose value depends upon α, the ratio between the normal and axial strains on the gage. When the gage is employed in a stress field corresponding to calibration conditions, $\alpha = -v_0$ and the modifying expression reverts to unity, since the indicated strain, for this case, represents the correct value.
3. Since the value of K will be small with respect to unity (less than about 4 percent for standard gages) for most gages, a precise knowledge of the exact value of α is not required. The ratio of the indicated normal and axial strains should be good enough without corrections; however, if a better value of the modifying factor is desired, then a further correction may be obtained by taking the ratio of the initially corrected values.

Some special cases

Strain relations	*Ratio, α*	*Correction factor,* $\left[\dfrac{1 - v_0 K}{1 + \alpha K}\right]$
1. Two equal and like principal strains: $\varepsilon_1 = \varepsilon_2 \;\; \varepsilon_n = \varepsilon_a$	$+1$	$\dfrac{1 - v_0 K}{1 + K}$
2. Two equal but unlike principal strains: $\varepsilon_2 = -\varepsilon_1 \;\; \varepsilon_n = -\varepsilon_a$	-1	$\dfrac{1 - v_0 K}{1 - K}$
3. Uniaxial stress with the gage axis in the direction of the stress axis: $\varepsilon_n = -v\varepsilon_a$; limits: $v = 0$ to $+\tfrac{1}{2}$	$0 \to -\tfrac{1}{2}$	$\dfrac{1 - v_0 K}{1 - vK}$
4. Uniaxial stress with the gage axis perpendicular to the stress axis: $\varepsilon_n = -\dfrac{\varepsilon_a}{v}$; limits: $v = 0$ to $+\tfrac{1}{2}$	$-\infty \to -2$	$\dfrac{1 - v_0 K}{1 - (K/v)}$

We can write an expression for the error that results when a single strain gage is used in a biaxial stress field. The actual strain along the gage axis is ε_a, while the actual strain normal to the gage axis is ε_n. From Eq. (7.27), the

248   THE BONDED ELECTRICAL RESISTANCE STRAIN GAGE

Table 7.4. Error in strains when using a uniaxial gage in a biaxial field

True strain, ε_a	True stain, ε_n	$\alpha = \varepsilon_n/\varepsilon_a$	η (%)
ε_a	$5\varepsilon_a$	5	18.7
ε_a	$3\varepsilon_a$	3	11.6
ε_a	$2\varepsilon_a$	2	8.1
ε_a	ε_a	1	4.5
ε_a	0	0	0.0
ε_a	$-\varepsilon_a$	-1	-2.5
ε_a	$-3\varepsilon_a$	-3	-9.6
ε_a	$-5\varepsilon_a$	-5	-16.7
ε_a	$-10\varepsilon_a$	-10	-34.3

strain indicator will read

$$\varepsilon'_a = \frac{\Delta R/R}{G_F} = \varepsilon_a \left[\frac{1 + K\alpha}{1 - v_0 K} \right] \quad (7.28)$$

The percent error, η, between the meter reading, ε'_a, and the actual strain, ε_a, is

$$\eta = \left[\frac{\varepsilon'_a - \varepsilon_a}{\varepsilon_a} \right] \times 100 = \left[\frac{1 + K\alpha}{1 - v_0 K} - 1 \right] \times 100$$

This reduces to

$$\eta = \left[\frac{K(\alpha + v_0)}{1 - v_0 K} \right] \times 100 \quad (7.29)$$

Table 7.4 shows the resulting error between the strain indicator reading, ε'_a, and the actual strain, ε_a, for a gage with a transverse sensitivity of $K = 0.035$.

7.5. Indication from a pair of like strain gages crossed at right angles

We assume that the strain gradient is so small that both gages are subjected to the same strain condition. Let us now examine the total unit change in resistance of both gages when they are connected in series. Since the gages are oriented at right angles, consider them to be aligned parallel, and perpendicular, to the reference axes, OA and ON, as shown in Fig. 7.8, and that they make any angle θ (or $\theta + 90°$) with respect to the directions of the principal axes.

The strain gages whose axes are parallel and perpendicular to OA and ON can now be examined. Subscripts a and n will refer the various quantities to these axes, respectively. When the two gages are connected in series, R_T

LATERAL EFFECTS IN STRAIN GAGES

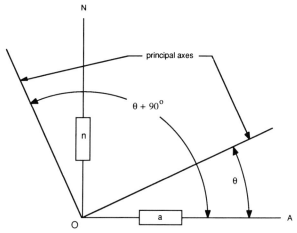

FIG. 7.8. Strain gages crossed at right angles.

is the total resistance of both gages, while ΔR_T is the change in resistance of both gages. Thus,

$$\frac{\Delta R_T}{R_T} = \frac{\Delta R_a + \Delta R_n}{R_a + R_n} \tag{7.30}$$

or

$$\frac{\Delta R_T}{R_T} = \frac{\Delta R_a + R_n\left(\dfrac{\Delta R_n}{R_n}\right)}{R_a\left(1 + \dfrac{R_n}{R_a}\right)} = \frac{\dfrac{\Delta R_a}{R_a} + \left(\dfrac{R_n}{R_a}\right)\left(\dfrac{\Delta R_n}{R_n}\right)}{1 + \dfrac{R_n}{R_a}}$$

This can be rewritten as

$$\frac{\Delta R_T}{R_T} = \frac{\dfrac{\Delta R_a}{R_a} + \beta\left(\dfrac{\Delta R_n}{R_n}\right)}{1 + \beta} \tag{7.31}$$

where $\beta = R_n/R_a$ = ratio of resistance of gage N to resistance of gage A.

From Eq. (7.11),

$$\frac{\Delta R}{R} = F_a\varepsilon_a + F_n\varepsilon_n = F_a(\varepsilon_a + K\varepsilon_n)$$

Also, from Eq. (7.21), for a single gage we have

$$F_a = \frac{G_F}{1 - v_0 K}$$

For each gage, then,

$$\frac{\Delta R_a}{R_a} = \frac{(G_F)_a}{1 - v_0 K_a}(\varepsilon_a + K_a \varepsilon_n) \tag{7.32}$$

$$\frac{\Delta R_n}{R_n} = \frac{(G_F)_n}{1 - v_0 K_n}(\varepsilon_n + K_n \varepsilon_a) \tag{7.33}$$

The unit change in resistance for the two gages in series, given by Eq. (7.31), can be rewritten by substituting the values of $\Delta R_a/R_a$ and $\Delta R_n/R_n$, given by Eqs. (7.32) and (7.33), respectively, into Eq. (7.31). This produces

$$\frac{\Delta R_T}{R_T} = \left(\frac{1}{1+\beta}\right)\left\{\frac{(G_F)_a}{1 - v_0 K_a}(\varepsilon_a + K_a \varepsilon_n) + \beta\left[\frac{(G_F)_n}{1 - v_0 K_n}(\varepsilon_n + K_n \varepsilon_a)\right]\right\} \tag{7.34}$$

Equation (7.34) is a general expression for the unit change in resistance of both gages in terms of variable strains, the gage factors, transverse sensitivity factors, and resistances, all of which may be different for each of the two gages.

In its present form, Eq. (7.34) is not very convenient; however, it can be reduced to workable conditions for certain special situations. For example, if the two crossed gages are alike in all respects, they will have the same gage factors, the same transverse sensitivity, and equal resistances. Thus,

$$(G_F)_a = (G_F)_n = G_F, \qquad K_a = K_n = K$$
$$R_a = R_n = R, \qquad \beta = R_n/R_a = 1$$

For these conditions, Eq. (7.34) reduces to

$$\frac{\Delta R_T}{R_T} = \frac{1}{2}\left\{\frac{G_F}{1 - v_0 K}(\varepsilon_a + K\varepsilon_n) + \frac{G_F}{1 - v_0 K}(\varepsilon_n + K\varepsilon_a)\right\} \tag{7.35}$$

This further simplies to

$$\frac{\Delta R_T}{R_T} = \frac{1}{2}\left(\frac{G_F}{1 - v_0 K}[(\varepsilon_a + \varepsilon_n)(1 + K)]\right) \tag{7.36}$$

An easy way to account for the transverse effect is to adjust the gage factor dial on the strain indicator (the scale factor) in order to correct for it. Also, a special situation of interest involving the solution of Eq. (7.34) will be shown when stress gages are discussed.

Problems

7.1. The following data are given for a thin-walled pressure vessel: diameter = 96 in, wall thickness = 2 in, internal pressure = 1000 psi, $v = 0.3$, $E = 30 \times 10^6$ psi. Two identical strain gages, with $G_F = 2.1$ and $K = 3.5$ percent, are bonded to the vessel, one in the longitudinal direction and one in the hoop direction.

  (a) Determine the actual strains.
  (b) Determine the strain indicator readings for each gage.
  (c) Determine the percent error in each reading.

7.2. At a point on a machine element the stresses are $\sigma_x = -8000$ psi, $\sigma_y = 4700$ psi, and $\tau_{xy} = 5500$ psi. A technician bonds two identical strain gages, with $G_F = 2.04$ and $K = -1.1$ percent, along what he believes are the principal stress directions. Gage a is located 55° CCW from the x axis, while gage b is located 90° CCW from gage a.

  (a) Have the gages been properly located?
  (b) Determine the actual strain at each gage if $v = 0.3$ and $E = 30 \times 10^6$ psi.
  (c) Determine the indicated strain for each gage.

7.3. A single strain gage, with $G_F = 1.9$ and $K = 2.5$ percent, is bonded to a member subjected to a uniaxial stress. The gage axis is aligned along the principal stress axis. If the maximum stress is 30 000 psi, $v = 0.3$, and $E = 30 \times 10^6$ psi, determine the value of the indicated strain.

7.4. In Problem 7.1, the two gages are wired in series in order to have the strain indicator read 1000 μin/in when the vessel is pressurized to 1000 psi. What gage factor setting must be used to accomplish this?

7.5. At a point on a machine element, $\sigma_x = 25\,000$ psi, $\sigma_y = -5000$ psi, and $\tau_{xy} = 12\,000$ psi. The member is loaded in such a manner that the principal stress axes remain fixed in direction. A strain gage is bonded along each principal stress axis, then they are connected in series. The gages are identical, with $G_F = 2.04$ and $K = -1.1$ percent. Determine the gage factor setting on the strain indicator so that the reading will be 500 μin/in when the sum of the principal stresses, $\sigma_1 + \sigma_2$, equals 20 000 psi.

REFERENCES

1. Baumberger, R. and F. Hines, "Practical Reduction Formulas for Use on Bonded Wire Strain Gages in Two-dimensional Stress Fields," *SESA Proceedings*, Vol. II, No. 1, 1944, pp. 113–127.
2. Bossart, K. J. and G. A. Brewer, "A Graphical Method of Rosette Analysis," *SESA Proceedings*, Vol. IV, No. 1, 1946, pp. 1–8.
3. Campbell, William R., "Performance Tests of Wire Strain Gages IV—Axial and Transverse Sensitivities," *NACA Technical Note No. 1042*, 1946.

4. *Handbook of Experimental Stress Analysis*, edited by M. Hetenyi, New York, Wiley, 1950, pp. 407–410.
5. Meier, J. H., "On the Transverse-strain Sensitivity of Foil Gages," *Experimental Mechanics*, Vol. 1, No. 7, July 1961, pp. 39–40.
6. Wu, Charles T., "Transverse Sensitivity of Bonded Strain Gages," *Experimental Mechanics*, Vol. 2, No. 11, Nov. 1962, pp. 338–344.
7. Meyer, M. L., "A Unified Rational Analysis for Gauge Factor and Cross-Sensitivity of Electric-Resistance Strain Gauges," Reprinted by permission of the Council of the Institution of Mechanical Engineers from *Journal of Strain Analysis*, Vol. 2, No. 4, 1967, pp. 324–331. On behalf of the Institution of Mechanical Engineers.
8. Meyer, M. L., "A Simple Estimate for the Effect of Cross Sensitivity on Evaluated Strain-gage Measurement," *Experimental Mechanics*, Vol. 7, No. 11, Nov. 1967, pp. 476–480.
9. "Errors Due to Transverse Sensitivity in Strain Gages," TN-509, Measurements Group, Inc., P.O. Box 27777, Raleigh, NC 27611, 1982.
10. Measurements Group, Inc., "Errors Due to Transverse Sensitivity in Strain Gages," *Experimental Techniques*, Vol. 7, No. 1, Jan. 1983, pp. 30–35.
11. *Handbook on Experimental Mechanics*, edited by Albert S. Kobayashi, Englewood Cliffs, Prentice-Hall, 1987, pp. 51–54.
12. *1986 Annual Book of ASTM Standards*, 1916 Race St., Philadelphia, PA 19103, "Performance Characteristics of Bonded Resistance Strain Gages," Vol. 03.01. Designation: E251-86, pp. 413–428. Copyright ASTM. Reprinted with permission.
13. "SR-4 Strain Gage Handbook," BLH Electronics, Inc., 75 Shawmut Rd., Canton, MA 02021, 1980.
14. Data furnished by Measurements Group, Inc., P.O. Box 27777, Raleigh, NC 27611, 1989.
15. *1986 Annual Book of ASTM Standards*, 1916 Race St., Philadelphia, PA 19103, "Verification and Classication of Extensometers," Vol. 03.01. Designation: E83-85, pp. 267–274. Copyright ASTM. Reprinted with permission.

8

STRAIN GAGE ROSETTES AND DATA ANALYSIS

8.1. Reason for rosette analysis

We saw in Chapter 2 that for any point on a free (unloaded) surface of a solid it is necessary to know three independent quantities in order to specify the state of stress completely. These quantities are the magnitudes of the two principal stresses, σ_1 and σ_2, and their directions, θ or $\theta + 90°$, with respect to some reference.

For isotropic elastic materials these values can be calculated from strains measured on the surface at the point in question, and since three independent quantities are to be determined, in general, it will be necessary to make three independent measurements of strain. There are, however, some special situations in which one or two observations of strain will suffice to provide the information necessary for completely establishing the state of stress.

It will be well, at this time, to draw attention to the fact that, although we refer to the stress condition at a point, the manner of measuring the strain gives the average over a small distance. Therefore, from the practical point of view, the results of a set of rosette observations will approximate the average conditions over a small area. This is not objectionable as long as the length over which the strain is measured is short enough that there is relatively little change from one end to the other. The gage length will therefore depend upon the strain gradient and may run from small ($\frac{1}{64}$ in to $\frac{1}{16}$ in) values to several inches or more.

8.2. Stress fields

Stress fields were examined in Chapter 2, where the several stress states were discussed. In general, the concern has been with plane stress, and transformation equations were developed to enable the determination of plane stress at a point in any direction relative to a chosen coordinate system. These concepts will be reexamined here in the development of rosette analysis.

Special case of uniaxial stress (simple tension or compression)

In the case of simple tension or compression, one knows that the directions of the principal stress axes will be parallel and perpendicular to the direction of the applied force, or load, and that the magnitude of the principal stress whose direction is at right angles to the load will be zero. This means that two of the three quantities are known from the prevailing physical conditions. On this account, it will therefore be necessary to make only a single observation of the strain along the direction of the load in order to determine the one remaining unknown quantity. For an elastic body, the stress is calculated as

$$\sigma = E\varepsilon \tag{8.1}$$

where σ = the stress intensity

E = the modulus of elasticity of the material

ε = the measured strain (positive for tension and negative for compression)

It should be noted here that if the stress is tension, σ represents σ_1, the algebraically larger principal stress, and $\sigma_2 = 0$. If the stress is compression, $\sigma_1 = 0$ and σ corresponds to σ_2, the algebraically smaller principal stress.

Special case of biaxial stress (principal stress directions known)

In a few special cases in which the directions of the principal stress axes (the angle θ) can be established by auxiliary means, such as conditions of symmetry or through a previous application of a brittle coat, there are only two unknowns, σ_1 and σ_2, the principal stress magnitudes, to be determined. These can be found by measuring the corresponding principal strains, ε_1 and ε_2, in the directions of the principal stress axes, and calculating the values from Eqs. (2.50a) and (2.50b). In these equations the subscripts, x and y, are changed to 1 and 2, respectively. Equations (2.50a) and (2.50b) can be rewritten as

$$\sigma_1 = \frac{E}{1 - v^2}(\varepsilon_1 + v\varepsilon_2) \tag{8.2}$$

$$\sigma_2 = \frac{E}{1 - v^2}(\varepsilon_2 + v\varepsilon_1) \tag{8.3}$$

where σ_1 = the algebraically larger principal stress

σ_2 = the algebraically smaller principal stress

ε_1 = the algebraically larger principal strain

ε_2 = the algebraically smaller principal strain

E = the modulus of elasticity of the material

v = Poisson's ratio

For later use it will be more convenient to express the principal stress values in the following form:

$$\sigma_1 = E\left[\frac{A}{1-v} + \frac{B}{1+v}\right] \quad (8.4)$$

$$\sigma_2 = E\left[\frac{A}{1-v} - \frac{B}{1+v}\right] \quad (8.5)$$

where $A = \frac{1}{2}(\varepsilon_1 + \varepsilon_2)$ = the hydrostatic component of strain and corresponds to the center of Mohr's circle
$B = \frac{1}{2}(\varepsilon_1 - \varepsilon_2)$ = the shear component of strain and corresponds to the radius of Mohr's circle

The general case

In many instances, neither the magnitudes of the principal stresses nor the directions of their axes will be known. This means that for a complete description of the state of stress, at any particular point, three independent quantities must be found. In consequence, it will be necessary to make three measurements of linear strain in different directions (see Section 2.6), and from these three observations, to compute the two principal stress magnitudes and the directions of the axes.

Figure 8.1 illustrates a pair of orthogonal reference axes, OX and OY, and three other axes, OA, OB, and OC, making angles θ_a, θ_b, and θ_c, respectively, with respect to the references axis OX. The axes OA, OB, and OC form what is described as a rosette, and if corresponding linear strains, ε_a, ε_b, and ε_c, are measured in their respective directions, the linear and shearing strains, ε_x, ε_y, and γ_{xy}, corresponding to the OX and OY axes of reference, can be calculated.

The values of ε_x, ε_y, and γ_{xy} are calculated in terms of the measured strains, ε_a, ε_b, and ε_c, by the use of Eq. (2.32). It is repeated here and

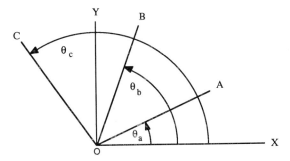

FIG. 8.1. Reference axes OX–OY with rosette axes.

renumbered as Eq. (8.6). Thus,

$$\varepsilon_{x'} = \frac{\varepsilon_x + \varepsilon_y}{2} + \frac{\varepsilon_x - \varepsilon_y}{2}\cos 2\theta + \frac{\gamma_{xy}}{2}\sin 2\theta \tag{8.6}$$

In Eq. (8.6), the subscript, x', takes on the values of the three measured strains in turn, and θ has the value associated with its particular strain. This gives three independent equations that can be solved simultaneously for ε_x, ε_y, and γ_{xy}. The three equations so formed are

$$\varepsilon_a = \frac{\varepsilon_x + \varepsilon_y}{2} + \frac{\varepsilon_x - \varepsilon_y}{2}\cos 2\theta_a + \frac{\gamma_{xy}}{2}\sin 2\theta_a \tag{8.7}$$

$$\varepsilon_b = \frac{\varepsilon_x + \varepsilon_y}{2} + \frac{\varepsilon_x - \varepsilon_y}{2}\cos 2\theta_b + \frac{\gamma_{xy}}{2}\sin 2\theta_b \tag{8.8}$$

$$\varepsilon_c = \frac{\varepsilon_x + \varepsilon_y}{2} + \frac{\varepsilon_x - \varepsilon_y}{2}\cos 2\theta_c + \frac{\gamma_{xy}}{2}\sin 2\theta_c \tag{8.9}$$

When ε_x, ε_y, and γ_{xy} have been determined by the simultaneous solution of Eqs. (8.7), (8.8), and (8.9), the principal strains may be found by using Eq. (2.37). Thus, the principal strains are

$$\varepsilon_{1,2} = \frac{\varepsilon_x + \varepsilon_y}{2} \pm \sqrt{\left(\frac{\varepsilon_x - \varepsilon_y}{2}\right)^2 + \left(\frac{\gamma_{xy}}{2}\right)^2} \tag{8.10}$$

or

$$\varepsilon_{1,2} = A \pm B \tag{8.11}$$

where $A = \dfrac{\varepsilon_x + \varepsilon_y}{2}$ (corresponds to the center of Mohr's circle)

$B = \sqrt{\left(\dfrac{\varepsilon_x - \varepsilon_y}{2}\right)^2 + \left(\dfrac{\gamma_{xy}}{2}\right)^2}$ (corresponds to the radius of Mohr's circle)

The orientation of the principel stresses relative to the reference axes is the same as the principal strain axes relative to the references axes. Thus, the orientation of the principal axes may be obtained from Mohr's circle for strain, or, analytically, by using Eq. 2.34 and either Eq. (2.35) or (2.36).

8.3. Rosette geometry

Theoretically, the relative directions of strain measurement (the angles θ_a, θ_b, and θ_c) are of no particular importance. However, from the practical consideration of solving the equations, one finds that certain preferred

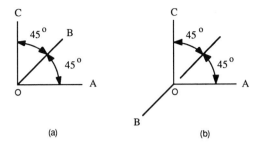

FIG. 8.2. Three-element rectangular rosette arrangements.

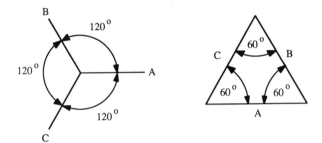

FIG. 8.3. Delta rosette arrangements.

orientations permit a much simpler reduction of the strains into terms of stress. At the present time there are four generally accepted arrangements of the gage axes for strain rosettes. Basically, there are just two arrangements, the rectangular and the equiangular, but each of these has a modification involving a redundant fourth observation of strain.

Basic arrangements involving three observations of strain

Figure 8.2 shows two arrangements of a three-element rectangular rosette. The three gage axes in arrangement (a) are laid out at 45° and 90° to each other. In arrangement (b), gage B forms a 135° angle with gages A and C.

The equiangular or delta rosette has the three gage axes laid out parallel to the sides of an equilateral triangle. This type of rosette has the most desirable orientations of the directions of strain observation, but the equations for computing stress values are not quite so simple as those of the rectangular rosette. For this reason, the rectangular rosette is preferred by many. Figure 8.3 illustrates two arrangements for the delta rosette.

Modified arrangements involving four observations of strain

The T-rectangular rosette has four gages with axes 45° apart, as indicated in Fig. 8.4. Although the fourth observation is theoretically unnecessary, it

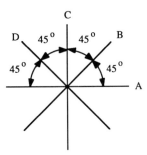

Fig. 8.4. T-rectangular rosette.

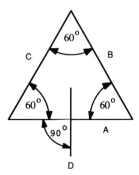

Fig. 8.5. T-delta rosette.

nevertheless provides a convenient means of checking the strain observations, since the sum of the strains in any two directions at right angles should be a constant for a given set of conditions. Thus,

$$\varepsilon_a + \varepsilon_c = \varepsilon_b + \varepsilon_d \tag{8.12}$$

The T-delta rosette is essentially the same as the equiangular arrangement with the addition of a fourth observation which is made at right angles to the direction of one of the other three. It is claimed that this form of rosette has all the desirable characteristics of the equiangular type plus the advantage of a little more precise determination of the hydrostatic component of strain at the reference point, if this coincides with the intersection of two perpendicular gage axes. The arrangement is shown in Fig. 8.5.

8.4. Analytical solution for the rectangular rosette

This analysis is for the three-element rectangular rosette, and is started by taking the OA axis of the rosette in Fig. 8.1 as the reference axis and making it coincident with the OX axis. Since the three-element rectangular rosette is being considered, the strain gage axes will be those shown in Fig. 8.2. For

STRAIN GAGE ROSETTES AND DATA ANALYSIS 259

this arrangement, then, one has the following angles:

$$\theta_a = 0°, \quad \theta_b = 45°, \quad \theta_c = 90°$$

Since the transformation equation given by Eq. (8.6) is written in terms of twice the angle, the required values of the trigonometric functions are

$$\cos 2\theta_a = 1, \quad \cos 2\theta_b = 0, \quad \cos 2\theta_c = -1$$
$$\sin 2\theta_a = 0, \quad \sin 2\theta_b = 1, \quad \sin 2\theta_c = 0$$

These values can be substituted into Eqs. (8.7), (8.8), and (8.9) to give the three simultaneous equations necessary to determine ε_x, ε_y, and γ_{xy}. Thus,

$$\varepsilon_a = \frac{\varepsilon_x + \varepsilon_y}{2} + \frac{\varepsilon_x - \varepsilon_y}{2}(1) + \frac{\gamma_{xy}}{2}(0) \tag{8.13}$$

$$\varepsilon_b = \frac{\varepsilon_x + \varepsilon_y}{2} + \frac{\varepsilon_x - \varepsilon_y}{2}(0) + \frac{\gamma_{xy}}{2}(1) \tag{8.14}$$

$$\varepsilon_c = \frac{\varepsilon_x + \varepsilon_y}{2} + \frac{\varepsilon_x - \varepsilon_y}{2}(-1) + \frac{\gamma_{xy}}{2}(0) \tag{8.15}$$

From Eq. (8.13), it is seen that

$$\varepsilon_x = \varepsilon_a \tag{8.16}$$

From Eq. (8.15), it is seen that

$$\varepsilon_y = \varepsilon_c \tag{8.17}$$

The shearing strain, γ_{xy}, can be determined in terms of the strain readings by substituting the values of ε_x and ε_y given by Eqs. (8.16) and (8.17), respectively, into Eq. (8.14). In doing this,

$$\varepsilon_b = \frac{\varepsilon_a + \varepsilon_c}{2} + \frac{\gamma_{xy}}{2}$$

Solving for γ_{xy},

$$\gamma_{xy} = 2\varepsilon_b - (\varepsilon_a + \varepsilon_c) \tag{8.18}$$

Thus, Eqs. (8.16), (8.17), and (8.18) give the values of ε_x, ε_y, and γ_{xy} in terms of the strain gage readings.

By substituting these values of ε_x, ε_y, and γ_{xy} into Eq. (8.10), the values of the principal strains are determined directly in terms of the strain readings from the rosette. Consequently,

$$\varepsilon_{1,2} = \frac{\varepsilon_a + \varepsilon_c}{2} \pm \tfrac{1}{2}\sqrt{(\varepsilon_a - \varepsilon_c)^2 + [2\varepsilon_b - (\varepsilon_a + \varepsilon_c)]^2} \quad (8.19)$$

Equation (8.19) can also be expressed as

$$\varepsilon_{1,2} = A \pm B \quad (8.20)$$

where

$$A = \frac{\varepsilon_a + \varepsilon_c}{2} \quad (8.20a)$$

$$B = \tfrac{1}{2}\sqrt{(\varepsilon_a - \varepsilon_c)^2 + [2\varepsilon_b - (\varepsilon_a + \varepsilon_c)]^2} \quad (8.20b)$$

The values of ε_1 and ε_2 given by Eq. (8.19) may now be substituted into Eqs. (8.2) and (8.3) in order to determine the principal stresses, σ_1 and σ_2. In most cases, however, the numerical values of the principal strains need not be known, since the values of A and B, given by Eqs. (8.20a) and (8.20b), respectively, can be substituted into Eqs. (8.4) and (8.5) in order to determine σ_1 and σ_2 directly in terms of the strain observations on the rosette. Carrying out this operation yields

$$\sigma_1 = E\left\{\frac{\varepsilon_a + \varepsilon_c}{2(1-v)} + \frac{1}{2(1+v)}\sqrt{(\varepsilon_a - \varepsilon_c)^2 + [2\varepsilon_b - (\varepsilon_a + \varepsilon_c)]^2}\right\} \quad (8.21)$$

$$\sigma_2 = E\left\{\frac{\varepsilon_a + \varepsilon_c}{2(1-v)} - \frac{1}{2(1+v)}\sqrt{(\varepsilon_a - \varepsilon_c)^2 + [2\varepsilon_b - (\varepsilon_a + \varepsilon_c)]^2}\right\} \quad (8.22)$$

Equations (8.21) and (8.22) are not in the simplest form but the form given lends itself better to the determination of the directions of the principal stress axes.

Determination of the principal stress axes directions

In order to determine the orientation of the principal stress axes, the orientation of the principal strain axes may be found instead, since the axes of each coincide. In Section 2.6, the orientation of the principal strain axes relative to the reference axes were found analytically by the use of Eqs. (2.34), (2.35), and (2.36). Note, also, that the angle θ is always measured in a counterclockwise direction from the positive OX axis to the positive O1 axis, which corresponds to the direction of ε_1, and therefore σ_1.

Equations (2.34), (2.35), and (2.36) can now be expressed in terms of the strain readings of the rosette. Thus,

$$\tan 2\theta = \frac{\gamma_{xy}}{\varepsilon_x - \varepsilon_y} = \frac{2\varepsilon_b - (\varepsilon_a + \varepsilon_c)}{\varepsilon_a - \varepsilon_c} \tag{8.23}$$

$$\sin 2\theta = \frac{\gamma_{xy}}{\sqrt{(\varepsilon_x - \varepsilon_y)^2 + (\gamma_{xy})^2}} = \frac{2\varepsilon_b - (\varepsilon_a + \varepsilon_c)}{\sqrt{(\varepsilon_a - \varepsilon_c)^2 + [2\varepsilon_b - (\varepsilon_a + \varepsilon_c)]^2}} \tag{8.24}$$

$$\cos 2\theta = \frac{\varepsilon_x - \varepsilon_y}{\sqrt{(\varepsilon_x - \varepsilon_y)^2 + (\gamma_{xy})^2}} = \frac{\varepsilon_a - \varepsilon_c}{\sqrt{(\varepsilon_a - \varepsilon_c)^2 + [2\varepsilon_b - (\varepsilon_a + \varepsilon_c)]^2}} \tag{8.25}$$

To establish the angular relationship, θ_1, between the $O1$ axis and the OA axis (OX and OA are coincident), two of the three equations are used. In fact, if Eq. (8.23) is chosen, then we need only to determine the sign of either $\sin 2\theta$ or $\cos 2\theta$ to obtain the matching quadrant. For example, if $\tan 2\theta$ is negative, then 2θ must be in either the second or fourth quadrant. If $\sin 2\theta$ is negative, then 2θ could be in either the third or fourth quadrant. The matching quadrants are the fourth, and so the angle 2θ must be in the fourth quadrant. From this, then, the orientation of axis $O1$ can be determined relative to axis OA.

Fortunately, a check can always be made by sketching a Mohr's circle. Three rules for determining the angle, θ_1, between the OA and the $O1$ axis will be stated.

1. $\varepsilon_b > \dfrac{\varepsilon_a + \varepsilon_c}{2}$

    θ_1 lies between $0°$ and $+90°$

2. $\varepsilon_b < \dfrac{\varepsilon_a + \varepsilon_c}{2}$

    θ_1 lies between $0°$ and $-90°$

3. $\varepsilon_b = \dfrac{\varepsilon_a + \varepsilon_c}{2}$

    (a) If $\varepsilon_a > \varepsilon_c$, $\varepsilon_a = \varepsilon_1$, $\theta_1 = 0°$
    (b) If $\varepsilon_a < \varepsilon_c$, $\varepsilon_a = \varepsilon_2$, $\theta_1 = \pm 90°$

Proof of rules

Figure 8.6 shows a Mohr's circle for a three-element rectangular strain rosette. The three strains, ε_a, ε_b, and ε_c, are represented by points A, B, and C, respectively, on the circumference of the circle and at the ends of the radial lines that are $90°$ apart and taken in the same sequence as the rosette axes, which are $45°$ apart.

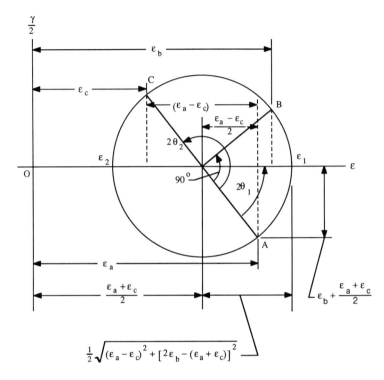

FIG. 8.6. Mohr's circle for the rectangular rosette with three observations of strain.

If point A lies anywhere along the semicircumference below the abscissa, then angle $2\theta_1$ will be positive and have values between $0°$ and $180°$, so that θ_1 will be between $0°$ and $90°$. If point A happens to lie on the semicircumference above the abscissa, then angle $2\theta_1$ will lie between 0 and $-180°$ and θ_1 will be between $0°$ and $-90°$.

How can we tell whether point A is above or below the abscissa on the Mohr diagram? A study of Fig. 8.6 shows that point A will lie below the abscissa whenever point B is to the right of the center of the circle; that is, when $\varepsilon_b > (\varepsilon_a + \varepsilon_c)/2$. Point A will be above the abscissa when $\varepsilon_b < (\varepsilon_a + \varepsilon_c)/2$. and will lie on the abscissa when $\varepsilon_b = (\varepsilon_a + \varepsilon_c)/2$. From this, the following rules can be set down:

1. The angle, θ_1, will lie between $0°$ and $+90°$ when $\varepsilon_b > (\varepsilon_a + \varepsilon_c)/2$. This is shown in Fig. 8.7.
2. The angle, θ_1, will lie between $0°$ and $-90°$ when $\varepsilon_b < (\varepsilon_a + \varepsilon_c)/2$. This is illustrated in Fig. 8.8.
3. Figure 8.9 shows that the angle, θ_1, will be zero when $\varepsilon_b = (\varepsilon_a + \varepsilon_c)/2$ and $\varepsilon_a > \varepsilon_c$. From the figure, it is evident that $\varepsilon_a = \varepsilon_1$, the maximum principal strain in the plane. Figure 8.10 shows that the angle, θ_1, will be $90°$ when $\varepsilon_b = (\varepsilon_a + \varepsilon_c)/2$ and $\varepsilon_a < \varepsilon_c$. It is apparent from the figure that $\varepsilon_a = \varepsilon_2$, the minimum principal strain in the plane.

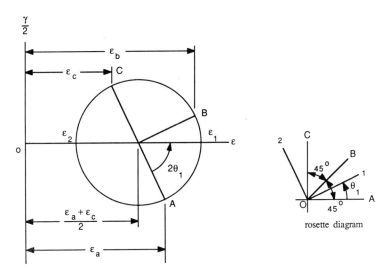

FIG. 8.7. Mohr diagram for $\varepsilon_b > \dfrac{\varepsilon_a + \varepsilon_c}{2}$.

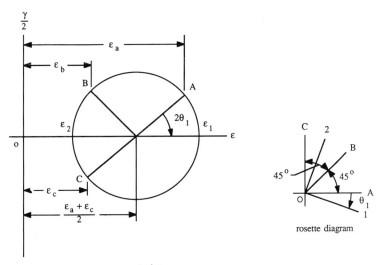

FIG. 8.8. Mohr diagram for $\varepsilon_b < \dfrac{\varepsilon_a + \varepsilon_c}{2}$.

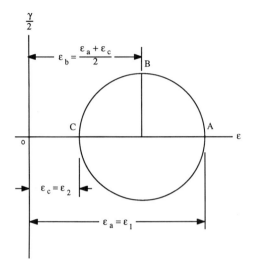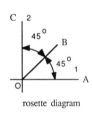

rosette diagram

FIG. 8.9. Mohr diagram for $\varepsilon_b = \dfrac{\varepsilon_a + \varepsilon_c}{2}$ and $\varepsilon_a > \varepsilon_c$.

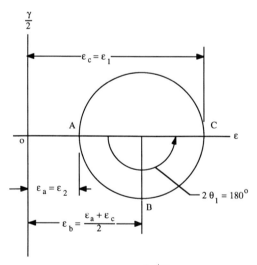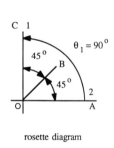

rosette diagram

FIG. 8.10. Mohr diagram for $\varepsilon_b = \dfrac{\varepsilon_a + \varepsilon_c}{2}$ and $\varepsilon_a < \varepsilon_c$.

STRAIN GAGE ROSETTES AND DATA ANALYSIS   265

The difference in the sign of the shearing strain, γ_{xy}, should again be reviewed in Section 2.6 in order to make the analytical solution for θ_1 compatible with the solution from Mohr's circle.

Example 8.1. A three-element rectangular strain rosette gives the following readings:

$$\varepsilon_a = 1350\ \mu\text{in/in}, \qquad \varepsilon_b = -500\ \mu\text{in/in}, \qquad \varepsilon_c = 560\ \mu\text{in/in}$$

(a) Determine the principal strains.
(b) Determine θ_1 analytically.
(c) Sketch the orientation of the principal axes relative to OA.
(d) Draw a Mohr's circle and check the position of θ_1.
(e) Determine σ_1 and σ_2 using $\nu = 0.3$ and $E = 30 \times 10^6$ psi.

Solution. (a) The principal strains are given by Eq. (8.19).

$$\varepsilon_{1,2} = \frac{\varepsilon_a + \varepsilon_c}{2} \pm \tfrac{1}{2}\sqrt{(\varepsilon_a - \varepsilon_c)^2 + [2\varepsilon_b - (\varepsilon_a + \varepsilon_c)]^2}$$

$$= \frac{1350 + 560}{2} \pm \tfrac{1}{2}\sqrt{(1350 - 560)^2 + [2(-500) - (1350 + 560)]^2}$$

$$= 955 \pm 1508$$

$$\varepsilon_1 = 2463\ \mu\text{in/in}, \qquad \varepsilon_2 = -553\ \mu\text{in/in}$$

(b) Equations (8.23) and (8.24) will be used to determine θ_1.

$$\tan 2\theta = \frac{2\varepsilon_b - (\varepsilon_a + \varepsilon_c)}{\varepsilon_a - \varepsilon_c} = \frac{2(-500) - (1350 + 560)}{1350 - 560} = -3.683\,54$$

From this, 2θ may be in either the second or fourth quadrant.

$$\sin 2\theta = \frac{2\varepsilon_b - (\varepsilon_a + \varepsilon_c)}{\sqrt{(\varepsilon_a - \varepsilon_c)^2 + [2\varepsilon_b - (\varepsilon_a + \varepsilon_c)]^2}}$$

The numerator of $\sin 2\theta$ is negative, as can be seen from $\tan 2\theta$, and so only the sign of $\sin 2\theta$ is needed. Since $\sin 2\theta$ is negative, 2θ may be in either the third or fourth quadrant. Since the fourth quadrant is the matching quadrant in each, 2θ is a fourth-quadrant angle. Thus,

$$2\theta_1 = 360 - \tan^{-1}|-3.683\,54| = 285.2°$$

$$\theta_1 = 142.6°$$

The angle, θ_1, is measured in a counterclockwise direction from the OA axis to the $O1$ axis.

(c) Figure 8.11 shows the orientation of the $O1$ axis relative to the OA axis.

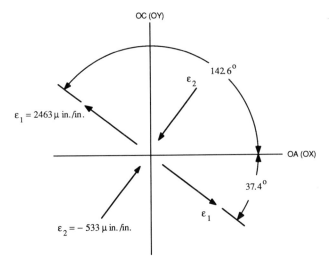

FIG. 8.11. Orientation of the principal strain (and stress) axes relative to the reference coordinates in Example 8.1.

This can also be verified by the use of Rule 2, since $\varepsilon_b < (\varepsilon_a + \varepsilon_c)/2$. In this case, θ_1 lies between $0°$ and $-90°$, which is the acute angle between the OA axis and ε_1, going in a clockwise (negative) direction.

(d) In order to draw Mohr's circle, ε_x, ε_y, and γ_{xy} are computed using Eqs. (8.16), (8.17), and (8.18).

$$\varepsilon_x = \varepsilon_a = 1350 \text{ μin/in}$$

$$\varepsilon_y = \varepsilon_c = 560 \text{ μin/in}$$

$$\gamma_{xy} = 2\varepsilon_b - (\varepsilon_a + \varepsilon_c) = 2(-500) - (1350 + 560) = -2910 \text{ μradians}$$

Figure 8.12 gives Mohr's circle. Note that γ_{xy} is negative from the transformation equation, but for the Mohr's circle it must be plotted as positive. This is in accordance with the sign convention established in Chapter 2. If the circle is traversed in a counterclockwise direction from the x axis, then $2\theta_1 = 285.2°$, which is the angle computed in part (b).

(e) Equations (8.2) and (8.3) can be used to determine σ_1 and σ_2, respectively.

$$\sigma_1 = \frac{E}{1 - v^2}(\varepsilon_1 + v\varepsilon_2) = \frac{30 \times 10^6}{[1 - (0.3)^2]}[2463 + 0.3(-553)] \times 10^{-6}$$

$$= 75\,729 \text{ psi}$$

$$\sigma_2 = \frac{E}{1 - v^2}(\varepsilon_2 + v\varepsilon_1) = \frac{30 \times 10^6}{[1 - (0.3)^2]}[-553 + 0.3(2463)] \times 10^{-6}$$

$$= 6129 \text{ psi}$$

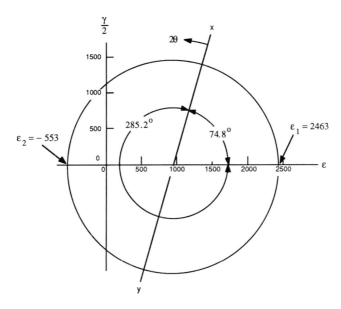

FIG. 8.12. Mohr's circle for Example 8.1.

8.5. Analytical solution for the equiangular or delta rosette

The procedure used for the rectangular rosette will also be used for the delta rosette. The OA axis of the rosette, Fig. 8.1, is taken coincident with the OX axis of reference. For this arrangement, then,

$$\theta_a = 0°, \quad \theta_b = 120°, \quad \theta_c = 240°$$

and

$$\cos 2\theta_a = 1, \quad \cos 2\theta_b = -1/2, \quad \cos 2\theta_c = -1/2$$
$$\sin 2\theta_a = 0, \quad \sin 2\theta_b = -\sqrt{3}/2, \quad \sin 2\theta_c = \sqrt{3}/2$$

Equations (8.7), (8.8), and (8.9) can be used with these trigonometric values to form the three simultaneous equations needed in order to determine ε_x, ε_y, and γ_{xy}. This results in

$$\varepsilon_a = \frac{\varepsilon_x + \varepsilon_y}{2} + \frac{\varepsilon_x - \varepsilon_y}{2}(1) + \frac{\gamma_{xy}}{2}(0) \tag{8.26}$$

$$\varepsilon_b = \frac{\varepsilon_x + \varepsilon_y}{2} + \frac{\varepsilon_x - \varepsilon_y}{2}\left(-\frac{1}{2}\right) + \frac{\gamma_{xy}}{2}\left(-\frac{\sqrt{3}}{2}\right) \tag{8.27}$$

$$\varepsilon_c = \frac{\varepsilon_x + \varepsilon_y}{2} + \frac{\varepsilon_x - \varepsilon_y}{2}\left(-\frac{1}{2}\right) + \frac{\gamma_{xy}}{2}\left(\frac{\sqrt{3}}{2}\right) \tag{8.28}$$

From Eq. (8.26), it is seen that

$$\varepsilon_x = \varepsilon_a \tag{8.29}$$

Substituting the value of ε_x given by Eq. (8.29) into Eqs. (8.27) and (8.28) yields

$$\varepsilon_b = \frac{\varepsilon_a}{4} + \frac{3\varepsilon_y}{4} - \frac{\sqrt{3}}{4}\gamma_{xy} \tag{a}$$

$$\varepsilon_c = \frac{\varepsilon_a}{4} + \frac{3\varepsilon_y}{4} + \frac{\sqrt{3}}{4}\gamma_{xy} \tag{b}$$

Equations (a) and (b) can be solved simultaneously for ε_y and γ_{xy}. Thus,

$$\varepsilon_y = \frac{1}{3}[2(\varepsilon_b + \varepsilon_c) - \varepsilon_a] \tag{8.30}$$

$$\gamma_{xy} = \frac{2}{\sqrt{3}}(\varepsilon_c - \varepsilon_b) \tag{8.31}$$

The principal strains in terms of the rosette readings may now be determined by substituting these values of ε_x, ε_y, and γ_{xy} into Eq. (8.10). Consequently,

$$\varepsilon_{1,2} = \frac{\varepsilon_a + \varepsilon_b + \varepsilon_c}{3} \pm \sqrt{\left[\frac{2\varepsilon_a - \varepsilon_b - \varepsilon_c}{3}\right]^2 + \left[\frac{\varepsilon_c - \varepsilon_b}{\sqrt{3}}\right]^2} \tag{8.32}$$

Equation (8.32) can also be expressed as

$$\varepsilon_{1,2} = A \pm B \tag{8.33}$$

where

$$A = \frac{\varepsilon_a + \varepsilon_b + \varepsilon_c}{3} \tag{8.33a}$$

$$B = \sqrt{\left[\frac{2\varepsilon_a - \varepsilon_b - \varepsilon_c}{3}\right]^2 + \left[\frac{\varepsilon_c - \varepsilon_b}{\sqrt{3}}\right]^2} \tag{8.33b}$$

The values of ε_1 and ε_2 given by Eq. (8.32) may now be substituted into Eqs. (8.2) and (8.3) in order to determine the principal stresses, σ_1 and σ_2, in terms of the rosette strain readings. This gives

$$\sigma_1 = E \left\{ \frac{\varepsilon_a + \varepsilon_b + \varepsilon_c}{3(1-v)} + \frac{1}{1+v} \sqrt{\left[\frac{2\varepsilon_a - \varepsilon_b - \varepsilon_c}{3}\right]^2 + \left[\frac{\varepsilon_c - \varepsilon_b}{\sqrt{3}}\right]^2} \right\} \quad (8.34)$$

$$\sigma_2 = E \left\{ \frac{\varepsilon_a + \varepsilon_b + \varepsilon_c}{3(1-v)} - \frac{1}{1+v} \sqrt{\left[\frac{2\varepsilon_a - \varepsilon_b - \varepsilon_c}{3}\right]^2 + \left[\frac{\varepsilon_c - \varepsilon_b}{\sqrt{3}}\right]^2} \right\} \quad (8.35)$$

Determination of the principal stress directions

As with any rectangular rosette, the orientation of the principal strain axes will be determined analytically through the use of Eqs. (2.34), (2.35), and (2.36). Once again, note that the principal strain axes and the principal stress axes coincide, and that the angle θ is measured in a counterclockwise direction from the positive OX axis to the positive $O1$ axis, which corresponds to the direction of ε_1, and therefore to σ_1.

Equations (2.34), (2.35), and (2.36) can now be expressed in terms of the strain readings of the rosette. They are

$$\tan 2\theta = \frac{\gamma_{xy}}{\varepsilon_x - \varepsilon_y} = \frac{\sqrt{3}(\varepsilon_c - \varepsilon_b)}{2\varepsilon_a - \varepsilon_b - \varepsilon_c} \quad (8.36)$$

$$\sin 2\theta = \frac{\gamma_{xy}}{\sqrt{(\varepsilon_x - \varepsilon_y)^2 + (\gamma_{xy})^2}} = \frac{\varepsilon_c - \varepsilon_b}{\sqrt{\frac{1}{3}(2\varepsilon_a - \varepsilon_b - \varepsilon_c)^2 + (\varepsilon_c - \varepsilon_b)^2}} \quad (8.37)$$

$$\cos 2\theta = \frac{\varepsilon_x - \varepsilon_y}{\sqrt{(\varepsilon_x - \varepsilon_y)^2 + (\gamma_{xy})^2}} = \frac{\frac{1}{\sqrt{3}}(2\varepsilon_a - \varepsilon_b - \varepsilon_c)}{\sqrt{\frac{1}{3}(2\varepsilon_a - \varepsilon_b - \varepsilon_c)^2 + (\varepsilon_c - \varepsilon_b)^2}} \quad (8.38)$$

As with the three-element rectangular rosette, any two of the three equations must be used in order to establish θ_1. If Eq. (8.36) is chosen, then only the sign of $\sin 2\theta$ or $\cos 2\theta$ need be determined in order to establish the matching quadrant.

As before, a check can always be made by sketching a Mohr's circle. Three rules for determining the angle, θ_1, between the OA axis and the $O1$ axis will be stated.

1. $\varepsilon_c > \varepsilon_b$: θ_1 lies between $0°$ and $+90°$
2. $\varepsilon_b > \varepsilon_c$: θ_1 lies between $0°$ and $-90°$
3. $\varepsilon_b = \varepsilon_c$:
   (a) If $\varepsilon_a > \varepsilon_b = \varepsilon_c$, $\varepsilon_a = \varepsilon_1$, $\theta_1 = 0°$
   (b) If $\varepsilon_a < \varepsilon_b = \varepsilon_c$, $\varepsilon_a = \varepsilon_2$, $\theta_1 = \pm 90°$.

Proof of rules

Figure 8.13 shows a Mohr's circle for the delta rosette. Since the gage axes of the equiangular rosette are inclined at 120° (or 60°) relative to each other, the points representing the corresponding strains on the circumference of Mohr's circle are located at the vertices of the equiangular triangle ABC, as indicated in the figure. A study of the diagram reveals that as the strains ε_a, ε_b, and ε_c vary, the triangle ABC will rotate about its centroid, which is located at the center of the circle.

Before continuing, however, attention is drawn particularly to the observation that if one starts at point A and follows around the circumference of Mohr's circle in the counterclockwise direction, the next station reached will be point C. On first thought, this might appear to be an error, since in going around the rosette axes in the same direction, axis B follows axis A, as shown in Fig. 8.13a. The apparent discrepancy is caused by the fact that the

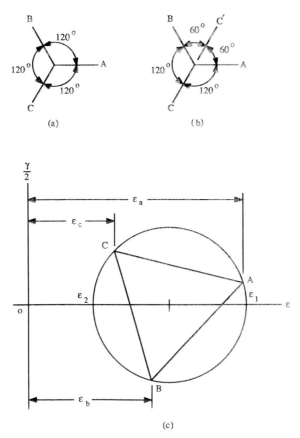

FIG. 8.13. Gage axes and Mohr diagram for equiangular rosette.

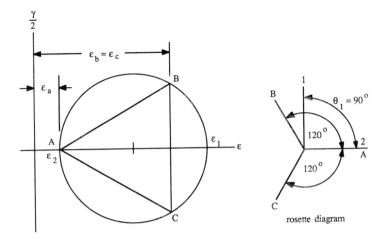

FIG. 8.14. Case in which $\varepsilon_b = \varepsilon_c > \varepsilon_a$, $\varepsilon_a = \varepsilon_2$, and $\theta_1 = \pm 90°$.

angular displacements are doubled in Mohr's diagram. If one extends the axis OC into the position OC' shown in Fig. 8.13b, then the reason for the relative positions of the points A, B, and C on the circumference of Mohr's circle should be clear.

If point A happens to fall at the extreme left of the circumference of the circle, Fig. 8.14, then, since the centroid of ABC lies on the abscissa, CB is at right angles to OA, which means that $\varepsilon_c = \varepsilon_b$. Also, because A is at the extreme left of the circle, $\varepsilon_a = \varepsilon_2$, which is the algebraically smaller principal strain. From the diagram it is seen that $2\theta_1 = \pm 180°$, and therefore $\theta_1 = \pm 90°$, which substantiates Rule 3(b).

If the relative values of ε_a, ε_b, and ε_c are now changed so that triangle ABC rotates in a counterclockwise direction from the position in Fig. 8.14, ε_b will become smaller than ε_c and point A will move on to the lower half of the circumference of the circle. Under these conditions the angle $2\theta_1$ will be between $0°$ and $+180°$, and θ_1 will be between $0°$ and $+90°$, as shown in Fig. 8.15 and stated in Rule 1.

When the triangle ABC has finally rotated through $180°$, point A will have moved along the entire lower semicircumference of the circle and taken up the position shown in Fig. 8.16, such that $2\theta_1 = 0°$, $\theta_1 = 0°$, $\varepsilon_a = \varepsilon_1$, and since A is again on the abscissa, $\varepsilon_c = \varepsilon_b$. This time $\varepsilon_a > \varepsilon_c = \varepsilon_b$ and Rule 3(a) is satisfied.

When the strains are further altered so that the continued rotation of the triangle causes point A to move on to the semicircumference above the abscissa, then, according to definition, $2\theta_1$ becomes negative and will lie between $0°$ and $-180°$. Strain ε_b will be larger than ε_c until A returns to the position corresponding to ε_2, where equality is again established between ε_b and ε_c. This establishes Rule 2 and is indicated in Fig. 8.17.

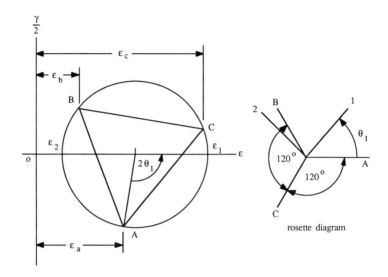

FIG. 8.15. Case in which $\varepsilon_c > \varepsilon_b$, $0° < 2\theta_1 < +180°$, and $0° < \theta_1 < +90°$.

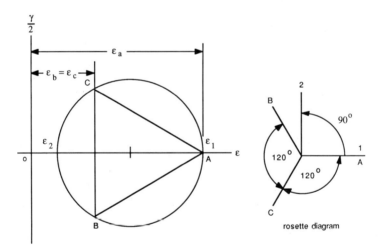

FIG. 8.16. Case in which $\varepsilon_a > \varepsilon_b = \varepsilon_c$, $\varepsilon_a = \varepsilon_1$, and $\theta_1 = 0°$.

STRAIN GAGE ROSETTES AND DATA ANALYSIS 273

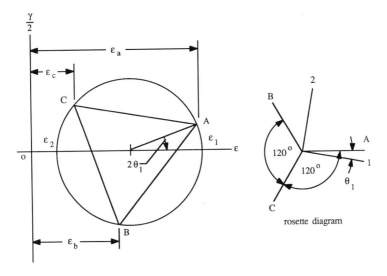

FIG. 8.17. Case in which $\varepsilon_b > \varepsilon_c$, $-180° < 2\theta_1 < 0°$, and $-90° < \theta_1 < 0°$.

Example 8.2. The following readings were obtained from a three-element delta rosette:

$$\varepsilon_a = -745 \text{ µin/in}, \quad \varepsilon_b = -165 \text{ µin/in}, \quad \varepsilon_c = 520 \text{ µin/in}$$

(a) Determine the principal strains.
(b) Determine θ_1 analytically and check using the rules listed.
(c) Determine σ_1 and σ_2, using $v = 0.3$ and $E = 30 \times 10^6$ psi.

Solution. (a) The principal strains are given by Eq. (8.32).

$$\varepsilon_{1,2} = \frac{\varepsilon_a + \varepsilon_b + \varepsilon_c}{3} \pm \sqrt{\left[\frac{2\varepsilon_a - \varepsilon_b - \varepsilon_c}{3}\right]^2 + \left[\frac{\varepsilon_c - \varepsilon_b}{\sqrt{3}}\right]^2}$$

$$= \frac{-745 - 165 + 520}{3} \pm \sqrt{\left[\frac{2(-745) - (-165) - 520}{3}\right]^2 + \left[\frac{520 - (-165)}{\sqrt{3}}\right]^2}$$

$$= -130 \pm 731$$

$$\varepsilon_1 = 601 \text{ µin/in}, \quad \varepsilon_2 = -861 \text{ µin/in}$$

(b) Equations (8.36) and (8.37) will be used to determine θ_1.

$$\tan 2\theta = \frac{\sqrt{3}(\varepsilon_c - \varepsilon_b)}{2\varepsilon_a - \varepsilon_b - \varepsilon_c} = \frac{\sqrt{3}[520 - (-165)]}{2(-745) - (-165) - 520} = -0.643\,06$$

This value of tan 2θ shows that the angle may be in either the second or fourth

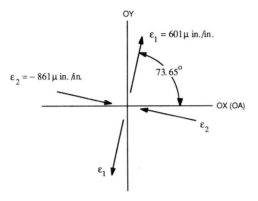

FIG. 8.18. Orientation of the principal strain (and stress) axes relative to the reference coordinates in Example 8.2.

quadrant.

$$\sin 2\theta = \frac{\varepsilon_c - \varepsilon_b}{\sqrt{\frac{1}{3}[2\varepsilon_a - \varepsilon_b - \varepsilon_c]^2 + [\varepsilon_c - \varepsilon_b]^2}}$$

The numerator of $\sin 2\theta$ is

$$\varepsilon_c - \varepsilon_b = 520 - (-165) = 685$$

Since the numerator is positive, $\sin 2\theta$ has a positive value, meaning that 2θ may be in either the first or second quadrant. The matching quadrant is the second, and so 2θ is a second-quadrant value. Thus,

$$2\theta_1 = 180 - \tan^{-1}|-0.643\,06| = 147.3°$$
$$\theta_1 = 73.65°$$

The angle, θ_1, is measured in a counterclockwise direction from the OA axis to the $O1$ axis, as shown in Fig. 8.18. To check the orientation of θ_1, Rule 1 applies, since $\varepsilon_c > \varepsilon_b$, and so θ_1 lies between zero and $+90°$. This checks with Fig. 8.18.

(c) Equations (8.2) and (8.3) may be used to determine σ_1 and σ_2, respectively.

$$\sigma_1 = \frac{E}{1-v^2}(\varepsilon_1 + v\varepsilon_2) = \frac{30 \times 10^6}{1-(0.3)^2}[601 + 0.3(-861)]10^{-6} = 11\,298 \text{ psi}$$

$$\sigma_2 = \frac{E}{1-v^2}(\varepsilon_2 + v\varepsilon_1) = \frac{30 \times 10^6}{1-(0.3)^2}[-861 + 0.3(601)]10^{-6} = -22\,441 \text{ psi}$$

Equations (8.34) and (8.35) could also have been used to compute σ_1 and σ_2.

8.6. Rosettes with four strain observations

These rosettes have been briefly described earlier. They are the T-rectangular rosette, which has four gages with axes 45° apart, and the T-delta rosette, which has a fourth gage at right angles to the axis of one of the equiangular gages. The equations for the principal strains, ε_1 and ε_2, and the principal stresses, σ_1 and σ_2, in terms of the four gage observations, will be given for each configuration.

The rectangular rosette with four observations

Figure 8.19 shows this arrangement. The fourth observation of strain is redundant, but it does provide a check since, within the limits of making the strain readings,

$$\varepsilon_a + \varepsilon_c = \varepsilon_b + \varepsilon_d \tag{8.39}$$

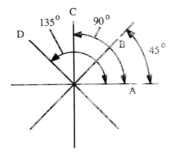

FIG. 8.19. Rectangular strain rosette with four gages.

In this case, it will be simpler to state the expressions for the principal strains, principal stresses, and the angle θ_1, and then to prove them graphically with Mohr's diagram. The principal strains can be written as

$$\varepsilon_{1,2} = \frac{\varepsilon_a + \varepsilon_b + \varepsilon_c + \varepsilon_d}{4} \pm \tfrac{1}{2}\sqrt{(\varepsilon_a - \varepsilon_c)^2 + (\varepsilon_b - \varepsilon_d)^2} \tag{8.40}$$

Equation (8.40) can also be expressed as

$$\varepsilon_{1,2} = A \pm B \tag{8.41}$$

where

$$A = \frac{\varepsilon_a + \varepsilon_b + \varepsilon_c + \varepsilon_d}{4} \tag{8.41a}$$

$$B = \tfrac{1}{2}\sqrt{(\varepsilon_a - \varepsilon_c)^2 + (\varepsilon_b - \varepsilon_d)^2} \tag{8.41b}$$

The direction of the principal axes may be found from the ratio of the quantities under the radical such that

$$\tan 2\theta = \frac{\varepsilon_b - \varepsilon_d}{\varepsilon_a - \varepsilon_c} \tag{8.42}$$

Insertion of the values of A and B given by Eqs. (8.41a) and (8.41b), respectively, into Eqs. (8.4) and (8.5) produces the expressions for the principal stresses. This gives

$$\sigma_1 = E\left[\frac{\varepsilon_a + \varepsilon_b + \varepsilon_c + \varepsilon_d}{4(1-v)} + \frac{1}{2(1+v)}\sqrt{(\varepsilon_a - \varepsilon_c)^2 + (\varepsilon_b - \varepsilon_d)^2}\right] \tag{8.43}$$

$$\sigma_2 = E\left[\frac{\varepsilon_a + \varepsilon_b + \varepsilon_c + \varepsilon_d}{4(1-v)} - \frac{1}{2(1+v)}\sqrt{(\varepsilon_a - \varepsilon_c)^2 + (\varepsilon_b - \varepsilon_d)^2}\right] \tag{8.44}$$

The rules for determining θ_1 are as follows:

1. If $\varepsilon_b > \varepsilon_d$: $2\theta_1$ lies between $0°$ and $+180°$
   θ_1 lies between $0°$ and $+90°$
2. If $\varepsilon_b < \varepsilon_d$: $2\theta_1$ lies between $0°$ and $-180°$
   θ_1 lies between $0°$ and $-90°$
3. If $\varepsilon_b = \varepsilon_d$:
   (a) If $\varepsilon_a > \varepsilon_c$: $\varepsilon_1 = \varepsilon_a$ and $\theta_1 = 0°$
   (b) If $\varepsilon_a < \varepsilon_c$: $\varepsilon_2 = \varepsilon_a$ and $\theta_1 = 90°$.

The above rules and Eqs. (8.39) through (8.44) may be proved by recourse to Fig. 8.20, which shows Mohr's diagram for this type of rosette. Since the directions of strain measurement in the rosette are inclined successively at $45°$, the radial lines to the points A, B, C, and D, which represent the strains of Mohr's circle, will be inclined successively at twice $45°$, or $90°$. Therefore, A, B, C, and D will be located at the corners of a square inscribed in a circle.

Since the intersection of the diagonals of the square will coincide with the center of the circle, and because the position of the center of the square corresponds to the average of the four corners, therefore

$$A = \frac{\varepsilon_a + \varepsilon_b + \varepsilon_c + \varepsilon_d}{4} \tag{8.41a}$$

Let us now determine B, the radius of the circle, in terms of ε_a, ε_b, ε_c, and ε_d, the horizontal distances from the ordinate through O to the corners of the square. This will require the following construction:

STRAIN GAGE ROSETTES AND DATA ANALYSIS 277

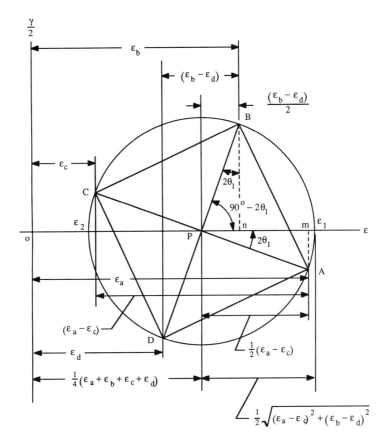

FIG. 8.20. Mohr's circle for rectangular rosette with four observations.

Let P be the center of the circle and drop perpendiculars Am and Bn, respectively, from A and B on to the abscissa at m and n. Then from the right-angled triangles APm and BPn,

$$AP = BP \qquad \text{(radius of the circle)}$$
$$\angle PmA = \angle PnB \qquad 90°$$

Since

$$\angle BPA = 90°$$
$$\angle BPn = \angle PAm \qquad (90° - 2\theta_1)$$

Therefore, triangles APm and BPn are equal, so that

$$Am = Pn = \frac{\varepsilon_b - \varepsilon_d}{2}$$

Also,

$$Pm = \frac{\varepsilon_a - \varepsilon_c}{2}$$

The hypotenuse, PA, is the radius of the circle, and so

$$B = \sqrt{\left(\frac{\varepsilon_a - \varepsilon_c}{2}\right)^2 + \left(\frac{\varepsilon_b - \varepsilon_d}{2}\right)^2}$$
$$= \tfrac{1}{2}\sqrt{(\varepsilon_a - \varepsilon_c)^2 + (\varepsilon_b - \varepsilon_d)^2} \qquad (8.41\text{b})$$

The value of $\tan 2\theta_1$ is

$$\tan 2\theta_1 = \frac{Am}{Pm} = \frac{\tfrac{1}{2}(\varepsilon_b - \varepsilon_d)}{\tfrac{1}{2}(\varepsilon_a - \varepsilon_c)} = \frac{\varepsilon_b - \varepsilon_d}{\varepsilon_a - \varepsilon_c} \qquad (8.42)$$

The T-delta rosette

If this rosette arrangement, shown in Fig. 8.21 is considered as containing a

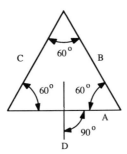

FIG. 8.21. T-delta rosette.

delta rosette with the addition of a fourth gage whose axis D is at right angles to the axis A, then, although the fourth observation is redundant, a variety of solutions can be obtained utilizing all four strain readings.

Meier (1) gives a solution based on the method of least squares, but its complexity is rather a disadvantage. The following simple solution is therefore presented, since its reduction of observed strains into terms of stress will be very much easier.

Since the average of any two strains measured at right angles gives the position of the center of Mohr's circle, we therefore have for the T-delta rosette the quantity

$$A = \frac{\varepsilon_a + \varepsilon_d}{2} \tag{8.45}$$

From Eq. (8.33a) for the delta rosette,

$$A = \frac{\varepsilon_a + \varepsilon_b + \varepsilon_c}{3} \tag{a}$$

Therefore, for the T-delta rosette,

$$A = \frac{\varepsilon_a + \varepsilon_d}{2} = \frac{\varepsilon_a + \varepsilon_b + \varepsilon_c}{3} \tag{b}$$

Again, from Eq. (8.33b) for the delta rosette,

$$B = \sqrt{\left[\frac{2\varepsilon_a - \varepsilon_b - \varepsilon_c}{3}\right]^2 + \left[\frac{\varepsilon_c - \varepsilon_b}{\sqrt{3}}\right]^2}$$

$$= \sqrt{\left[\varepsilon_a - \frac{\varepsilon_a + \varepsilon_b + \varepsilon_c}{3}\right]^2 + \left[\frac{\varepsilon_c - \varepsilon_b}{\sqrt{3}}\right]^2}$$

If $(\varepsilon_a + \varepsilon_d)/2$ is substituted for $(\varepsilon_a + \varepsilon_b + \varepsilon_c)/3$ from Eq. (b), then

$$\varepsilon_a - \frac{\varepsilon_a + \varepsilon_b + \varepsilon_c}{3} = \varepsilon_a - \frac{\varepsilon_a + \varepsilon_d}{2} = \frac{\varepsilon_a - \varepsilon_d}{2} \tag{c}$$

The expression for B can now be written as

$$B = \sqrt{\left[\frac{\varepsilon_a - \varepsilon_d}{2}\right]^2 + \left[\frac{\varepsilon_c - \varepsilon_b}{\sqrt{3}}\right]^2} \tag{8.46}$$

Again, from the delta rosette, the value of $\tan 2\theta$ is given by Eq. (8.36). Thus,

$$\tan 2\theta = \frac{\sqrt{3}(\varepsilon_c - \varepsilon_b)}{2\varepsilon_a - \varepsilon_b - \varepsilon_c} = \frac{\sqrt{3}(\varepsilon_c - \varepsilon_b)}{3\left[\varepsilon_a - \frac{\varepsilon_a + \varepsilon_b + \varepsilon_c}{3}\right]} \tag{d}$$

The value of $\varepsilon_a - (\varepsilon_a + \varepsilon_b + \varepsilon_c)/3$ given by Eq. (c) can be substituted into

Eq. (d) to produce

$$\tan 2\theta = \frac{\sqrt{3}(\varepsilon_c - \varepsilon_b)}{3\left(\dfrac{\varepsilon_a - \varepsilon_d}{2}\right)} = \frac{(\varepsilon_c - \varepsilon_b)/\sqrt{3}}{\tfrac{1}{2}(\varepsilon_a - \varepsilon_d)} \qquad (8.47)$$

The value of tan 2θ given by Eq. (8.47) is the ratio of the quantities under the radical in Eq. (8.46). The rules for assigning the two values of θ, obtained through the use of Eq. (8.47), to the correct principal axes is exactly the same as in the case of the equiangular (delta) rosette.

For the T-delta rosette, the values of the principal strains may be expressed as

$$\varepsilon_{1,2} = A \pm B \qquad (8.48)$$

Substituting the values of A and B, given by Eqs. (8.45) and (8.46), respectively, into Eq. (8.48) yields

$$\varepsilon_{1,2} = \frac{\varepsilon_a + \varepsilon_d}{2} \pm \sqrt{\left[\frac{\varepsilon_a - \varepsilon_d}{2}\right]^2 + \left[\frac{\varepsilon_c - \varepsilon_b}{\sqrt{3}}\right]^2} \qquad (8.49)$$

Insertion of the values of A and B, Eqs. (8.45) and (8.46) respectively, into Eqs. (8.4) and (8.5) gives the expressions for the principal stresses. Thus,

$$\sigma_1 = E\left[\frac{\varepsilon_a + \varepsilon_d}{2(1-v)} + \frac{1}{1+v}\sqrt{\left[\frac{\varepsilon_a - \varepsilon_d}{2}\right]^2 + \left[\frac{\varepsilon_c - \varepsilon_b}{\sqrt{3}}\right]^2}\right] \qquad (8.50)$$

$$\sigma_2 = E\left[\frac{\varepsilon_a + \varepsilon_d}{2(1-v)} - \frac{1}{1+v}\sqrt{\left[\frac{\varepsilon_a - \varepsilon_d}{2}\right]^2 + \left[\frac{\varepsilon_c - \varepsilon_b}{\sqrt{3}}\right]^2}\right] \qquad (8.51)$$

Summary of equations

Three-element rectangular rosette:

$$\sigma_{1,2} = \left[\frac{E}{1-v}\right]\left[\frac{\varepsilon_a + \varepsilon_d}{2}\right] \pm \left[\frac{E}{2(1+v)}\right]\sqrt{(\varepsilon_a - \varepsilon_c)^2 + [2\varepsilon_b - (\varepsilon_a + \varepsilon_c)]^2}$$

When $\varepsilon_b > (\varepsilon_a + \varepsilon_c)/2$, θ lies between $0°$ and $+90°$.

Three-element delta rosette:

$$\sigma_{1,2} = \left[\frac{E}{1-v}\right]\left[\frac{\varepsilon_a + \varepsilon_b + \varepsilon_c}{3}\right] \pm \left[\frac{E}{1+v}\right]\sqrt{\left(\frac{2\varepsilon_a - \varepsilon_b - \varepsilon_c}{3}\right)^2 + \left(\frac{\varepsilon_c - \varepsilon_b}{\sqrt{3}}\right)^2}$$

When $\varepsilon_c > \varepsilon_b$, θ lies between 0° and +90°.
Four-element rectangular rosette:

$$\sigma_{1,2} = \left[\frac{E}{1-v}\right]\left[\frac{\varepsilon_a + \varepsilon_b + \varepsilon_c + \varepsilon_d}{4}\right] \pm \left[\frac{E}{2(1+v)}\right]\sqrt{(\varepsilon_a - \varepsilon_c)^2 + (\varepsilon_b - \varepsilon_d)^2}$$

or

$$\sigma_{1,2} = \left[\frac{E}{1-v}\right]\left[\frac{\varepsilon_a + \varepsilon_c}{2}\right] \pm \left[\frac{E}{2(1+v)}\right]\sqrt{(\varepsilon_a - \varepsilon_c)^2 + (\varepsilon_b - \varepsilon_d)^2}$$

When $\varepsilon_b > \varepsilon_d$, θ lies between 0° and +90°.
Four-element delta rosette:

$$\sigma_{1,2} = \left[\frac{E}{1-v}\right]\left[\frac{\varepsilon_a + \varepsilon_d}{2}\right]\left[\frac{E}{1+v}\right] \pm \sqrt{\left(\frac{\varepsilon_a - \varepsilon_d}{2}\right)^2 + \left(\frac{\varepsilon_c - \varepsilon_b}{\sqrt{3}}\right)^2}$$

When $\varepsilon_c > \varepsilon_b$, θ lies between 0° and +90°.

Directions of principal axes for all the summary equations are given by

$$\tan 2\theta = \frac{\text{2nd quantity under radical}}{\text{1st quantity under radical}}$$

8.7. Graphical solutions

If a number of rosette observations are to be analyzed, the task can be time-consuming and tedious. The data, however, can be reduced rapidly and easily with the use of a programmable calculator or a small computer. There are times, though, when graphical solutions may be desirable, either for the purpose of several people checking each other, or if a computer is not available.

For our purpose, the discussion of graphical methods of solving the rosette equations will be confined to the general case. This method, which has been put forward by McClintock (2), applies to the general case in which the rosette axes may have any arbitrarily chosen axes, θ_{ab}, and θ_{bc}, between them. A rosette can always be represented diagrammatically so that $\theta_{ab} + \theta_{bc}$ is always less than 180°, as indicated in Fig. 8.22.

The objective is to establish Mohr's circle for strain by a very simple procedure. The following steps are employed for finding the strain circle:

1. The rosette axes are rearranged, by extending them if necessary, so that they are arranged in sequence in order of ascending or descending strain magnitudes (algebraic order). The included angle between the axes of

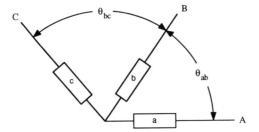

FIG. 8.22. Arbitrary rosette axes.

minimum and maximum strain must be less than 180°. For the rearranged rosette, the angle between the maximum and intermediate strain axes is designated as α, while the angle between the intermediate and minimum strain axes is designated as β. For the rearranged rosette, compute α and β, then place the intermediate axis in the vertical position and lay off, on either side, the maximum and minimum axes. The possible arrangements and the values of α and β are shown in Figs. 8.23 and 8.24. Note, in Fig. 8.24, that the maximum and minimum axes have been extended and the angles α and β are also shown below the crossover point. The reason for this will be explained in a subsequent step.

2. Lay out a strain scale parallel to the direction of the abscissa (which will be established later). Next, draw in ordinates at locations corresponding to zero strain, ε_a, ε_b, and ε_c. This procedure is shown in Fig. 8.25. While the strain values shown in Fig. 8.26 are positive, they might all be negative or some positive and some negative. Furthermore, the measured strains, ε_a, ε_b, and ε_c may have any relation with each other. The strains in Fig. 8.25 have been plotted in sequence according to magnitude.

3. When the diagram corresponding to Fig. 8.25 has been drawn, choose any point, D, on the ordinate corresponding to the intermediate strain value. From point D draw straight lines DE and EF, making angles α and β, respectively, with the ordinate of intermediate strain, to meet the ordinates of ε_{max} and ε_{min} at E and F, respectively. Notice that there are two possibilities for drawing the lines DE and DF, since the angles of α and β can be measured from either the upward or the downward direction of the ordinate of intermediate strain, as shown in Fig. 8.26. The choice is governed as follows:

(a) In Fig. 8.23, the right-hand diagrams show the strain axes in sequence. Here it can be seen that they go in a counterclockwise direction from ε_{max} to ε_{int} to ε_{min}. In this case, the axis of maximum strain falls to the right of the intermediate strain axis, and so α and β are measured from the upward direction.

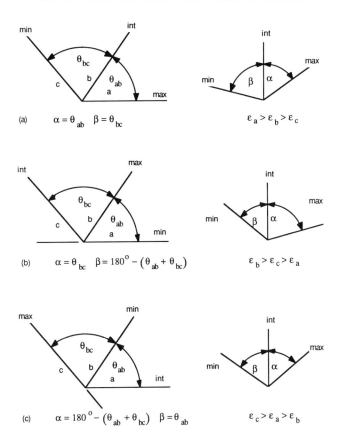

FIG. 8.23.

(b) In Fig. 8.24, the right-hand diagrams show the strain axes in sequence. Here it can be seen that they go in a counterclockwise direction from ε_{min} to ε_{int} to ε_{max}. In this case, the axis of maximum strain falls to the left of the intermediate strain axis, and so α and β are measured from the downward direction. This is shown by the extended lines in Fig. 8.24.

4. The final step is to draw a circle through points D, E, and F. This will be Mohr's circle for strain. The abscissa, which can now be drawn in, will pass through the center of the circle, and the extreme right-hand and left-hand positions of the circumference will represent the principal strains ε_1 and ε_2. Case (a) from Fig. 8.23 and Case (b) from Fig. 8.24 are plotted as Figs. 8.27 and 8.28, respectively.

The points A, B, and C, which represent the strains along the rosette axes, can now be located on the circumference of the circle according to the following two requirements:

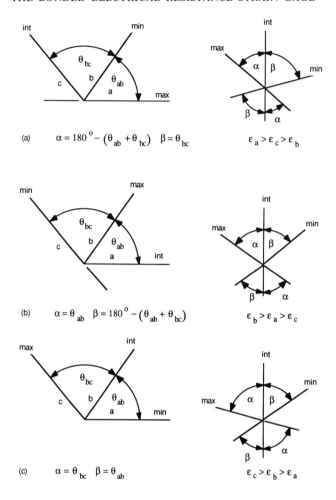

FIG. 8.24.

1. The magnitudes of the strains ε_a, ε_b, and ε_c.
2. The sequence as we go along the circumference of the circle. This must correspond to the sequence in the physical layout of the rosette. For example, if the rosette axes follow the sequence A, B, and C when one proceeds in the counterclockwise direction, the same order must prevail as one goes around Mohr's circle in the same sense.

Although there are two possible positions for each of points A, B, and C that will satisfy the first requirement, the second requirement eliminates half of them. This means that there is only one arrangement for the points A, B, and C on the circumference of the circle.

STRAIN GAGE ROSETTES AND DATA ANALYSIS

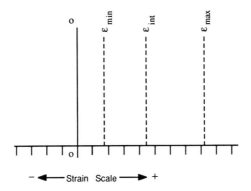

FIG. 8.25.

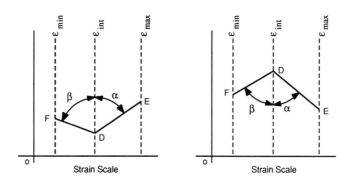

FIG. 8.26.

Angle of reference, θ_1

As soon as point A has been located on the circumference of the circle, the angle between the radial lines to point A and to ε_1 will establish the angle $2\theta_1$, as shown in Figs. 8.27 and 8.28. From this we can determine the angle θ_1 and locate the axis of ε_1, the algebraically larger principal strain, relative to the A axis of the rosette.

Principal stress determination

Once the magnitudes of the principal strains, ε_1 and ε_2, have been determined, then the principal stress values can be computed from Eqs. (8.2) and (8.3).

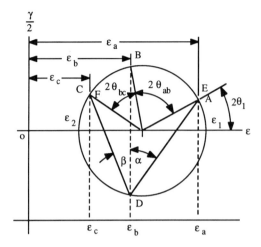

FIG. 8.27. Case in which $\varepsilon_a > \varepsilon_b > \varepsilon_c$.

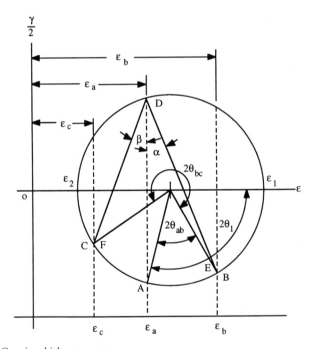

FIG. 8.28. Case in which $\varepsilon_b > \varepsilon_a > \varepsilon_c$.

STRAIN GAGE ROSETTES AND DATA ANALYSIS

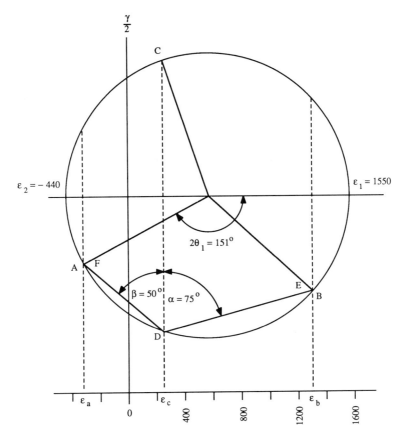

FIG. 8.29. Mohr's circle for Example 8.3.

Thus,

$$\sigma_1 = \frac{E}{1-v^2}(\varepsilon_1 + v\varepsilon_2) \qquad (8.2)$$

$$\sigma_2 = \frac{E}{1-v^2}(\varepsilon_2 + v\varepsilon_1) \qquad (8.3)$$

Example 8.3. Three strain gages are arranged into a rosette as shown in Fig. 8.22. The following data are given: $\varepsilon_a = -325$ μstrain; $\varepsilon_b = 1300$ μstrain; $\varepsilon_c = 250$ μstrain; $\theta_{ab} = 55°$; $\theta_{bc} = 75°$. Construct a Mohr's circle and determine ε_1, ε_2, and θ_1.

Solution. Since $\varepsilon_b > \varepsilon_c > \varepsilon_a$, a rearrangement of the rosette axes will produce the configuration shown in Fig. 8.23b. The angles, α and β, are

$$\alpha = \theta_{bc} = 75°, \qquad \beta = 180° - (\theta_{ab} + \theta_{bc}) = 180° - (55° + 75°) = 50°$$

FIG. 8.30. Computer-based data-aquisition system. (Courtesy of Measurements Group, Inc.)

Lay out a horizontal strain axis. On this axis, erect vertical lines representing ε_a, ε_b, and ε_c. Since the maximum strain axis falls to the right of the vertical line representing the intermediate axis, the angles, α and β, will be measured from the upward vertical, as shown in Fig. 8.23b. The construction of Mohr's circle is shown in Fig. 8.29. From the circle, the following values are obtained:

$$\varepsilon_1 = 1550 \text{ µstrain}, \quad \varepsilon_2 = -440 \text{ µstrain}, \quad 2\theta_1 = 151°$$

Machine solutions

In situations involving the solution of large numbers of rosette equations, the employment of machines can be very advantageous for economy of both time and cost. A number of special-purpose computers have been developed over the years in order to evaluate rosette data (3–7). Today, however, many hand-held programmable calculators, some with graphics display, are available at small cost. For reduction of data for a few rosettes at a time, these are quite convenient. Small desk-top computers are also now available at reasonable prices and are found in nearly every organization. These can reduce and print out large quantities of data in a short period of time once the raw data have been entered.

The ultimate aim, however, has been to develop a combined computer–plotter–tabulator for direct connection to strain gages. Such systems (8) are now available that are dedicated solely to the acquisition of strain gage data (also transducers, thermocouples, etc.). No programming is necessary; the operator enters the required constants and the machine automatically scans the test points and reduces the data. Such a system is shown in Fig. 8.30.

Problems

8.1. A tensile specimen has two gages bonded to its surface, one aligned along the longitudinal axis and the other perpendicular to it. Show that only the longitudinal gage is required in order to determine the longitudinal stress.

8.2. In a long, thin-walled pressure vessel, the hoop stress is twice the longitudinal stress. If the vessel is made of steel, determine the ratio of the hoop strain to the longitudinal strain.

The following rectangular rosettes, illustrated in Fig. 8.2, are bonded to steel with gage A aligned along the x axis. For the readings shown, in μin/in, compute the principal strains, the principal stresses, their orientation relative to the x axis, and the maximum shear stress at the point. Sketch the principal stress element and its relation to the xy coordinate system. Check the analytical results by using Mohr's circle.

	ε_a	ε_b	ε_c
8.3.	1225	115	905
8.4.	395	-760	985
8.5.	1000	1000	1000
8.6.	-725	-285	530
8.7.	-940	545	210

The following delta rosettes, illustrated in Fig. 8.3, are bonded to steel with gage A aligned along the x axis. For the readings shown, in μin/in, compute the principal strains, the principal stresses, their orientation relative to the x axis, and the maximum shear stress at the point. Sketch the principal stress element and its relation to the xy coordinate system. Check the analytical results by using Mohr's circle.

	ε_a	ε_b	ε_c
8.8.	-880	0	-880
8.9.	455	-205	110
8.10.	-610	235	-105
8.11.	975	435	435
8.12.	-720	-610	-185

8.13. A four-element rectangular rosette, illustrated in Fig. 8.4, is bonded to aluminium with gage A aligned along the x axis. The strain observations, given in μin/in, are the following: $\varepsilon_a = -295$, $\varepsilon_b = -350$, $\varepsilon_c = 550$, $\varepsilon_d = 605$. Using $E = 10.5 \times 10^6$ psi and $v = 0.33$, determine σ_1 and σ_2.

8.14. If a T-delta rosette, illustrated in Fig. 8.5, is applied at a point whose strain field is identical to that of Problem 8.13, determine the rosette readings.

8.15. Solve Problem 8.3 by graphical methods.

8.16. Solve Problem 8.7 by graphical methods.

8.17. Solve Problem 8.9 by graphical methods.

8.18. Solve Problem 8.11 by graphical methods.

REFERENCES

1. Meier, J. H., "Improvements in Rosette Computer," *SESA Proceedings*, Vol. III, No. 2, 1946, pp. 1–3.
2. McClintock, F. A., "On Determining Principal Strains from Strain Rosettes with Arbitrary Angles," Letter to the Editor, *SESA Proceedings*, Vol. IX, No. 1, 1951, pp. 209–210.

3. Hoskins, E. E. and R. C. Olesen, "An Electrical Computer for the Evaluation of Strain Rosette Data," *SESA Proceedings*, Vol. II, No. 1, 1944, pp. 67–77.
4. Meier, J. H. and W. R. Mehaffey, "Electronic Computing Apparatus for Rectangular and Equiangular Strain Rosettes," *SESA Proceedings*, Vol. II, No. 1, 1944, pp. 78–101.
5. Murray, W. M., "Machine Solution of the Strain Rosette Equations," *SESA Proceedings*, Vol. II, No. 1, 1944, pp. 106–112.
6. Bassett, W. V., Helen Cromwell, and W. E. Wooster, "Improved Techniques and Devices for Stress Analysis with Resistance Wire Gages," *SESA Proceedings*, Vol. III, No. 2, 1946, pp. 76–88.
7. Williams, S. B., "Geometry in the Design of Stress Measurement Circuits; Improved Methods Through Simpler Concepts," *SESA Proceedings*, Vol. XVII, No. 2, 1960, pp. 161–178.
8. "System 4000," Bulletin 235-B, Measurements Group, Inc., P.O. Box 27777, Raleigh, NC 27611, 1985.

9

STRAIN GAGE ROSETTES AND TRANSVERSE SENSITIVITY EFFECT

9.1. Introduction

In Chapter 7 the effect of transverse sensitivity on a strain gage measurement was considered. It was pointed out that the total unit resistance change in a gage was made up of two parts: namely, (1) the unit resistance change in the gage's axial direction, and (2) the unit resistance change normal (transverse) to the gage axis. Furthermore, the axial strain sensitivity, F_a, and the normal strain sensitivity, F_n, are defined by Eqs. (7.3) and (7.4), respectively. The transverse sensitivity of the gage is then taken as the ratio of the normal sensitivity to the axial sensitivity, or $K = F_n/F_a$.

It was also stated that if a strain gage is used under conditions differing from those of calibration, an error will exist in the indicated value of axial strain. Thus, if the strain is measured by a single gage under biaxial conditions, the error will depend on both the value of the transverse sensitivity factor, K, and the ratio of the normal strain to the axial strain, $\varepsilon_n/\varepsilon_a$. Fortunately, this error is usually rather small and can be neglected. For instance, if the normal strain does not exceed the axial strain and the value of K is 3 percent or less, then the maximum error will not exceed 4 percent. This is easily verified by using Eq. (7.29) to compute the error.

When strain gage rosettes were examined in Chapter 8, the effect of transverse sensitivity was not taken into account. In general, though, the effect of transverse sensitivity should be considered when using strain gages in a biaxial stress field (1–4). If it can be demonstrated that the transverse effect is negligible, then the expressions in Chapter 8 may be used; if, on the other hand, the effect is not negligible, then the expressions for determining the actual strain that will be developed here should be used.

9.2. Two identical orthogonal gages

Figure 9.1 shows two identical gages mounted at 90° to each other. The longitudinal axis of gage a is aligned along axis OX, while the longitudinal axis of gage b is aligned along axis OY. The strain in the axial, or longitudinal, direction of a gage is represented by ε_a, while the strain normal (transverse) to the gage axis is represented by ε_n. In order to identify the gage that is

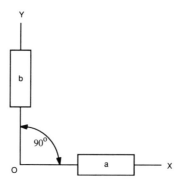

FIG. 9.1. Two identical strain gages aligned along the OX and OY axes.

subjected to strain and the strain direction, a double subscript will be used. The first subscript denotes the strain direction while the second subscript identifies the gage. For instance, if a strain is designated ε_{aa}, the first subscript shows the strain is in the axial direction of the gage, and the second subscript identifies the gage as gage a. The strain ε_{nb} is the transverse strain on gage b.

Since the gages are identical, they have equal axial strain sensitivities, F_a, equal manufacturer's gage factor, G_F, and equal transverse sensitivities, K. When the gages are subjected to an unknown biaxial stress field, the unit change in resistance for each gage is

$$\left(\frac{\Delta R}{R}\right)_a = G_F \varepsilon'_{aa} \tag{9.1}$$

$$\left(\frac{\Delta R}{R}\right)_b = G_F \varepsilon'_{ab} \tag{9.2}$$

where ε'_{aa} = indicated strain for gage a
ε'_{ab} = indicated strain for gage b

Using Eq. (7.13), we can also write for each gage

$$\left(\frac{\Delta R}{R}\right)_a = F_a(\varepsilon_{aa} + K\varepsilon_{na}) \tag{9.3}$$

$$\left(\frac{\Delta R}{R}\right)_b = F_a(\varepsilon_{ab} + K\varepsilon_{nb}) \tag{9.4}$$

The right-hand sides of Eqs. (9.1) and (9.3) may be equated, and also the

STRAIN GAGE ROSETTES AND TRANSVERSE SENSITIVITY EFFECT 293

right-hand sides of Eqs. (9.2) and (9.4). This gives

$$G_F \varepsilon'_{aa} = F_a(\varepsilon_{aa} + K\varepsilon_{na})$$

$$G_F \varepsilon'_{ab} = F_a(\varepsilon_{ab} + K\varepsilon_{nb})$$

Dividing both sides of each equation by G_F produces

$$\varepsilon'_{aa} = \frac{F_a}{G_F}(\varepsilon_{aa} + K\varepsilon_{na}) \tag{9.5}$$

$$\varepsilon'_{ab} = \frac{F_a}{G_F}(\varepsilon_{ab} + K\varepsilon_{nb}) \tag{9.6}$$

From Eq. (7.21) it is seen that

$$\frac{F_a}{G_F} = \frac{1}{1 - v_0 K} \tag{9.7}$$

Substituting the value of F_a/G_F given by Eq. (9.7) into Eqs. (9.5) and (9.6) yields

$$\varepsilon'_{aa} = \frac{1}{1 - v_0 K}(\varepsilon_{aa} + K\varepsilon_{na}) \tag{9.8}$$

$$\varepsilon'_{ab} = \frac{1}{1 - v_0 K}(\varepsilon_{ab} + K\varepsilon_{nb}) \tag{9.9}$$

Since the gages are orthogonal, we know that

$$\varepsilon_{na} = \varepsilon_{ab} \tag{a}$$

$$\varepsilon_{nb} = \varepsilon_{aa} \tag{b}$$

Substituting the values of the transverse strains given by Eqs. (a) and (b) into Eqs. (9.8) and (9.9), respectively, results in

$$\varepsilon'_{aa} = \frac{1}{1 - v_0 K}(\varepsilon_{aa} + K\varepsilon_{ab}) \tag{9.10}$$

$$\varepsilon'_{ab} = \frac{1}{1 - v_0 K}(\varepsilon_{ab} + K\varepsilon_{aa}) \tag{9.11}$$

Equations (9.10) and (9.11) are now expressed in terms of the strains in the axial direction of each gage, and so the first subscript, a, can be dropped. The apparent, or indicated, strains are now expressed in terms of the actual strains in the axial directions of the gages. Thus,

$$\varepsilon'_a = \frac{1}{1 - v_0 K}(\varepsilon_a + K\varepsilon_b) \tag{9.12}$$

$$\varepsilon'_b = \frac{1}{1 - v_0 K}(\varepsilon_b + K\varepsilon_a) \tag{9.13}$$

If Eqs. (9.12) and (9.13) are solved simultaneously, the actual strains, ε_a and ε_b, will be determined in terms of the apparent (indicated) strains. This operation gives

$$\varepsilon_a = \frac{(1 - v_0 K)(\varepsilon'_a - K\varepsilon'_b)}{1 - K^2} \tag{9.14}$$

$$\varepsilon_b = \frac{(1 - v_0 K)(\varepsilon'_b - K\varepsilon'_a)}{1 - K^2} \tag{9.15}$$

Equations (9.14) and (9.15) show that, in order to determine the actual strain in a desired direction, two gages must be used. One gage is aligned in the desired direction; the other gage is mounted normal to the direction of the desired strain. If one chooses to ignore the transverse sensitivity ($K = 0$), then Eqs. (9.14) and (9.15) reduce to $\varepsilon_a = \varepsilon'_a$ and $\varepsilon_b = \varepsilon'_b$.

9.3. Two different orthogonal gages

The case can now be considered in which there are two orthogonal gages, each with a different F_a, G_F, and K. Again the gages are arranged as shown in Fig. 9.1. For gage a we have the axial strain sensitivity, F_{aa}, the manufacturer's gage factor, G_{Fa}, and the transverse sensitivity factor, K_a. The corresponding values for gage b are F_{ab}, G_{Fb}, and K_b. We can use Eqs. (9.5) and (9.6) to write the apparent strains, ε'_{aa} and ε'_{ab}, in terms of the actual strains, ε_{aa} and ε_{ab}, and the individual gage factors. Thus,

$$\varepsilon'_{aa} = \frac{F_{aa}}{G_{Fa}}(\varepsilon_{aa} + K_a \varepsilon_{na}) \tag{9.16}$$

$$\varepsilon'_{ab} = \frac{F_{ab}}{G_{Fb}}(\varepsilon_{ab} + K_b \varepsilon_{nb}) \tag{9.17}$$

Equation (7.21) also shows that, for each gage,

$$\frac{F_{aa}}{G_{Fa}} = \frac{1}{1 - v_0 K_a} \tag{9.18}$$

$$\frac{F_{ab}}{G_{Fb}} = \frac{1}{1 - v_0 K_b} \tag{9.19}$$

Substituting the values of F_{aa}/G_{Fa} and F_{ab}/G_{Fb} given by Eqs. (9.18) and (9.19) into Eqs. (9.16) and (9.17) results in

$$\varepsilon'_{aa} = \frac{1}{1 - v_0 K_a}(\varepsilon_{aa} + K_a \varepsilon_{na}) \tag{9.20}$$

$$\varepsilon'_{ab} = \frac{1}{1 - v_0 K_b}(\varepsilon_{ab} + K_b \varepsilon_{nb}) \tag{9.21}$$

From Eqs. (a) and (b) in Section 9.2, the normal strains can be expressed in terms of the axial strains; that is, $\varepsilon_{na} = \varepsilon_{ab}$ and $\varepsilon_{nb} = \varepsilon_{aa}$. Replacing the normal strains with axial strains in Eqs. (9.20) and (9.21), we obtain

$$\varepsilon'_{aa} = \frac{1}{1 - v_0 K_a}(\varepsilon_{aa} + K_a \varepsilon_{ab}) \tag{9.22}$$

$$\varepsilon'_{ab} = \frac{1}{1 - v_0 K_b}(\varepsilon_{ab} + K_a \varepsilon_{aa}) \tag{9.23}$$

Again the first subscript, a, for each strain can be dropped since the strains are in the axial direction of each gage. The apparent strains, ε'_a and ε'_b, in terms of the actual strains, ε_a and ε_b, will now be

$$\varepsilon'_a = \frac{1}{1 - v_0 K_a}(\varepsilon_a + K_a \varepsilon_b) \tag{9.24}$$

$$\varepsilon'_b = \frac{1}{1 - v_0 K_b}(\varepsilon_b + K_b \varepsilon_a) \tag{9.25}$$

If Eqs. (9.24) and (9.25) are solved simultaneously, the true (actual) strains will be expressed in terms of the apparent (indicated) strains. This operation gives

$$\varepsilon_a = \frac{(1 - v_0 K_a)\varepsilon'_a - K_a(1 - v_0 K_b)\varepsilon'_b}{1 - K_a K_b} \qquad (9.26)$$

$$\varepsilon_b = \frac{(1 - v_0 K_b)\varepsilon'_b - K_b(1 - v_0 K_a)\varepsilon'_a}{1 - K_a K_b} \qquad (9.27)$$

If $K_a = K_b = K$, then Eqs. (9.26) and (9.27) reduce to Eqs. (9.14) and (9.15). Furthermore, if the transverse sensitivity factor, K, is taken as zero, then $\varepsilon_a = \varepsilon'_a$ and $\varepsilon_b = \varepsilon'_b$.

9.4. Three-element rectangular rosette

A three-element rectangular rosette, with all gages different, will be examined next. The rosette is shown in Fig. 9.2. The apparent strains for each gage

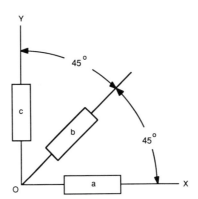

FIG. 9.2. Three-element rectangular rosette.

can be developed and expressed in the same manner as those leading to Eqs. (9.20) and (9.21). The three equations are

$$\varepsilon'_{aa} = \frac{1}{1 - v_0 K_a}(\varepsilon_{aa} + K_a \varepsilon_{na}) \qquad (9.28)$$

$$\varepsilon'_{ab} = \frac{1}{1 - v_0 K_b}(\varepsilon_{ab} + K_b \varepsilon_{nb}) \qquad (9.29)$$

$$\varepsilon'_{ac} = \frac{1}{1 - v_0 K_c}(\varepsilon_{ac} + K_c \varepsilon_{nc}) \qquad (9.30)$$

STRAIN GAGE ROSETTES AND TRANSVERSE SENSITIVITY EFFECT 297

The normal strains for each gage must be expressed in terms of the axial strains. For gages a and c we have $\varepsilon_{na} = \varepsilon_{ac}$ and $\varepsilon_{nc} = \varepsilon_{aa}$, since these gages are 90° to each other. For gage b, however, a Mohr's circle or the transformation equation given by Eq. (2.32) must be used in order to determine the normal strain. The transformation equation is

$$\varepsilon_{x'} = \frac{\varepsilon_x + \varepsilon_y}{2} + \frac{\varepsilon_x - \varepsilon_y}{2}\cos 2\theta + \frac{\gamma_{xy}}{2}\sin 2\theta \tag{9.31}$$

Before proceeding to determine ε_{nb}, which is 90° from gage b and 135° from gage a, the shearing strain in the plane must be determined. In order to do this, Eq. (9.31) is used with $\varepsilon_{x'} = \varepsilon_{ab}$, $\varepsilon_x = \varepsilon_{aa}$, $\varepsilon_y = \varepsilon_{ac}$, and $\theta = 45°$. Using these values,

$$\varepsilon_{ab} = \frac{\varepsilon_{aa} + \varepsilon_{ac}}{2} + \frac{\varepsilon_{aa} - \varepsilon_{ac}}{2}\cos 2(45°) + \frac{\gamma_{xy}}{2}\sin 2(45°) \tag{a}$$

From Eq. (a),

$$\frac{\gamma_{xy}}{2} = \varepsilon_{ab} - \frac{\varepsilon_{aa} + \varepsilon_{ac}}{2} \tag{b}$$

Equation (9.31) is once again used with $\varepsilon_{x'} = \varepsilon_{nb}$, $\theta = 135°$, and the value of $\gamma_{xy}/2$ given by Eq. (b). Thus,

$$\varepsilon_{nb} = \frac{\varepsilon_{aa} + \varepsilon_{ac}}{2} + \frac{\varepsilon_{aa} - \varepsilon_{ac}}{2}\cos 2(135°) + \left(\varepsilon_{ab} - \frac{\varepsilon_{aa} + \varepsilon_{ac}}{2}\right)\sin 2(135°) \tag{c}$$

Equation (c) reduces to

$$\varepsilon_{nb} = \varepsilon_{aa} + \varepsilon_{ac} - \varepsilon_{ab} \tag{9.32}$$

The normal strains are now in terms of the axial strains.

Equation (9.32) could also have been obtained by considering the first strain invariant. That is,

$$\varepsilon_{aa} + \varepsilon_{ac} = \varepsilon_{ab} + \varepsilon_{nb} = \text{constant}$$

or

$$\varepsilon_{nb} = \varepsilon_{aa} + \varepsilon_{ac} - \varepsilon_{ab}$$

The values of the normal strains can now be substituted into their respective equations; that is, Eqs. (9.28), (9.29), and (9.30). This gives

$$\varepsilon'_{aa} = \frac{1}{1 - v_0 K_a} [\varepsilon_{aa} + K_a \varepsilon_{ac}] \tag{d}$$

$$\varepsilon'_{ab} = \frac{1}{1 - v_0 K_b} [\varepsilon_{ab} + K_b(\varepsilon_{aa} + \varepsilon_{ac} - \varepsilon_{ab})] \tag{e}$$

$$\varepsilon'_{ac} = \frac{1}{1 - v_0 K_c} [\varepsilon_{ac} + K_c \varepsilon_{aa}] \tag{f}$$

The first subscript, a, for each strain can now be dropped, and so Eqs. (d), (e), and (f) become

$$\varepsilon'_a = \frac{\varepsilon_a + K_a \varepsilon_c}{1 - v_0 K_a} \tag{9.33}$$

$$\varepsilon'_b = \frac{(1 - K_b)\varepsilon_b + K_b(\varepsilon_a + \varepsilon_c)}{1 - v_0 K_b} \tag{9.34}$$

$$\varepsilon'_c = \frac{\varepsilon_c + K_c \varepsilon_a}{1 - v_0 K_c} \tag{9.35}$$

Equations (9.33), (9.34), and (9.35) may be solved simultaneously for the actual strains, ε_a, ε_b, and ε_c. This operation results in

$$\varepsilon_a = \frac{(1 - v_0 K_a)\varepsilon'_a - K_a(1 - v_0 K_c)\varepsilon'_c}{1 - K_a K_c} \tag{9.36}$$

$$\varepsilon_b = \frac{(1 - v_0 K_b)\varepsilon'_b}{1 - K_b} - \frac{K_b[(1 - v_0 K_a)(1 - K_c)\varepsilon'_a + (1 - v_0 K_c)(1 - K_a)\varepsilon'_c]}{(1 - K_b)(1 - K_a K_c)} \tag{9.37}$$

$$\varepsilon_c = \frac{(1 - v_0 K_c)\varepsilon'_c - K_c(1 - v_0 K_a)\varepsilon'_a}{1 - K_a K_c} \tag{9.38}$$

When the transverse sensitivities of gages a and c are the same, then

STRAIN GAGE ROSETTES AND TRANSVERSE SENSITIVITY EFFECT   299

$K_a = K_c = K_{ac}$. For this condition Eqs. (9.36), (9.37), and (9.38) become

$$\varepsilon_a = \frac{(1 - v_0 K_{ac})(\varepsilon_a' - K_{ac}\varepsilon_c')}{1 - K_{ac}^2} \quad (9.39)$$

$$\varepsilon_b = \frac{(1 - v_0 K_b)(1 + K_{ac})\varepsilon_b' - K_b(1 - v_0 K_{ac})(\varepsilon_a' + \varepsilon_c')}{(1 - K_b)(1 + K_{ac})} \quad (9.40)$$

$$\varepsilon_c = \frac{(1 - v_0 K_{ac})(\varepsilon_c' - K_{ac}\varepsilon_a')}{1 - K_{ac}^2} \quad (9.41)$$

If $K_a = K_b = K_c = K$, then Eqs. (9.39), (9.40), and (9.41) reduce to

$$\varepsilon_a = \frac{(1 - v_0 K)(\varepsilon_a' - K\varepsilon_c')}{1 - K^2} \quad (9.42)$$

$$\varepsilon_b = \frac{(1 - v_0 K)[(1 + K)\varepsilon_b' - K(\varepsilon_a' + \varepsilon_c')]}{1 - K^2} \quad (9.43)$$

$$\varepsilon_c = \frac{(1 - v_0 K)(\varepsilon_c' - K\varepsilon_a')}{1 - K^2} \quad (9.44)$$

The actual strains, ε_a, ε_b, and ε_c, have been determined by taking into account the transverse sensitivities of the gages making up the three-element rectangular rosette. In order to determine the principal strains, the principal stresses, and the directions of the principal stress (or strain) axes, the equations developed in Section 8.4 can be used. The appropriate equations from that section will be identified and renumbered here.

The principal strains in terms of the gage values are given by Eq. (8.19). The expression is

$$\varepsilon_{1,2} = \frac{\varepsilon_a + \varepsilon_c}{2} \pm \tfrac{1}{2}\sqrt{(\varepsilon_a - \varepsilon_c)^2 + [2\varepsilon_b - (\varepsilon_a + \varepsilon_c)]^2} \quad (9.45)$$

The principal stresses are given by Eqs. (8.21) and (8.22). These are

$$\sigma_1 = E\left\{\frac{\varepsilon_a + \varepsilon_c}{2(1 - v)} + \frac{1}{2(1 + v)}\sqrt{(\varepsilon_a - \varepsilon_c)^2 + [2\varepsilon_b + (\varepsilon_a + \varepsilon_c)]^2}\right\} \quad (9.46)$$

$$\sigma_2 = E\left\{\frac{\varepsilon_a + \varepsilon_c}{2(1 - v)} - \frac{1}{2(1 + v)}\sqrt{(\varepsilon_a - \varepsilon_c)^2 + [2\varepsilon_b + (\varepsilon_a + \varepsilon_c)]^2}\right\} \quad (9.47)$$

In order to establish the angular relationship, θ_1, between the $O1$ axis and the OX axis, Eqs. (8.23), (8.24), and (8.25) were given. Any two of the three equations are needed in order to establish θ_1. The three equations are

$$\tan 2\theta = \frac{2\varepsilon_b - (\varepsilon_a + \varepsilon_c)}{\varepsilon_a - \varepsilon_c} \tag{9.48}$$

$$\sin 2\theta = \frac{2\varepsilon_b - (\varepsilon_a + \varepsilon_c)}{\sqrt{(\varepsilon_a - \varepsilon_c)^2 + [2\varepsilon_b - (\varepsilon_a + \varepsilon_c)]^2}} \tag{9.49}$$

$$\cos 2\theta = \frac{\varepsilon_a - \varepsilon_c}{\sqrt{(\varepsilon_a - \varepsilon_c)^2 + [2\varepsilon_b - (\varepsilon_a + \varepsilon_c)]^2}} \tag{9.50}$$

Graphical methods could, of course, also be used.

Example 9.1. The following data are given for a three-element rectangular rosette:

$$\varepsilon'_a = 1450 \; \mu\text{in/in}, \quad K_a = -6.0 \text{ percent}$$
$$\varepsilon'_b = -960 \; \mu\text{in/in}, \quad K_b = 2.5 \text{ percent}$$
$$\varepsilon'_c = 870 \; \mu\text{in/in}, \quad K_c = -5.0 \text{ per cent}$$

(a) Determine the actual strains, ε_a, ε_b, and ε_c.
(b) Determine the principal strains, ε_1 and ε_2.
(c) What error exists if the principal strains are computed using apparent strains rather than actual strains?

Solution. (a) The actual strains may be computed using Eqs. (9.36), (9.37), and (9.38).

$$\varepsilon_a = \frac{(1 - \nu_0 K_a)\varepsilon'_a - K_a(1 - \nu_0 K_c)\varepsilon'_c}{1 - K_a K_c}$$

$$= \frac{[1 - 0.285(-0.06)](1450) - (-0.06)[1 - 0.285(-0.05)](870)}{1 - (-0.06)(-0.05)}$$

$$\varepsilon_a = 1532 \; \mu\text{in/in}$$

$$\varepsilon_b = \frac{(1 - \nu_0 K_b)\varepsilon'_b}{1 - K_b} - \frac{K_b[(1 - \nu_0 K_a)(1 - K_c)\varepsilon'_a + (1 - \nu_0 K_c)(1 - K_a)\varepsilon'_c]}{(1 - K_b)(1 - K_a K_c)}$$

$$= \frac{[1 - 0.285(0.025)](-960)}{1 - 0.025}$$

$$- \frac{(0.025)[1 - 0.285(-0.06)][1 - (-0.05)](1450)}{(1 - 0.025)[1 - (-0.06)(-0.05)]}$$

$$- \frac{(0.025)[1 - 0.285(-0.05)][1 - (-0.06)](870)}{(1 - 0.025)[1 - (-0.06)(-0.05)]}$$

$$\varepsilon_b = -1041 \; \mu\text{in/in}$$

$$\varepsilon_c = \frac{(1 - v_0 K_c)\varepsilon_c' - K_c(1 - v_0 K_a)\varepsilon_a'}{1 - K_a K_c}$$

$$= \frac{[1 - 0.285(-0.05)](870) - (-0.05)[1 - 0.285(-0.06)](1450)}{1 - (-0.06)(-0.05)}$$

$$\varepsilon_c = 959 \text{ μin/in}$$

(b) The principal strains are given by Eq. (9.45).

$$\varepsilon_{1,2} = \frac{\varepsilon_a + \varepsilon_c}{2} \pm \tfrac{1}{2}\sqrt{(\varepsilon_a - \varepsilon_c)^2 + [2\varepsilon_b - (\varepsilon_a + \varepsilon_c)]^2}$$

$$= \frac{1532 + 959}{2} \pm \tfrac{1}{2}\sqrt{(1532 - 959)^2 + [2(-1041) - (1532 + 959)]^2}$$

$$= 1246 \pm 2304$$

$$\varepsilon_1 = 3550 \text{ μin/in}, \qquad \varepsilon_2 = -1058 \text{ μin/in}$$

(c) Equation (9.45) is again used, but the strains will be the apparent strains.

$$\varepsilon_{1,2}' = \frac{1450 + 870}{2} \pm \tfrac{1}{2}\sqrt{(1450 - 870)^2 + [2(-960) - (1450 + 870)]^2}$$

$$= 1160 \pm 2140$$

$$\varepsilon_1' = 3300 \text{ μin/in}, \qquad \varepsilon_2' = -980 \text{ μin/in}$$

The principal strains computed by using the apparent strain readings are slightly more than 7 percent lower than the actual principal strains.

9.5. The equiangular or delta rosette

The equiangular or delta rosette is shown in Fig. 9.3. The apparent strains for each gage are given by Eqs. (9.28), (9.29), and (9.30). The transformation equation, Eq. (9.31), must be used to determine ε_{na}, ε_{nb}, and ε_{nc}, the strains normal to gages a, b, and c.

The X axis is established along gage a, and since these two axes coincide, $\varepsilon_x = \varepsilon_{aa}$. In order to determine the strains normal to the gages, the values of ε_y and $\gamma_{xy}/2$ must first be computed. Two expressions are obtained through the use of the transformation equation; these are solved simultaneously for ε_y and $\gamma_{xy}/2$.

The first equation uses gage b. Here $\varepsilon_x' = \varepsilon_{ab}$, $\theta = 120°$, and $\varepsilon_x = \varepsilon_{aa}$. Substituting these values into Eq. (9.31) gives

$$\varepsilon_{ab} = \frac{\varepsilon_{aa} + \varepsilon_y}{2} + \frac{\varepsilon_{aa} - \varepsilon_y}{2}\cos 2(120°) + \frac{\gamma_{xy}}{2}\sin 2(120°) \qquad (a)$$

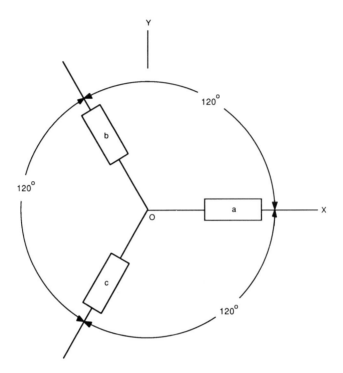

FIG. 9.3. The equiangular or delta rosette.

The second equation uses gage c. Here $\varepsilon'_x = \varepsilon_{ac}$, $\theta = 240°$, and $\varepsilon_x = \varepsilon_{aa}$. Substituting these values into Eq. (9.31) gives the second independent equation that is needed.

$$\varepsilon_{ac} = \frac{\varepsilon_{aa} + \varepsilon_y}{2} + \frac{\varepsilon_{aa} - \varepsilon_y}{2} \cos 2(240°) + \frac{\gamma_{xy}}{2} \sin 2(240°) \qquad (b)$$

Solving Eqs. (a) and (b) simultaneously for ε_y and $\gamma_{xy}/2$ produces

$$\varepsilon_y = \frac{1}{3}[2(\varepsilon_{ab} + \varepsilon_{ac}) - \varepsilon_{aa}] \qquad (9.51)$$

$$\frac{\gamma_{xy}}{2} = \frac{1}{\sqrt{3}}(\varepsilon_{ac} - \varepsilon_{ab}) \qquad (9.52)$$

Since the values of ε_x, ε_y, and $\gamma_{xy}/2$ are now known in terms of the actual strains along each gage axis, the strains normal to each gage may now be determined through the use of the transformation equation, Eq. (9.31). Because ε_{na} is along the Y axis, its value is the same as ε_y. Thus, ε_{na} is written

as
$$\varepsilon_{na} = \tfrac{1}{3}[2(\varepsilon_{ab} + \varepsilon_{ac}) - \varepsilon_{aa}] \tag{9.53}$$

The transverse strain, ε_{nb}, has an angle of 210° relative to the X axis. Letting $\varepsilon_{x'} = \varepsilon_{nb}$ and $\theta = 210°$, the transformation equation is

$$\varepsilon_{nb} = \frac{\varepsilon_x + \varepsilon_y}{2} + \frac{\varepsilon_x - \varepsilon_y}{2}\cos 2(210°) + \frac{\gamma_{xy}}{2}\sin 2(210°) \tag{c}$$

If $\varepsilon_x = \varepsilon_{aa}$ and the values of ε_y and $\gamma_{xy}/2$, given by Eqs. (9.51) and (9.52), respectively, are substituted into Eq. (c), then

$$\varepsilon_{nb} = \tfrac{2}{3}(\varepsilon_{aa} + \varepsilon_{ac}) - \tfrac{1}{3}\varepsilon_{ab} \tag{9.54}$$

Finally, the transverse strain, ε_{nc}, has an angle of 330° relative to the X axis. Letting $\varepsilon_{x'} = \varepsilon_{nc}$ and $\theta = 330°$, the transformation equation is

$$\varepsilon_{nc} = \frac{\varepsilon_x + \varepsilon_y}{2} + \frac{\varepsilon_x - \varepsilon_y}{2}\cos 2(330°) + \frac{\gamma_{xy}}{2}\sin 2(330°) \tag{d}$$

If $\varepsilon_x = \varepsilon_{aa}$ and the values of ε_y and $\gamma_{xy}/2$, given by Eqs. (9.51) and (9.52), respectively, are substituted into Eq. (d), then

$$\varepsilon_{nc} = \tfrac{2}{3}(\varepsilon_{aa} + \varepsilon_{ab}) - \tfrac{1}{3}\varepsilon_{ac} \tag{9.55}$$

The required normal strains are given by Eqs. (9.53), (9.54), and (9.55). As pointed out in Section 9.4, once ε_{na} was established the first strain invariant could be used to determine ε_{nb} and ε_{nc}. Thus,

$$\varepsilon_{ab} + \varepsilon_{nb} = \varepsilon_{aa} + \varepsilon_{na}$$

or

$$\varepsilon_{nb} = \varepsilon_{aa} + \varepsilon_{na} - \varepsilon_{ab} = \tfrac{2}{3}(\varepsilon_{aa} + \varepsilon_{ac}) - \tfrac{1}{3}\varepsilon_{ab}$$

Also,

$$\varepsilon_{ac} + \varepsilon_{nc} = \varepsilon_{aa} + \varepsilon_{na}$$

or

$$\varepsilon_{nc} = \varepsilon_{aa} + \varepsilon_{na} - \varepsilon_{ac} = \tfrac{2}{3}(\varepsilon_{aa} + \varepsilon_{ab}) - \tfrac{1}{3}\varepsilon_{ac}$$

The expressions for the normal strains are now given in terms of the axial strains. Substituting the values of ε_{na}, ε_{nb}, and ε_{nc}, given by Eqs. (9.53), (9.54), and (9.55), respectively, into Eqs. (9.28), (9.29), and (9.30) will give the indicated strains in terms of the axial strains at each gage location. Also, since only axial strains are involved, the first subscript, a, for each strain can now be dropped. Carrying out these substitutions gives

$$\varepsilon'_a = \frac{1}{1 - v_0 K_a}\left[\left(1 - \frac{K_a}{3}\right)\varepsilon_a + \frac{2K_a}{3}(\varepsilon_b + \varepsilon_c)\right] \tag{9.56}$$

$$\varepsilon'_b = \frac{1}{1 - v_0 K_b}\left[\left(1 - \frac{K_b}{3}\right)\varepsilon_b + \frac{2K_b}{3}(\varepsilon_a + \varepsilon_c)\right] \tag{9.57}$$

$$\varepsilon'_c = \frac{1}{1 - v_0 K_c}\left[\left(1 - \frac{K_c}{3}\right)\varepsilon_c + \frac{2K_c}{3}(\varepsilon_a + \varepsilon_b)\right] \tag{9.58}$$

Equations (9.56), (9.57), and (9.58) may now be solved simultaneously for ε_a, ε_b, and ε_c. This yields

$$\varepsilon_a = \frac{(1 - v_0 K_a)(3 - K_b - K_c - K_b K_c)\varepsilon'_a}{3 - K_a - K_b - K_c - K_a K_b - K_a K_c - K_b K_c + 3K_a K_b K_c}$$
$$- \frac{2K_a[(1 - v_0 K_b)(1 - K_c)\varepsilon'_b + (1 - v_0 K_c)(1 - K_b)\varepsilon'_c]}{3 - K_a - K_b - K_c - K_a K_b - K_a K_c - K_b K_c + 3K_a K_b K_c} \tag{9.59}$$

$$\varepsilon_b = \frac{(1 - v_0 K_b)(3 - K_a - K_c - K_a K_c)\varepsilon'_b}{3 - K_a - K_b - K_c - K_a K_b - K_a K_c - K_b K_c + 3K_a K_b K_c}$$
$$- \frac{2K_b[(1 - v_0 K_a)(1 - K_c)\varepsilon'_a + (1 - v_0 K_c)(1 - K_a)\varepsilon'_c]}{3 - K_a - K_b - K_c - K_a K_b - K_a K_c - K_b K_c + 3K_a K_b K_c} \tag{9.60}$$

$$\varepsilon_c = \frac{(1 - v_0 K_c)(3 - K_a - K_b - K_a K_b)\varepsilon'_c}{3 - K_a - K_b - K_c - K_a K_b - K_a K_c - K_b K_c + 3K_a K_b K_c}$$
$$- \frac{2K_c[(1 - v_0 K_a)(1 - K_b)\varepsilon'_a + (1 - v_0 K_b)(1 - K_a)\varepsilon'_b]}{3 - K_a - K_b - K_c - K_a K_b - K_a K_c - K_b K_c + 3K_a K_b K_c} \tag{9.61}$$

If two gages have the same transverse sensitivity factor, then these expressions can be simplified accordingly. If gages a and c have the same transverse sensitivity factor, then $K_a = K_c = K_{ac}$. Under these conditions Eqs. (9.59), (9.60), and (9.61) reduce to

$$\varepsilon_a = \frac{(1 - v_0 K_{ac})(3 - K_b - K_{ac} - K_b K_{ac})\varepsilon'_a}{3 - K_b - 2K_{ac} - 2K_b K_{ac} - K_{ac}^2 + 3K_b K_{ac}^2}$$
$$- \frac{2K_{ac}[(1 - v_0 K_b)(1 - K_{ac})\varepsilon'_b + (1 - v_0 K_{ac})(1 - K_b)\varepsilon'_c]}{3 - K_b - 2K_{ac} - 2K_b K_{ac} - K_{ac}^2 + 3K_b K_{ac}^2} \tag{9.62}$$

$$\varepsilon_b = \frac{(1 - v_0 K_b)(3 - 2K_{ac} - K_{ac}^2)\varepsilon_b'}{3 - K_b - 2K_{ac} - 2K_b K_{ac} - K_{ac}^2 + 3K_b K_{ac}^2}$$

$$- \frac{2K_b[(1 - v_0 K_{ac})(1 - K_{ac})(\varepsilon_a' + \varepsilon_c')]}{3 - K_b - 2K_{ac} - 2K_b K_{ac} - K_{ac}^2 + 3K_b K_{ac}^2} \quad (9.63)$$

$$\varepsilon_c = \frac{(1 - v_0 K_{ac})(3 - K_b - K_{ac} - K_b K_{ac})\varepsilon_c'}{3 - K_b - 2K_{ac} - 2K_b K_{ac} - K_{ac}^2 + 3K_b K_{ac}^2}$$

$$- \frac{2K_{ac}[(1 - v_0 K_{ac})(1 - K_b)\varepsilon_a' + (1 - v_0 K_b)(1 - K_{ac})\varepsilon_b']}{3 - K_b - 2K_{ac} - 2K_b K_{ac} - K_{ac}^2 + 3K_b K_{ac}^2} \quad (9.64)$$

The denominators of Eqs. (9.62), (9.63), and (9.64) are alike. However, Eq. (9.63) can be simplified further by finding common factors in the numerator and denominator (1).

If all gages have the same transverse sensitivity factor, then the expressions simplify further. Thus, for $K_a = K_b = K_c = K$, we have

$$\varepsilon_a = \frac{(1 - v_0 K)[(3 + K)\varepsilon_a' - 2K(\varepsilon_b' + \varepsilon_c')]}{3(1 - K^2)} \quad (9.65)$$

$$\varepsilon_b = \frac{(1 - v_0 K)[(3 + K)\varepsilon_b' - 2K(\varepsilon_a' + \varepsilon_c')]}{3(1 - K^2)} \quad (9.66)$$

$$\varepsilon_c = \frac{(1 - v_0 K)[(3 + K)\varepsilon_c' - 2K(\varepsilon_a' + \varepsilon_b')]}{3(1 - K^2)} \quad (9.67)$$

The actual strains, ε_a, ε_b, and ε_c, have been determined by accounting for the transverse sensitivities of the gages making up the delta rosette. The equations developed in Section 8.5 can now be used to determine (1) the principal strains, (2) the principal stresses, and (3) the orientation of the principal axes relative to the original coordinate system. For ease of use, the pertinent equations will be repeated and renumbered here.

The principal strains, given by Eq. (8.32), are

$$\varepsilon_{1,2} = \frac{\varepsilon_a + \varepsilon_b + \varepsilon_c}{3} \pm \sqrt{\left[\frac{2\varepsilon_a - \varepsilon_b - \varepsilon_c}{3}\right]^2 + \left[\frac{\varepsilon_c - \varepsilon_b}{\sqrt{3}}\right]^2} \quad (9.68)$$

The principal stresses, given by Eqs. (8.34) and (8.35), are

$$\sigma_1 = E\left\{\frac{\varepsilon_a + \varepsilon_b + \varepsilon_c}{3(1 - v)} + \frac{1}{1 + v}\sqrt{\left[\frac{2\varepsilon_a - \varepsilon_b - \varepsilon_c}{3}\right]^2 + \left[\frac{\varepsilon_c - \varepsilon_b}{\sqrt{3}}\right]^2}\right\} \quad (9.69)$$

$$\sigma_2 = E\left\{\frac{\varepsilon_a + \varepsilon_b + \varepsilon_c}{3(1-v)} - \frac{1}{1+v}\sqrt{\left[\frac{2\varepsilon_a - \varepsilon_b - \varepsilon_c}{3}\right]^2 + \left[\frac{\varepsilon_c - \varepsilon_b}{\sqrt{3}}\right]^2}\right\} \quad (9.70)$$

In order to establish the angle, θ_1, between the $O1$ axis and the OX axis, Eqs. (8.36), (8.37), and (8.38) are used. Any two of the three equations are needed in order to establish θ_1.

$$\tan 2\theta = \frac{\sqrt{3}(\varepsilon_c - \varepsilon_b)}{2\varepsilon_a - \varepsilon_b - \varepsilon_c} \quad (9.71)$$

$$\sin 2\theta = \frac{\varepsilon_c - \varepsilon_b}{\sqrt{\frac{1}{3}(2\varepsilon_a - \varepsilon_b - \varepsilon_c)^2 + (\varepsilon_c - \varepsilon_b)^2}} \quad (9.72)$$

$$\cos 2\theta = \frac{(2\varepsilon_a - \varepsilon_b - \varepsilon_c)/\sqrt{3}}{\sqrt{\frac{1}{3}(2\varepsilon_a - \varepsilon_b - \varepsilon_c)^2 + (\varepsilon_c - \varepsilon_b)^2}} \quad (9.73)$$

Example 9.2. If the strain gages used in Example 9.1 are arranged in a delta rosette, as shown in Fig. 9.3, determine the apparent strains indicated by each gage when subjected to the stress field of Example 9.1.

Solution. The following actual strains have been determined in Example 9.1:

$$\varepsilon_a = 1532 \text{ μin/in}, \qquad \theta_a = 0°$$
$$\varepsilon_b = -1041 \text{ μin/in}, \qquad \theta_b = 45°$$
$$\varepsilon_c = 959 \text{ μin/in}, \qquad \theta_c = 90°$$

Since the gages in the delta rosette are arranged at $\theta_a = 0°$, $\theta_b = 120°$, and $\theta_c = 240°$, the actual strains in these directions are required before the apparent strains can be computed. In order to compute the actual strains, the shearing strain must first be determined through the use of Eq. (9.31), the transformation equation. For this purpose $\varepsilon_x = 1532$ μin/in, $\varepsilon_y = 959$ μin/in, $\varepsilon_{x'} = \varepsilon_{45°} = -1041$ μin/in, and $\theta = 45°$. Equation (9.31) is

$$\varepsilon_{x'} = \frac{\varepsilon_x + \varepsilon_y}{2} + \frac{\varepsilon_x - \varepsilon_y}{2}\cos 2\theta + \frac{\gamma_{xy}}{2}\sin 2\theta$$

Thus,

$$-1041 = \frac{1532 + 959}{2} + \frac{1532 - 959}{2}\cos 2(45°) + \frac{\gamma_{xy}}{2}\sin 2(45°)$$

From this,

$$\frac{\gamma_{xy}}{2} = -2287 \text{ μradians}$$

STRAIN GAGE ROSETTES AND TRANSVERSE SENSITIVITY EFFECT 307

The actual strain in the axial direction of gage b can be determined by using the transformation equation with $\theta = 120°$.

$$\varepsilon_{x'} = \varepsilon_b = \frac{1532 + 959}{2} + \frac{1532 - 959}{2} \cos 2(120°) + (-2287) \sin 2(120°)$$

$$= 3083 \text{ μin/in}$$

The actual strain in the axial direction of gage c may be determined by using the transformation equation with $\theta = 240°$.

$$\varepsilon_{x'} = \varepsilon_c = \frac{1532 + 959}{2} + \frac{1532 - 959}{2} \cos 2(240°) + (-2287) \sin 2(240°)$$

$$= -878 \text{ μin/in}$$

The apparent strains, ε'_a, ε'_b, and ε'_c are given by Eqs. (9.56), (9.57), and (9.58), respectively. The value of ε'_a, however, must be the same for both rosettes, since both are aligned along the same axis. Thus,

$$\varepsilon'_a = 1450 \text{ μin/in}$$

$$\varepsilon'_b = \frac{1}{1 - v_0 K_b} \left[\left(1 - \frac{K_b}{3}\right) \varepsilon_b + \frac{2K_b}{3}(\varepsilon_a + \varepsilon_c) \right]$$

$$= \frac{1}{1 - 0.285(0.025)} \left[\left(1 - \frac{0.025}{3}\right)(3083) + \frac{2(0.025)}{3}(1450 - 878) \right]$$

$$= 3089 \text{ μin/in}$$

$$\varepsilon'_c = \frac{1}{1 - v_0 K_c} \left[\left(1 - \frac{K_c}{3}\right) \varepsilon_c + \frac{2K_c}{3}(\varepsilon_a + \varepsilon_b) \right]$$

$$= \frac{1}{1 - 0.285(-0.05)} \left[\left(1 - \frac{-0.05}{3}\right)(-878) + \frac{2(-0.05)}{3}(1450 + 3083) \right]$$

$$= -1029 \text{ μin/in}$$

The apparent strain readings on gages b and c for this rosette are quite different from those for the rectangular rosette. Whether or not these values are correct can be verified by using Eqs. (9.59), (9.60), and (9.61) to compute the actual strains at each gage, which are already known.

Problems

9.1. Two identical strain gages are arranged as shown in Fig. 9.1. The transverse sensitivity factor is $K = -0.026$. If the indicated strains are $\varepsilon'_a = 765$ μin/in and $\varepsilon'_b = 255$ μin/in, determine (a) the true strain in each direction, and (b) the error if the transverse sensitivity factor is ignored.

9.2. Two different gages are bonded to a thin-walled pressure vessel. Gage a with $K_a = 3.0$ percent is aligned in the longitudinal direction, while gage b with $K_b = -3.9$ percent is aligned in the hoop direction. The following data are available for the vessel: internal pressure = 800 psi, diameter = 60 in, and wall thickness = 1.25 in. Determine the indicated strains.

9.3. Two like gages with $K = -1.7$ percent are bonded along the principal stress axes of a round shaft subjected to pure torsion. Determine the percent error if the transverse sensitivity factor is ignored.

The given data for Problems 9.4 through 9.8, with all strains in µin/in are for three-element rectangular rosettes. Determine the true strains for each rostte and then compute the error if the transverse sensitivity factor had been ignored.

	ε'_a	K_a, percent	ε'_b	K_b, percent	ε'_c	K_c, percent
9.4.	960	1.3	150	0.7	445	1.3
9.5.	−565	1.5	−760	−0.5	315	1.5
9.6.	135	2.0	−820	1.0	865	2.0
9.7.	−355	−2.0	460	1.5	−715	2.0
9.8.	800	1.8	800	1.8	800	1.8

The given data for Problems 9.9 through 9.13, with all strains in µin/in, are for three element-delta rosettes. Determine the true strains for each rosette and then compute the error if the transverse sensitivity factor has been ignored.

	ε'_a	K_a, percent	ε'_b	K_b, percent	ε'_c	K_c, percent
9.9.	445	3.0	−225	1.0	−565	−3.0
9.10.	810	3.0	405	1.0	−195	3.0
9.11.	1000	1.8	1000	1.8	1000	1.8
9.12.	800	−1.3	0	0.7	800	−1.3
9.13.	−565	2.0	260	2.0	695	2.0

9.14. A three-element rectangular rosette is bonded to a steel specimen, a gage factor of 2.0 is set on the strain indicator, and the recorded data are as follows:

	Gage a	Gage b	Gage c
Gage factor	2.15	2.05	2.15
K, percent	1.8	1.0	1.8
Strain, µin/in	200	1608	850

(a) Correct for the gage factor setting.
(b) Determine the true strains.
(c) Compute the principal strains.
(d) Compute the principal stresses and their orientation relative to the axis of gage a. Sketch the element.
(e) Compute the maximum shearing stress at the point.

REFERENCES

1. "Errors Due to Transverse Sensitivity in Strain Gages," TN-509, Measurements Group, Inc., P.O. Box 27777, Raleigh, NC 27611, 1982.

2. Dove, Richard C. and Paul H. Adams, *Experimental Stress Analysis and Motion Measurement*, Columbus, OH, Charles E. Merrill Books, Inc., 1964, pp. 243–251. From *Experimental Stress Analysis and Motion Measurement* by Richard C. Dove and Paul H. Adams. Copyright © 1964. Reprinted by permission of Merrill, an imprint of Macmillan Publishing Company.
3. Dally, James W. and William F. Riley, *Experimental Stress Analysis*, 2nd edition, New York, McGraw-Hill, 1978, pp. 328–329. Material is reproduced with permission of McGraw-Hill, Inc.
4. *Handbook on Experimental Mechanics*, edited by Albert S. Kobayashi, Englewood Cliffs, Prentice-Hall, 1987, pp. 52–54.

10

STRESS GAGES

10.1. Introduction

There are a number of situations in which one wishes to determine either the normal or shearing stress in some particular direction without being required to establish the complete state of stress at any particular point. For example, if it is desired to evaluate the radial force at a given cross section of an aircraft propeller blade, this can be accomplished by multiplying the average radial stress by the area of cross section of the blade. This sounds simple, but it may involve the use of a great deal of equipment, especially under dynamic conditions when all strain observations, at all gage locations, may have to be made simultaneously.

The standard procedure for approaching this problem would be to mount rosette gages at each of the desired stations around the blade, and then calculate, from the three strains indicated by each rosette, the corresponding stress in the radial direction, and hence the radial force at this section. This involves considerable computation, and so one can appreciate that a gage whose response is directly proportional to normal stress will not only reduce the amount of instrumentation required, but in addition will reduce the amount of calculation involved in determining the final result. Thus, using a stress gage rather than a three-element rosette at each station reduces the number of channels from three to one.

10.2. The normal stress gage (1)

A much simpler method, however, involves the use of the stress gage, which has the capacity to measure two strains at right angles and to combine them in the proper proportions so that its indication, when multiplied by the proper constant, gives the value of the stress in the given direction. The use of a stress gage reduces the amount of instrumentation required by two-thirds, and the time involved in data reduction by even more than that.

Theory of the normal stress gage

Let us consider the references axes, OA and ON, which are at right angles on a free surface in a two-dimensional stress system (Fig. 10.1). The following

STRESS GAGES

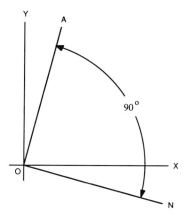

FIG. 10.1. References axes OA and ON.

relations exist between normal stress and linear strain:

$$\varepsilon_a = \frac{1}{E}(\sigma_a - \nu\sigma_n) \tag{10.1}$$

$$\varepsilon_n = \frac{1}{E}(\sigma_n - \nu\sigma_a) \tag{10.2}$$

Simultaneous solution of Eqs. (10.1) and (10.2) for σ_a in terms of the strains gives

$$\sigma_a = \frac{E}{1-\nu^2}(\varepsilon_a + \nu\varepsilon_n) \tag{10.3}$$

In passing, one should observe that the directions of the axes, OA and ON, although 90° apart, have no particular inclinations with respect to the directions of the principal axes.

Let us now examine the expression for the indication from a strain gage, which will be a dimensionless quantity in terms of $\Delta R/R$. Since Eq. (10.3) involves strains in two perpendicular directions, one can refer to Chapter 7 on lateral effects in strain gages for a general expression for the unit change in resistance of a strain gage. From Eq. (7.13) we have the general relation that

$$\frac{\Delta R}{R} = F_a(\varepsilon_a + K\varepsilon_n) \tag{7.13}$$

where F_a and K are constants for the gage. Furthermore, from Eq. (7.21),

$$F_a = \frac{G_F}{1 - v_0 K} \tag{7.21}$$

where v_0 is the Poisson ratio of the material upon which the gage factor was determined. Substituting the value of F_a into Eq. (7.13) produces

$$\frac{\Delta R}{R} = \left(\frac{G_F}{1 - v_0 K}\right)(\varepsilon_a + K\varepsilon_n) \tag{10.4}$$

Equations (7.13) and (10.3) indicate that $\Delta R/R$ is proportional to $(\varepsilon_a + K\varepsilon_n)$ and σ_a is proportional to $(\varepsilon_a + v\varepsilon_n)$. Therefore, if $K = v$, then

$$\frac{\Delta R}{R} \propto (\varepsilon_a + K\varepsilon_n) = (\varepsilon_a + v\varepsilon_n) \propto \sigma_a \tag{10.5}$$

This means that, in order for the gage to respond directly in proportion to the normal stress in the direction of OA, K must be equal to v.

From Eq. (10.4), we can find the value of $(\varepsilon_a + K\varepsilon_n)$ and then substitute this value into Eq. (10.3) for $(\varepsilon_a + v\varepsilon_n)$. This gives

$$\sigma_a = \frac{E}{1 - v^2}\left[\left(\frac{\Delta R}{R}\right)\left(\frac{1 - v_0 K}{G_F}\right)\right] \tag{10.6}$$

We will now consider certain grid configurations, for both wire gages and foil gages, which possess characteristics suitable for stress gages. Fortunately, strain gages with metal sensing elements lend themselves rather well to fulfilling the requirements for stress gages.

Single round wire in an L configuration

The L is the simplest configuration, as shown in Fig. 10.2. It consists of two straight parts of round wire, at right angles, so proportioned that the following ratio exists:

$$\frac{\text{Length of the short piece}}{\text{Length of the long piece}} = v$$

where v is the Poisson ratio of the material upon which the gage is to be used as a stress indicator.

The following assumptions will be made in the analysis for the transverse sensitivity factor:

1. The change in direction from the longer piece of wire to the shorter piece of wire takes place very abruptly.

STRESS GAGES

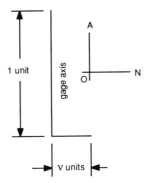

FIG. 10.2. Single round wire in an L configuration.

2. The lateral effect of the wire is zero, due to the lack of efficiency of the bonding agent in this direction. (The reader should appreciate that this may not be true for a slender strip of foil, in which the width of the element may be several times the thickness.)

If a gage of this configuration is subjected to strains ε_a and ε_n, the change in gage resistance, ΔR, is

$$\Delta R = kvS_t\varepsilon_n + k(1)S_t\varepsilon_a = kS_t[v\varepsilon_n + (1)\varepsilon_n] \tag{a}$$

where k = resistance per unit length of the wire
S_t = strain sensitivity of the wire

The value of $\Delta R/R$ is then

$$\frac{\Delta R}{R} = \frac{kS_t}{R}[v\varepsilon_n + (1)\varepsilon_a] \tag{b}$$

The transverse sensitivity factor, K, for a gage of the L configuration may be computed by using Eq. (7.14), which is

$$K = \frac{\left(\dfrac{\Delta R/R}{\varepsilon_n}\right)_{\varepsilon_a=0}}{\left(\dfrac{\Delta R/R}{\varepsilon_a}\right)_{\varepsilon_n=0}} \tag{7.14}$$

The value of $\Delta R/R$ given by Eq. (b), but subject to the restrictions on the strains of Eq. (7.14), may be substituted into Eq. (7.14). Therefore,

$$K = \frac{\left(\dfrac{kS_t}{R}[v\varepsilon_n + 1(0)]\right)\varepsilon_n}{\left(\dfrac{kS_t}{R}[v(0) + (1)\varepsilon_a]\right)\varepsilon_a} = v \qquad (c)$$

Equation (c) shows that $\Delta R/R$ will be proportional to σ_a when this configuration is used as a stress gage for the normal stress in the OA direction. However, due to the amount of wire required to make a practical gage, this form will usually occupy too much space. On this account, it is customary to arrange the wire, or foil, in a more compact grid form.

Two orthogonal gages of different resistances

Let us imagine that two strain gages with resistances R_a and R_n have been installed in directions parallel to the reference axes, OA and ON, which are at right angles. These two orthogonal gages, connected in series, have a combined output expressed by Eq. (7.34). It is

$$\frac{\Delta R_T}{R_T} = \left(\frac{1}{1+\beta}\right)\left\{\frac{(G_F)_a}{1-v_0 K_a}(\varepsilon_a + K_a\varepsilon_n) + \beta\left[\frac{(G_F)_n}{1-v_0 K_n}(\varepsilon_n + K_n\varepsilon_a)\right]\right\} \qquad (7.34)$$

where the subscripts a and n refer to the gages which are parallel to the OA and ON axes, respectively, and $\beta = R_n/R_a$.

It will be assumed that the gage factors for both gages have been determined on the same calibrating device so that v_0 is the same for the two gages. Equation (7.34) can be simplied to

$$\frac{\Delta R_T}{R_T} = \left(\frac{1}{1+\beta}\right)[M_a(\varepsilon_a + K_a\varepsilon_n) + \beta M_n(\varepsilon_n + K_n\varepsilon_a)] \qquad (10.7)$$

where

$$M_a = \frac{(G_F)_a}{1-v_0 K_a}, \qquad M_n = \frac{(G_F)_n}{1-v_0 K_n}$$

Rearrangement of Eq. (10.7) gives

$$\frac{\Delta R_T}{R_T} = \left(\frac{1}{1+\beta}\right)[(M_a + \beta M_n K_n)\varepsilon_a + (M_a K_a + \beta M_n)\varepsilon_n] \qquad (10.8)$$

or

$$\frac{\Delta R_T}{R_T} = \left(\frac{1}{1+\beta}\right)(M_a + \beta M_n K_n)\left[\varepsilon_a + \left(\frac{M_a K_a + \beta M_n}{M_a + \beta M_n K_n}\right)\varepsilon_n\right] \quad (10.9)$$

It is seen that

$$\left(\frac{1}{1+\beta}\right)(M_a + \beta M_n K_n) = \text{a constant} = C$$

so

$$\frac{\Delta R_T}{R_T} = C\left[\varepsilon_a + \left(\frac{M_a K_a + \beta M_n}{M_a + \beta M_n K_n}\right)\varepsilon_n\right] \quad (10.10)$$

Furthermore, if

$$\frac{M_a K_a + \beta M_n}{M_a + \beta M_n K_n} = v \quad (10.11)$$

then

$$\frac{\Delta R_T}{R_T} \propto (\varepsilon_a + v\varepsilon_n) \propto \sigma_a$$

This concept covers all values of gage factor and transverse sensitivity factor, which may be different for both gages, for any particular value of β. However, since commercially available gages may not be obtainable to satisfy the required values of v and β, it may be necessary to seek a compromise, or, possibly, some other method.

The above relations are somewhat complicated, so a first approximation may be examined. If the two gage factors are nearly equal and the two transverse sensitivity factors are also nearly equal, one can make the approximation of equality without causing very much error (possibly less than the error in the value of the modulus of elasticity) by using the values of the gage factor and the transverse sensitivity factor for the gage in the direction of the OA axis, especially if the strain, ε_a, is somewhat larger numerically than ε_n. According to this approximation,

$$(G_F)_a = (G_F)_n = G_F \quad \text{and} \quad K_a = K_n = K$$

Consequently,

$$M_a = M_n = M$$

Using these values, Eq. (7.34) becomes

$$\frac{\Delta R_T}{R_T} = \left(\frac{1}{1+\beta}\right)\left(\frac{G_F}{1-v_0 K}\right)[(\varepsilon_a + K\varepsilon_n) + \beta(\varepsilon_n + K\varepsilon_a)]$$

Rearranging gives

$$\frac{\Delta R_T}{R_T} = \left(\frac{1}{1+\beta}\right)\left(\frac{G_F}{1-v_0 K}\right)(1+K\beta)\left[\varepsilon_a + \left(\frac{K+\beta}{1+K\beta}\right)\varepsilon_n\right] \quad (10.12)$$

From Eq. (10.12) one can see that if

$$\frac{K+\beta}{1+K\beta} = v$$

then

$$\frac{\Delta R_T}{R_T} \propto (\varepsilon_a + v\varepsilon_n) \propto \sigma_a$$

Two special cases of interest can now be examined. The first case takes $K = 0$ for both gages and $(G_F)_a = (G_F)_n = G_F$. Using these values in Eq. (10.12) produces

$$\frac{\Delta R_T}{R_T} = \left(\frac{G_F}{1+\beta}\right)(\varepsilon_a + \beta\varepsilon_n) \quad (10.13)$$

where $\beta = R_n/R_a$ can have any value between 0 and 1. However, if $\beta = v$, the two gages in series represent a single stress gage. For the second case when $(G_F)_a = (G_F)_n = G_F$ and $K_a = K_n = K = v$, the gage in the direction of the OA axis is a stress gage by itself and the second gage is not required. In this case $\beta = 0$. Consequently, using these values in Eq. (10.12) gives

$$\frac{\Delta R_T}{R_T} = \left(\frac{G_F}{1-v_0 K}\right)(\varepsilon_a + v\varepsilon_n) \propto \sigma_a$$

The problem is to select gages with appropriate gage factors and transverse sensitivity factors and then to establish a suitable value of the ratio $\beta = R_n/R_a$. It will be best to commence by choosing a pair of gages with equal gage factors and equal transverse sensitivity factors. If gages are chosen such that the transverse sensitivity factors are not quite equal, an average value might be used without causing serious error, since these factors represent a secondary effect.

10.3. The SR-4 stress–strain gage

The stress–strain gage shown in Fig. (10.3) was produced as a special item, but has since been discontinued (2). It is, however, an interesting concept and worth examining. The gage consists of a pair of foil strain gages mounted

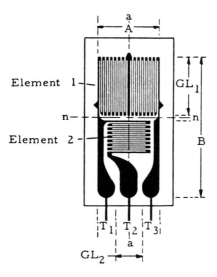

FIG. 10.3. Stress–strain gage. (From ref. 2.)

at right angles on a common carrier and possessing a ratio of resistances such that

$$\frac{R_2}{R_1} = v \qquad (10.14)$$

where v is the Poisson ratio of the material upon which the stress gage is to be used. The two gages are arranged with three-lead connections so that either grid can be used independently to measure the strains in the two perpendicular directions, a–a and n–n. If the two gages are connected in series, however, they can be used together to indicate stress in the a–a direction.

Since the Poisson ratio of the material on which the gage might be used could have many different values, the gages were limited to two particular values, namely, 0.28 for steel and 0.33 for aluminum and its alloys. These gages were also furnished in temperature compensations for use on mild steel, stainless steel, and aluminum. The resistances of all large grids were fixed at 350 ohms, while the resistances of the smaller grids were either 98 or 115 ohms, to correspond with the different Poisson ratio values for steel and aluminum.

The user was supplied with the following three factors:

G_{Fa} = factor for sensing strain along the a–a axis

G_{Fn} = factor for sensing strain along the n–n axis

G_{Fs} = factor for sensing stress along the a–a axis

The resulting strains are given by

$$\varepsilon_a = \frac{1}{G_{Fa}}\left(\frac{\Delta R_1}{R_1}\right) \qquad (10.15)$$

$$\varepsilon_n = \frac{1}{G_{Fn}}\left(\frac{\Delta R_2}{R_2}\right) \qquad (10.16)$$

The resulting stress along the a–a axis is given by

$$\sigma_a = \frac{1}{G_{Fs}}\left(\frac{\Delta R}{R}\right) \qquad (10.17)$$

where $\Delta R = \Delta R_1 + \Delta R_2$
$R = R_1 + R_2$.

In discussing the stress–strain gage, the description by Hines (3) outlines its essential characteristics, and his argument will be followed here. From previous work, we know that the relationship between stress and strain is

$$\varepsilon_a = \frac{1}{E}(\sigma_a - \nu\sigma_n) \qquad (10.18)$$

$$\varepsilon_n = \frac{1}{E}(\sigma_n - \nu\sigma_a) \qquad (10.19)$$

where σ_a = stress along the a–a axis
σ_n = stress along the n–n axis

From Chapter 7, the unit change in resistance of a gage is given by Eq. (7.11). For the stress–strain gage, this is

$$\frac{\Delta R}{R} = F_a\varepsilon_a + F_n\varepsilon_n \qquad (10.20)$$

where F_a = strain sensitivity of the gage elements for uniaxial strain along axis a–a with zero strain along axis n–n
F_n = strain sensitivity of the gage elements for uniaxial strain along axis n–n with zero strain along axis a–a

If the values of ε_a and ε_n given by Eqs. (10.18) and (10.19), respectively, are substituted into Eq. (10.20), the result is

$$\frac{\Delta R}{R} = \frac{F_a}{E}(\sigma_a - \nu\sigma_n) + \frac{F_n}{E}(\sigma_n - \nu\sigma_a)$$

Rearranging,

$$\frac{\Delta R}{R} = \frac{\sigma_a}{E}(F_a - vF_n) + \frac{\sigma_n}{E}(F_n - vF_a) \tag{10.21}$$

Since $F_n = KF_a$, Eq. (10.21) can be rewritten as

$$\frac{\Delta R}{R} = \frac{F_a}{E}[\sigma_a(1 - vK) + \sigma_n(K - v)] \tag{10.22}$$

When the gage is being calibrated, the following two conditions must be met:

1. When $\Delta R/R = 0$, $\sigma_a = 0$. This means the change in gage resistance must be independent of the transverse stress, σ_n.
2. When $\sigma_n = 0$, then $\Delta R/R = (\sigma_a G_{Fs})/E$. This means the change in gage resistance must be proportional to the stress, σ_a, applied along the a–a axis.

From the first condition, Eq. (10.22) gives

$$\frac{\Delta R}{R} = \frac{F_a \sigma_n}{E}(K - v_0) = 0 \tag{10.23}$$

where $v_0 =$ Poisson ratio of the test material. Thus, Eq. (10.23) shows that

$$K = v_0 \tag{10.24}$$

For the second condition, Eq. (10.22) gives

$$\frac{\Delta R}{R} = \frac{F_a \sigma_a}{E}(1 - v_0 K) = \frac{G_{Fs} \sigma_a}{E} \tag{10.25}$$

where

$$G_{Fs} = F_a(1 - v_0 K) \tag{10.26}$$

Equation (10.24) shows that in order for the gage to have a unit change in resistance proportional to σ_a, the stress gage must have a transverse sensitivity factor equal to the Poisson ratio of the material to which it is bonded. Equation (10.26) shows how the stress-gage factor, G_{Fs}, is related to the other gage constants.

In Fig. 10.3, element 1 is the principal strain-measuring grid, while element 2 provides the necessary transverse sensitivity when the two grids are connected in series. Therefore, the transverse sensitivity of the entire gage may be controlled by the ratio of the resistance of element 2 to the resistance of element 1; that is, R_2/R_1. The ratio $R_2/R_1 = v_0 = K$ is only approximate, since each element has a small but measurable transverse sensitivity factor.

This could be corrected, however, through calibration or by computation from the known characteristics of each element.

10.4. Electrical circuit for two ordinary gages to indicate normal stress

The circuit shown in Fig. 10.4 was developed in 1945 by S. B. Williams in order to produce indications which are directly proportional to the normal stresses in the directions of the gage axes. The circuit was first reported by Kern (4) and then appeared in later papers (5, 6). If the directions of the principal axes coincide with the gage axes, this provides a method for observing the principal stresses directly.

The value of the resistor, R_c, is given by the expression

$$R_c = \frac{R_0 R_g}{R_0 + R_g} \left[\frac{(1+K)(1-v)}{(v-K)} \right] \quad (10.27)$$

where K is the transverse sensitivity factor of the strain gages. If $K = 0$, then

$$R_c = \frac{R_0 R_g}{R_0 + R_g} \left[\frac{1-v}{v} \right] \quad (10.28)$$

If $K = v$, then R_c becomes infinite, and in this case the two gages, aligned in the X and Y directions, are themselves stress indicators, as each will respond in direct proportion of the normal stresses, σ_x and σ_y, respectively.

For further details on the circuit, the reader should consult the references cited.

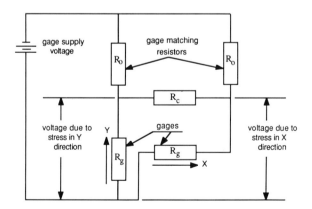

FIG. 10.4. Circuit for conversion of T-strain gage rosette into two equivalent stress gage circuits. (From ref. 4.)

10.5. The V-type stress gage (4, 7)

Schematic diagrams of two forms of wire grids of a V-type stress gage are shown in Fig. 10.5. Although these are depicted as wire gages, foil gages are also manufactured as a single unit with this configuration. The great advantage of this shape, which is formed by two like gages, is that it can easily be made up by connecting two ordinary strain gages in series, providing, however, that the angle between their axes corresponds to the value of the Poisson ratio of the material upon which they will be used to indicate stress.

For commercially manufactured gages of this nature, the angle between the axes of the two grids can be determined under controlled factory conditions. If two separate gages are to be installed in the field, the engineer in charge will need to be particularly careful to see that the two gages are mounted with the correct relative inclination, 2ϕ, between their center lines. The direction in which the stress is to be determined will be established by the direction of the bisector of the angle between the grid axes. There are, therefore, two considerations about which the installer of the gages should be meticulous. They are

1. The angle between the gage axes.
2. The direction of the bisector of the angle between the gage axes.

The following two considerations should also be kept in mind:

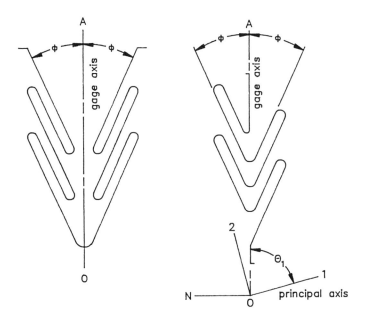

FIG. 10.5. Schematic diagrams of two forms of V gage.

1. Foil and wire strain gages respond essentially to linear strains, primarily in the axial direction, but frequently and to a lesser extent, to the transverse strain as well.
2. For two like grids in series, the combined output will correspond to the average output from each of the two gages.

Mathematical analysis of the V-type configuration

The development that Kern (4) used will be followed but extended to take in the lateral response of the strain gages to strains in the direction perpendicular to the gage axes. Let us consider a stress gage of the V type as consisting of two like strain gages connected in series, as shown in Fig. 10.5. For each gage, Eq. (7.13) can be used to give

$$\frac{\Delta R}{R} = F_a(\varepsilon_a + K\varepsilon_n) \tag{10.29}$$

where F_a is the axial strain sensitivity and K is the transverse sensitivity factor.

The stress, σ_a, in the direction of the axis OA may now be stated in terms of the principal strains. Note, in Fig. 10.5, that the principal stress axis, 1, makes an angle of θ_1 with respect to the OA axis. Again, σ_a is

$$\sigma_a = \frac{E}{1 - v^2}(\varepsilon_a + v\varepsilon_n) \tag{a}$$

where ε_a and ε_n are the strains along and transverse, respectively, to the OA axis. The two strains, ε_a and ε_n, are to be written in terms of the principal strains, ε_1 and ε_2. In order to do this, the transformation equation, Eq. (2.32), can be written in terms of the principal strains by taking $\varepsilon_x = \varepsilon_1$, $\varepsilon_y = \varepsilon_2$, and $\gamma_{xy} = 0$. Thus,

$$\varepsilon_{x'} = \frac{\varepsilon_1 + \varepsilon_2}{2} + \frac{\varepsilon_1 - \varepsilon_2}{2} \cos 2\theta \tag{10.30}$$

The strain, ε_a, in the OA direction becomes

$$\varepsilon_a = \frac{\varepsilon_1 + \varepsilon_2}{2} + \frac{\varepsilon_1 - \varepsilon_2}{2} \cos 2\theta_1 \tag{b}$$

The strain, ε_n, normal to OA becomes

$$\varepsilon_n = \frac{\varepsilon_1 + \varepsilon_2}{2} + \frac{\varepsilon_1 - \varepsilon_2}{2} \cos 2(\theta_1 + 90) = \frac{\varepsilon_1 + \varepsilon_2}{2} - \frac{\varepsilon_1 - \varepsilon_2}{2} \cos 2\theta_1 \tag{c}$$

Substituting the values of ε_a and ε_n given by Eqs. (b) and (c), respectively, into Eq. (a) results in

$$\sigma_a = E\left[\left(\frac{1}{1-v}\right)\left(\frac{\varepsilon_1 + \varepsilon_2}{2}\right) + \left(\frac{1}{1+v}\right)\left(\frac{\varepsilon_1 - \varepsilon_2}{2}\right)\cos 2\theta_1\right] \quad (10.31)$$

Equation (10.31) contains two terms. The first term is

$$\varepsilon_H = \tfrac{1}{2}(\varepsilon_1 + \varepsilon_2)$$

which represents the hydrostatic component of the principal strains and is the center of a Mohr's strain circle. The second term is

$$\varepsilon_S = \tfrac{1}{2}(\varepsilon_1 - \varepsilon_2)$$

which represents the pure shear component of the principal strains and is the radius of a Mohr's strain circle. Since the hydrostatic strain is the same in all directions, the axial and normal strains acting on the gage due to this component are the same. Using Eq. (10.29), the hydrostatic component gives a unit resistance change of

$$\frac{\Delta R}{R} = F_a(\varepsilon_H + K\varepsilon_H) = F_a\varepsilon_H(1 + K)$$

In terms of principal strains, $\Delta R/R$ is

$$\frac{\Delta R}{R} = F_a\left(\frac{\varepsilon_1 + \varepsilon_2}{2}\right)(1 + K) \quad (10.32)$$

In a similar manner, since the pure shear component corresponds to two equal strains of unlike sign, the unit resistance change due to this component is

$$\frac{\Delta R}{R} = F_a(\varepsilon_S - K\varepsilon_S) = F_a\varepsilon_S(1 - K)$$

In terms of principal strains, $\Delta R/R$ is

$$\frac{\Delta R}{R} = F_a\left(\frac{\varepsilon_1 - \varepsilon_2}{2}\right)(1 - K) \quad (10.33)$$

The average unit resistance change for the V-type gage (two gages in series), is $\Delta R/(2R_g)$, where R_g is the resistance of one half of the two gages, or grids. This can be written as

$$\frac{\Delta R}{R} \propto \frac{(\varepsilon_{\theta_1+\phi} + \varepsilon_{\theta_1-\phi})}{2} \tag{d}$$

where $\varepsilon_{\theta_1+\phi}$ and $\varepsilon_{\theta_1-\phi}$ are the strains the grids are subjected to. These strains are

$$\varepsilon_{\theta_1+\phi} = \left(\frac{\varepsilon_1 + \varepsilon_2}{2}\right)(1+K) + \left(\frac{\varepsilon_1 - \varepsilon_2}{2}\right)(1-K)\cos 2(\theta_1 + \phi) \tag{e}$$

$$\varepsilon_{\theta_1-\phi} = \left(\frac{\varepsilon_1 + \varepsilon_2}{2}\right)(1+K) + \left(\frac{\varepsilon_1 - \varepsilon_2}{2}\right)(1-K)\cos 2(\theta_1 - \phi) \tag{f}$$

If the right-hand sides of Eqs. (e) and (f) are multiplied and divided by $(1-v)$ and these values are then substituted into Eq. (d), we have

$$\frac{\Delta R}{R} \propto \left(\frac{1-v}{2}\right)\left[2\left(\frac{1+K}{1-v}\right)\left(\frac{\varepsilon_1+\varepsilon_2}{2}\right) + \left(\frac{1-K}{1-v}\right)\left(\frac{\varepsilon_1-\varepsilon_2}{2}\right)[\cos A + \cos B]\right] \tag{g}$$

where

$$\cos A + \cos B = \cos 2(\theta_1 + \phi) + \cos 2(\theta_1 - \phi) = 2\cos 2\theta_1 \cos 2\phi \tag{h}$$

Substituting the value of the sum of the cosine terms given by Eq. (h) into Eq. (g), then multiplying and dividing the coefficient of $2\cos 2\theta_1 \cos 2\phi$ by $(1+K)$, the final expression for $\Delta R/R$ is

$$\frac{\Delta R}{R} \propto (1-v)(1+K)\left[\left(\frac{1}{1-v}\right)\left(\frac{\varepsilon_1+\varepsilon_2}{2}\right) + \left(\frac{\cos 2\phi}{1-v}\right)\left(\frac{1-K}{1+K}\right)\left(\frac{\varepsilon_1-\varepsilon_2}{2}\right)\cos 2\theta_1\right]$$

(10.34)

Examination of Eqs. (10.31) and (10.34) tells us that when

$$\left(\frac{1-K}{1+K}\right)\left(\frac{\cos 2\phi}{1-v}\right) = \frac{1}{1+v} \tag{i}$$

the quantities in the square bracket of each equation will be identical, so that $\Delta R/R$ will be directly proportional to σ_a. This means that when one computes the correct angle between the two gages, or grids, one will have a stress gage. From Eq. (i), the relative angle of inclination between the two gages, or grids, can be established. Thus,

$$\cos 2\phi = \frac{(1+K)(1-v)}{(1-K)(1+v)} \tag{10.35}$$

This expression can also be written as

$$\cos 2\phi = \frac{(1-vK)-(v-K)}{(1-vK)+(v-K)} \tag{10.36}$$

Since v will be between 0 and 0.5, and because K can be expected to be less than 0.04, the product vK will be very small with respect to unity, and so by neglecting the product, vK, the value of $\cos 2\phi$ becomes

$$\cos 2\phi = \frac{1-(v-K)}{1+(v-K)} \tag{10.37}$$

which is a very close approximation, especially when K tends towards zero.
The term, $\cos 2\phi$, can also be written as

$$\cos 2\phi = \frac{1-\tan^2 \phi}{1+\tan^2 \phi} \tag{10.38}$$

Using the value of $\cos 2\phi$ given by Eq. (10.36), the value of $\tan \phi$ becomes

$$\tan \phi = \sqrt{\frac{v-K}{1-vK}} \tag{10.39}$$

If the product vK is ignored, then

$$\tan \phi = \sqrt{v-K} \tag{10.40}$$

Taking the transverse sensitivity factor, K, equal to zero gives

$$\tan \phi = \sqrt{v} \tag{10.41}$$

Stress gages may be made using either wire or foil. A typical foil stress gage is shown in Fig. 10.6.

FIG. 10.6. Foil stress gage in V configuration. (Courtesy of Measurements Group, Inc.)

10.6. Application of a single strain gage to indicate principal stress (8)

There are certain situations in which the directions of the principal stresses are known. This may be through the conditions of symmetry, a preliminary study with a brittle lacquer coat, or by some other method. Under these conditions, if two strain gages are installed, one in the direction of each principal axis, the two strain readings thus obtained will provide sufficient information to enable the computation of the two principal stresses. At each point to be investigated, this saves the use of at least one set of instrumentation when all readings are required to be made simultaneously. Also, the time involved with calculations will be much reduced in comparison with that required for standard rosette analysis, which might employ equipment for three or four observations at each station for each load level.

If one is only interested in one of the two principal stress magnitudes, however, and knows to which axis this corresponds, a single strain gage indication for each load at each station is all that is required to provide the necessary information. Such a gage is shown in Fig. 10.7.

The amount of required instrumentation is reduced by two-thirds in comparison with the general rosette method. The corresponding calculation time can also be reduced by an even greater proportion, especially if the equipment can be calibrated to give a direct readout in terms of the desired data.

Due to the fact that the strain conditions are symmetrical with respect to a principal axis, if a V-type stress gage is to be used and lined up with the principal axis, both halves of the grid will be subjected to exactly the same strain conditions and will show the same unit change in resistance. This means that one half of the V-type gage will be redundant. Because of this only one half of the gage needs to be used, which means a single strain gage inclined at the correct angle ϕ with respect to the principal axis will

STRESS GAGES

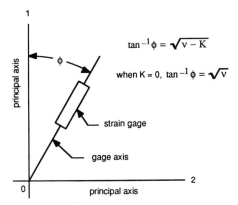

FIG. 10.7. Single strain gage inclined at an angle ϕ with respect to the principal axis.

yield all the necessary information to enable the evaluation of the principal stress magnitude. As shown in Section 10.5, $\tan \phi = \sqrt{v - K}$. If $K = 0$, then $\tan \phi = \sqrt{v}$.

10.7. Determination of plane shearing stress

Wire and foil gages have little response to shearing strains so, if this quantity is to be determined, it is necessary to make the approach through the measurement of linear strains, which can be converted into the equivalent of plane shear strain, and then into terms of shearing stress by means of the shearing modulus of elasticity. In Chapter 8 on rosette analysis, it was shown that linear strains in certain given directions can be converted into their equivalent values in terms of the hydrostatic component, ε_H, and the pure shear component, ε_S. These two values are written in terms of the principal strains as follows:

$$\varepsilon_H = \frac{\varepsilon_1 + \varepsilon_2}{2} \quad \text{(position of the center of Mohr's circle)}$$

$$\varepsilon_S = \frac{\varepsilon_1 - \varepsilon_2}{2} \quad \text{(radius of Mohr's circle)}$$

This is shown in Fig. 10.8.

As shearing stresses are currently being considered, the hydrostatic component, which is equivalent to two principal strains of equal magnitude and like sign, will not concern us. The pure shear component, which is equivalent to two principal strains of equal magnitude but of opposite sign, will be examined rather carefully. It is from the pure shear component that the shearing condition in any particular direction can be determined, as

328 THE BONDED ELECTRICAL RESISTANCE STRAIN GAGE

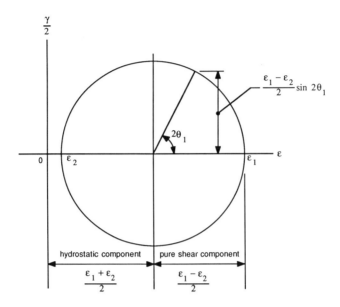

FIG. 10.8. Shear strain in terms of pure shear component.

shown in Fig. 10.8. Since the data produced with any rosette arrangement can be converted into its hydrostatic and pure shear components, it can be seen that rosettes might, in general, be used to find the shearing strain in any particular direction. However, except in two special cases, the data reduction involved in this procedure is too complicated and time-consuming.

Let us now look into the simpler procedures which can be accomplished with just two strain indications. Consider the rectangular rosette with four strain observations, which, for this special case, can be reduced to two measurements. This rosette consists of four strain gages mounted in the directions OA, OB, OC, and OD, as indicated in Fig. 10.9. For this particular rosette, the values of ε_H, ε_S, and $\tan 2\theta$ are given by Eqs. (8.41a), (8.41b), and (8.42). These are

$$\varepsilon_H = \frac{\varepsilon_a + \varepsilon_b + \varepsilon_c + \varepsilon_d}{4} \tag{8.41a}$$

$$\varepsilon_S = \tfrac{1}{2}\sqrt{(\varepsilon_a - \varepsilon_c)^2 + (\varepsilon_b - \varepsilon_d)^2} \tag{8.41b}$$

$$\tan 2\theta = \frac{\varepsilon_b - \varepsilon_d}{\varepsilon_a - \varepsilon_c} \tag{8.42}$$

Furthermore, if gages A and C and gages B and D can be connected in adjacent arms of the Wheatstone bridge, the differences, $(\varepsilon_a - \varepsilon_c)$ and

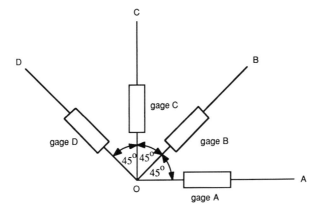

FIG. 10.9. Arrangement of gage axes for four-element rectangular rosette.

$(\varepsilon_b - \varepsilon_d)$, can be read out directly so that the determination of the maximum shear strain and the directions of the principal axes may be found from just two strain observations. This saves both time and equipment. It should also be noted that with this form of rosette it is also possible to determine the magnitudes of the two principal stresses as well as the directions of the two principal axes.

When $K \neq 0$, the gages respond to lateral strain in addition to axial strain. Thus,

$$\varepsilon_H = (1 + K)\left(\frac{\varepsilon_a + \varepsilon_b + \varepsilon_c + \varepsilon_d}{4}\right) \tag{10.42}$$

$$\varepsilon_S = \left(\frac{1 - K}{2}\right)\sqrt{(\varepsilon_a - \varepsilon_c)^2 + (\varepsilon_b - \varepsilon_d)^2} \tag{10.43}$$

There is no change in the directions of the principal axes, since the value of $\tan 2\theta$ is independent of K. Figure 10.10 shows an available stacked four-element rectangular rosette of the configuration shown in Fig. 10.9.

If stress gages rather than strain gages were used, the change in these relations would be due essentially to the difference in the numerical value of K, which would then take on the particular value of K that would be

FIG. 10.10. Four-element rectangular rosette. (Courtesy of Measurements Group, Inc.)

equal to the value of Poisson's ratio of the material upon which the gages were installed. Hence,

$$\sigma_{1,2} = \frac{E}{1-v}(1+K)\varepsilon_H \pm \frac{E}{1+v}(1-K)\varepsilon_S \qquad (10.44)$$

When $K = v$,

$$\sigma_{1,2} = E\left[\left(\frac{1+v}{1-v}\right)\varepsilon_H \pm \left(\frac{1-v}{1+v}\right)\varepsilon_S\right]$$

This can be rewritten as

$$\sigma_{1,2} = E\left(\frac{1+v}{1-v}\right)\varepsilon_H \pm \frac{E(1-v)}{2(1+v)}(\varepsilon_1 - \varepsilon_2)$$

or

$$\sigma_{1,2} = E\left(\frac{1+v}{1-v}\right)\varepsilon_H \pm G(1-v)(\varepsilon_1 - \varepsilon_2) \qquad (10.45)$$

where G, the shear modulus, is

$$G = \frac{E}{2(1+v)}$$

The shear gage

In order to determine the linear strain equivalent of shearing strain, two gages can be used that are mounted with their axes having any arbitrary angle between them. In order to show this, the argument given by Perry (9) will be followed. The two gages are bonded to the test material as illustrated by Fig. 10.11.

The expressions for the strains, ε_a and ε_b, can be written as

$$\varepsilon_a = \frac{\varepsilon_x + \varepsilon_y}{2} + \frac{\varepsilon_x - \varepsilon_y}{2}\cos 2\theta_A + \frac{\gamma_{xy}}{2}\sin 2\theta_A \qquad (10.46)$$

$$\varepsilon_b = \frac{\varepsilon_x + \varepsilon_y}{2} + \frac{\varepsilon_x - \varepsilon_y}{2}\cos 2\theta_B + \frac{\gamma_{xy}}{2}\sin 2\theta_B \qquad (10.47)$$

Solving Eqs. (10.46) and (10.47) simultaneously for γ_{xy} produces

$$\gamma_{xy} = \frac{2(\varepsilon_a - \varepsilon_b) - (\varepsilon_x - \varepsilon_y)(\cos 2\theta_A - \cos 2\theta_B)}{\sin 2\theta_A - \sin 2\theta_B} \qquad (10.48)$$

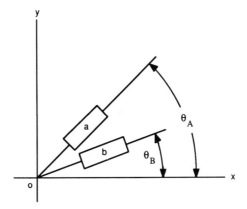

FIG. 10.11. Arbitrary gage arrangement.

If gages a and b are bisected by the x axis, then $\theta_A = -\theta_B$ and so $\cos\theta_A = \cos\theta_B$. In that case Eq. (10.48) is rewritten as

$$\gamma_{xy} = \frac{2(\varepsilon_a - \varepsilon_b)}{\sin 2\theta_A - \sin 2\theta_B} \tag{10.49}$$

Furthermore, $\sin 2\theta_A = -\sin 2\theta_B$, which further reduces Eq. (10.49) to

$$\gamma_{xy} = \frac{\varepsilon_a - \varepsilon_b}{\sin 2\theta_A} = -\frac{\varepsilon_a - \varepsilon_b}{\sin 2\theta_B} \tag{10.50}$$

Perry (9) generalizes these results as follows: *The difference in normal strain sensed by any two arbitrarily oriented strain gages in a uniform field is proportional to the shear strain along an axis bisecting the strain gage axes, irrespective of the included angle between the gages.*

An examination of Eq. (10.50) shows that if the two gages are 90° apart the denominator becomes unity, since $\theta_A = 45°$ (or $\theta_B = -45°$). Thus,

$$\gamma_{xy} = \varepsilon_a - \varepsilon_b \tag{10.51}$$

Equation (10.51) tells us that the shearing strain along the bisector of the gages' axes is equal to the difference in the normal strains. It can be seen, then, that a two-element rectangular rosette makes an ideal shear gage when the two gages are arranged in adjacent arms of a Wheatstone bridge. Figure 10.12 shows the gage and bridge arrangement.

Unless the two gages happen to be lined up with the principal axes (when the individual strain indications will correspond to each of the principal strains), no information about the principal strain magnitudes, or the directions of the principal axes, is available from the two gages. However,

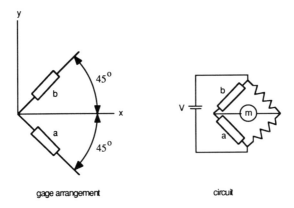

FIG. 10.12. Two-element rectangular rosette arranged to measure γ_{xy}.

if one is only interested in shear in a given direction, the two gages will provide the necessary data.

Since two strain gages only occupy half of the Wheatstone bridge, the bridge output can be doubled by adding two more gages with their axes parallel and perpendicular to the first two. If the bisector of the gage axes happens to line up with the principal axes, then twice the maximum shear is indicated by the output from the complete Wheatstone bridge. In particular, it should be noted that, since the four gages in the Wheatstone bridge only measure strains in two directions (at right angles), they will do nothing to determine the directions of the principal strain axes. Figure 10.13 shows several configurations of commercially available four-element gages for determining shear strain.

Care must be taken in order to avoid confusion between the four-element shear gage and the four-element rectangular rosette. Both employ four active strain gages. The shear gage, which measures strain in two perpendicular directions, involves a simple procedure for establishing the shearing strain, and thus the shearing stress, but it will only permit one to find this in a particular direction. A rectangular rosette, on the other hand, is much more general in nature, permitting strain observations in four different directions spaced successively at 45°. With the rosette, one can determine the two principal strains, the directions of the two principal axes, and the shearing strain in any direction, including the maximum value. However, the corresponding computations are somewhat more elaborate.

Equations (10.50) and (10.51) were developed by considering the transverse sensitivity factor, K, to be zero. If K is to be accounted for, then ε_a and ε_b in Eqs. (10.50) and (10.51) will have to be modified. This can be accomplished by returning to Eqs. (9.14) and (9.15). These are

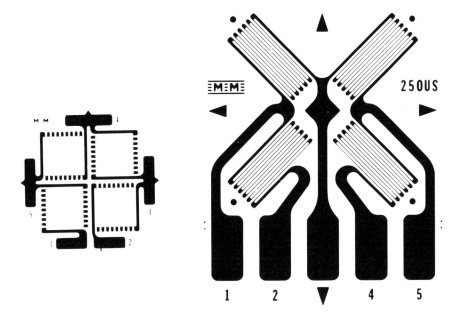

FIG. 10.13. Full bridges for shear measurement. (Courtesy of Measurements Group, Inc.)

$$\varepsilon_a = \frac{(1 - v_0 K)(\varepsilon_a' - K\varepsilon_b')}{1 - K^2} \tag{9.14}$$

$$\varepsilon_b = \frac{(1 - v_0 K)(\varepsilon_b' - K\varepsilon_a')}{1 - K^2} \tag{9.15}$$

These equations represent strains for a pair of orthogonal gages. Since the gages used to indicate shear strain are arbitrarily oriented, these two expressions will be rewritten. Thus,

$$\varepsilon_a = \frac{(1 - v_0 K)(\varepsilon_a' - K\varepsilon_a'')}{1 - K^2} \tag{10.52}$$

$$\varepsilon_b = \frac{(1 - v_0 K)(\varepsilon_b' - K\varepsilon_b'')}{1 - K^2} \tag{10.53}$$

where ε_a'' and ε_b'' are strains perpendicular to ε_a' and ε_b', respectively. The values of ε_a'' and ε_b'' can be determined by using the first strain invariant, so that

$$\varepsilon_a'' = \varepsilon_x + \varepsilon_y - \varepsilon_a' \tag{a}$$

$$\varepsilon_b'' = \varepsilon_x + \varepsilon_y - \varepsilon_b' \tag{b}$$

Substituting the values of ε''_a given by Eq. (a) into Eq. (10.52),

$$\varepsilon_a = \frac{1 - v_0 K}{1 - K^2}[(1 + K)\varepsilon'_a - K(\varepsilon_x + \varepsilon_y)] \tag{c}$$

Substituting the value of ε''_b given by Eq. (b) into Eq. (10.53),

$$\varepsilon_b = \frac{1 - v_0 K}{1 - K^2}[(1 + K)\varepsilon'_b - K(\varepsilon_x + \varepsilon_y)] \tag{d}$$

Since the shearing strain is proportional to $(\varepsilon_a - \varepsilon_b)$, we have from Eqs. (c) and (d)

$$\varepsilon_a - \varepsilon_b = \frac{1 - v_0 K}{1 - K^2}[(1 + K)\varepsilon'_a - K(\varepsilon_x + \varepsilon_y) - (1 + K)\varepsilon'_b + K(\varepsilon_x + \varepsilon_y)]$$

This reduces to

$$\varepsilon_a - \varepsilon_b = \frac{1 - v_0 K}{1 - K}(\varepsilon'_a - \varepsilon'_b) \tag{10.54}$$

Note that ε'_a and ε'_b are indicated strains.

Equation (10.54) shows that Eqs. (10.50) and (10.51) can be corrected for transverse sensitivity by multiplying the shearing strain by $(1 - v_0 K)/(1 - K)$. Therefore, Eq. (10.50) becomes

$$\gamma_{xy} = \left(\frac{1 - v_0 K}{1 - K}\right)\left(\frac{\varepsilon'_a - \varepsilon'_b}{\sin 2\theta_A}\right) = -\left(\frac{1 - v_0 K}{1 - K}\right)\left(\frac{\varepsilon'_a - \varepsilon'_b}{\sin 2\theta_b}\right) \tag{10.55}$$

Likewise, Eq. (10.51) for the two-element rectangular rosette becomes

$$\gamma_{xy} = \left(\frac{1 - v_0 K}{1 - K}\right)(\varepsilon'_a - \varepsilon'_b) \tag{10.56}$$

Problems

10.1. Two strain gages with their axes perpendicular to each other are to be used as a stress gage. The following data are available for the gages: $R_a = 350$ ohms, $(G_F)_a = 2.15$, $K_a = 0.007$, $R_n = 120$ ohms, $(G_F)_n = 2.05$, $K_n = 0.009$. Will this arrangement be suitable for a stress gage? If so, specify the material on which it may be used.

10.2. For the V stress gage shown in Fig. 10.6, determine the included angle, using $K = 0$, when designed for use on materials having the following Poisson ratios:

(a) $v = 0.25$; (b) $v = 0.30$; (c) $v = 0.34$.

10.3. A single strain gage is used to measure the longitudinal stress at a point on an aluminum cantilever beam. Using $v = 0.33$ and $E = 10 \times 10^6$ psi, determine the following:

(a) The angle ϕ between the beam axis and the gage axis
(b) The longitudinal stress for a recorded strain of 884 microstrain.

10.4. Thin-walled pressure vessels are to be made from several different materials. They have an internal pressure p, a diameter d, a wall thickness t, Poisson ratio v, and a modulus of elasticity E. A single strain gage is to be bonded to each vessel so the hoop stress, σ_H, may be monitored and the vessel automatically shut down if a specified stress is exceeded. Develop an expression for σ_H in terms of the vessel dimensions, the material properties and the strain.

10.5. Check the expressions developed in Problem 10.4 by using $p = 900$ psi, $d = 40$ in, $t = 1.5$ in, $v = 0.29$, and $E = 28 \times 10^6$ psi.

10.6. On the vessel in Problem 10.5, a line 45° from the longitudinal axis is drawn from the origin of the longitudinal and circumferential axes. Two identical strain gages are bonded to the vessel at 15° on either side of this line. For the conditions in Problem 10.5, determine the strains at each gage and show they measure the maximum shearing strain.

10.7. A two-element rectangular rosette is to be bonded to the web of a beam, whose cross section is shown in Fig. 10.14, in order to determine the maximum shear strain at that section and thereby the maximum shear stress. The material properties of the beam are $v = 0.3$ and $E = 30 \times 10^6$ psi.

(a) Determine the point where the maximum shear stress occurs.
(b) Sketch the gage arrangement.
(c) Determine the strain at each gage when the total vertical shearing force is 48 000 lb.

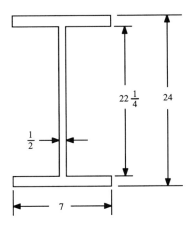

Fig. 10.14.

REFERENCES

1. Williams, Sidney B., "The Dyadic Gage," *SESA Proceedings*, Vol. I, No. 2, 1944, pp. 43–55.
2. "SR-4 Stress-Strain-Gage," Product Data 4323, BLH Electronics, Inc., 75 Shawmut Rd., Canton, MA 02021, May, 1961. (Now out of print.)
3. Hines, Frank F., "The Stress-Strain Gage," *Proc. 1st International Congress on Experimental Mechanics*, 1963, pp. 237–253.
4. Kern, Robert E., "The Stress Gage," *SESA Proceedings*, Vol. IV, No. 1, 1946, pp. 124–129.
5. Kern, Robert E. and Sidney B. Williams, "Stress Measurement by Electrical Means," *Electrical Engineering, Transactions*, Vol. 65, March 1946, pp. 100–107.
6. Williams, Sidney B., "Geometry in the Design of Stress Measurement Circuits; Improved Methods Through Simpler Concepts," *SESA Proceedings*, Vol. XVII, No. 2, 1960, pp. 161–178.
7. Sevenhuijsen, Pieter J., "Stress Gages," *Experimental Techniques*, Vol. 8, No. 3, March 1984, pp. 26–27.
8. Lissner, H. R. and C. C. Perry, "Conventional Wire Strain Gage Used as a Principal Stress Gage," *SESA Proceedings*, Vol. XIII, No. 1, 1955, pp. 25–34.
9. Perry, C. C., "Plane-shear Measurement with Strain Gages," *Experimental Mechanics*, Vol. 9, No. 1, Jan. 1969, pp. 19N–22N.

11

TEMPERATURE EFFECTS ON STRAIN GAGES

11.1. Introduction

When using strain gages, the engineer wants to measure strains produced only by the loading on the structure and to eliminate strains produced by other variables, particularly temperature. Since metals change their resistance with temperature as well as with strain, the purpose of the electrical resistance strain gage is to measure the strain-induced resistance change independently of the temperature-induced resistance change. Therefore, we want to account, or to compensate automatically, for the effects of temperature on the strain observations.

The physical phenomena occurring in a strain gage bonded to a test specimen are complex when a change of temperature takes place (1). Among them are the following:

1. The base material expands or contracts.
2. The strain-sensitive filament of the gage expands or contracts.
3. The resistance of the filament changes.
4. The gage factor of the gage is subject to variation.
5. The bond between the gage and the base material may be affected.
6. Due to the transverse sensitivity, dimensional changes which take place in the lateral direction, either in the gage or in the base material, will show an indicated change in resistance.
7. The carrier on which the sensitive filament is mounted may change its properties.

11.2. Basic considerations of temperature-induced strain (2–4)

In Chapter 1, in the discussion of the strain sensitivity of a wire, Eq. (1.18) was developed. From this, the expression for the unit change in resistance can be written as

$$\frac{dR}{R} = (1 + 2v)\frac{dL}{L} + \frac{d\rho}{\rho} \tag{11.1}$$

where v = Poisson's ratio
 L = conductor length
 ρ = resistivity of the conductor material

Equation (11.1) shows that the unit change in resistance is dependent on the unit changes in length and resistivity of the conductor. If no mechanical strain takes place, a unit resistance change can still occur when the conductor is subjected to a temperature change.

Consider a strain gage bonded to a base material and connected to a strain indicator. If the base material is unrestrained and then undergoes a temperature change, the strain indicator will show an indicated strain consisting of the algebraic sum of three components.

1. The base material to which the gage is bonded expands or contracts in the direction of the gage axis. This unit change in length, or strain, is

$$\varepsilon_m = \frac{\Delta L_m}{L_m} = \alpha_m \Delta T \tag{11.2}$$

where α_m = coefficient of thermal expansion of the base material
ΔT = temperature change from a reference temperature.

2. The strain gage grid material expands or contracts due to the temperature change. This unit change in length, or strain, is

$$\varepsilon_g = \frac{\Delta L_g}{L_g} = \alpha_g \Delta T \tag{11.3}$$

where α_g = coefficient of thermal expansion of the grid material.

3. Since the resistivity of the strain gage grid material changes with temperature, the gage resistance will change. The unit resistance change of the gage is

$$\frac{\Delta R}{R} = \beta \Delta T \tag{11.4}$$

where β = resistance–temperature coefficient of the strain gage grid material. Equation (11.4) can be expressed in terms of strain by dividing both sides of the equation by G_F, the gage factor. Thus,

$$\varepsilon_\rho = \frac{\beta \Delta T}{G_F} \tag{11.5}$$

The strain, called thermal output (sometimes referred to as apparent strain), that will be registered on the strain indicator may be expressed as the algebraic sum of the three strains. Hence,

$$\varepsilon_{TO} = \varepsilon_m - \varepsilon_g + \varepsilon_\rho$$

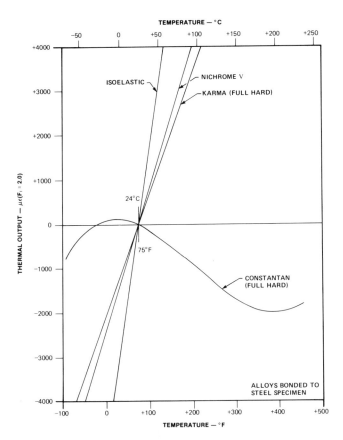

FIG. 11.1. Apparent strain vs. temperature for strain gage alloys bonded to steel. (From ref. 2.)

or

$$\varepsilon_{TO} = \left[\alpha_m - \alpha_g + \frac{\beta}{G_F} \right] \Delta T \qquad (11.6)$$

We are not to assume that the strain given by Eq. (11.6) is a linear function of temperature. It is not, since the coefficients α_m, α_g, and β are also functions of temperature. We must know, therefore, the temperature characteristics of each gage used as well as the temperature characteristics of the material on which the gage is bonded.

For illustration only, Fig. 11.1 shows the variation of strain with temperature for several strain gage materials bonded to steel. The figure shows that large errors can occur when the strain gage and the material to which it is bonded are subjected to temperatures differing from the reference, or bonding, temperature. This illustrates the need for correction when the strain gage system is subjected to temperature fluctuations. Corrections may be accomplished by computation using a temperature–strain calibration curve, or by using a compensating, or dummy, gage in an adjacent arm of the bridge that is subject to certain restrictions. If the correction is made by computation, then a temperature record must be kept during the test.

Gage factor variation with temperature

The gage factor, G_F, also varies with temperature. If the temperature range is small and the variation in G_F is slight, then a correction may be ignored. If, however, the test temperature range is large and the variation in G_F with temperature cannot be disregarded, then, depending on the required accuracy of the strain measurement, a gage factor correction may be necessary. This is illustrated in Fig. 11.2, which shows the variation in gage factor with temperature for several strain gage alloys. Several of the alloys are linear over a considerable temperature range and show quite a variation in slopes. Constantan shows an increase of less than 1 percent per 100°F with increasing temperature, while Nichrome V shows a decrease of over 2 percent per 100°F with increasing temperature. Isoelastic has a very slight change in gage factor between room temperature and 200°F, but changes quite perceptibly outside of this region. This latter material, however, is used for dynamic measurements rather than static measurements. Under dynamic conditions, other errors may be considerably greater than the change in gage factor, and so correcting the gage factor may be inessential.

In order to correct the gage factor from its value at the reference temperature to its value at the test temperature, a simple procedure is followed. If Fig. 11.2 is examined, it is seen that the percent change in gage factor is plotted versus temperature. The gage factor, G_{FT}, at some temperature different from the reference temperature is

$$G_{FT} = G_{FR}\left[1 + \frac{\Delta G_F(\%)}{100}\right] \quad (11.7)$$

where G_{FR} = gage factor at the reference temperature, generally at room temperature where $G_{FR} = G_F$
$\Delta G_F(\%)$ = percentage change in the gage factor from the reference temperature to the test temperature

The proper sign of $\Delta G_F(\%)$ must, of course, be used in Eq. (11.7).

Method of determining gage factor variation with temperature

The method of determining the variation in the gage factor with temperature for resistance strain gages is given by ASTM(5). Two methods, static and dynamic, are discussed, but only the static method will be outlined here. The test apparatus, shown in Fig. 11.3, consists of a beam having a uniform stress

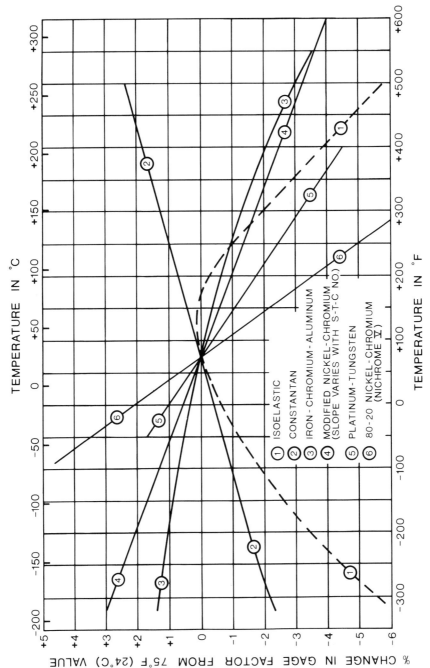

Fig. 11.2. Gage factor variation with temperature for several strain gage alloys. (From ref. 3.)

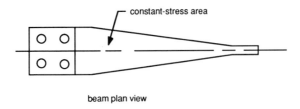

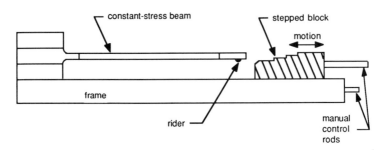

Fig. 11.3. Apparatus for static determination of gage-factor variation vs. temperature. (From ref. 5 with permission. © ASTM.)

area that is directly proportional to the deflection of the end point containing the rider, which is located at the apex of the angle formed by the beam sides.

The fixture holding the beam must be much more rigid than the beam in order to prevent errors due to its own deformation. The sliding stepped block has four surfaces that must be of nearly equal steps, with the surfaces parallel to each other and to the sliding surfaces. The steps are such that a maximum surface strain of 1000 ± 100 μin/in is produced on the beam. When the rider is resting on the lowest surface of the sliding block, the end of the beam should be deflected about 2 percent of its total planned deflection in order to insure positive contact.

The gages to be tested are symmetrically mounted in the constant-stress area of the beam and aligned with the longitudinal axis. Thermocouples are mounted as near the gages as possible and at each end of the constant stress area. The entire test unit is then placed in a temperature chamber, the gages are connected to the instrumentation, and the fixture and beam are allowed to come to equilibrium at the reference temperature, which is usually room temperature. With the rider resting on the lowest step of the sliding block, the instrumentation is balanced, then gage output is recorded as the rider is displaced to subsequent steps on the sliding block. Readings, taken three times, are recorded for both increasing and decreasing deflections.

The test chamber is brought to previously selected temperatures and the process is repeated after the temperature has stabilized. The temperature difference over the constant stress area shall not exceed 5°F (3°C) or 1 percent of the temperature of the gage area, whichever is greater. Neither shall the temperature change more than 5°F (3°C) during a test at any temperature.

The change in gage factor is computed as the difference between the gage output due to the strain for a given temperature and that at the reference temperature. This is expressed as a percentage change. Thus, the percent change in gage factor is

$$\Delta G_F(\%) = \left[\frac{E_t}{E_r} - 1\right] \times 100 \qquad (11.8)$$

where E_t = gage output at test temperature
E_r = gage output at reference temperature

If more accuracy is desired, corrections can be made for the thermal expansion of the beam and the stepped block. This gives

$$\Delta G_F(\%) = \left\{\left[\frac{E_t}{E_r}\right]\left[\frac{1 + \alpha_b \Delta T}{1 + \alpha_s \Delta T}\right] - 1\right\} \times 100 \qquad (11.9)$$

where α_b = coefficient of thermal expansion of the beam
α_s = coefficient of thermal expansion of the sliding block
ΔT = difference between the test and reference temperature

11.3. Self-temperature-compensated strain gages (2, 4)

The manufacturer of strain gage alloys can control temperature–resistance coefficients within reasonable limits. With careful selection of particular melts, followed by judicious process control, the alloy will exhibit a minimum temperature response over a given temperature range when bonded to a test specimen whose coefficient of thermal expansion matches that of the strain gage alloy. By choosing a gage that is temperature compensated for the material being tested, a three-wire, quarter-bridge circuit may be used rather than using a half-bridge circuit with a matching dummy, or compensating, gage. In the case of the quarter-bridge circuit, the Wheatstone bridge can be completed by using a stable precision resistor in the adjacent arm at the instrument, or by using an instrument that accommodates the three-wire, quarter-bridge circuit. This circuit (with lead-line resistance) was discussed in Chapter 5. Figure 11.4 shows the three-wire circuit (without lead-line resistance) with the bridge-completion resistor.

Self-temperature-compensated gages are readily available from strain gage manufacturers. In the gage designation code, a number usually appears that indicates the material for which the gage is temperature-compensated. For instance, 6 or 06 indicates a gage compensated for mild steel, where the

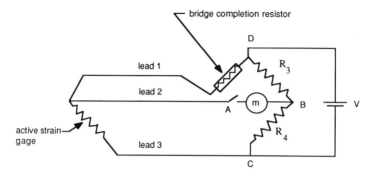

FIG. 11.4. Quarter bridge with three lead wires and bridge completion resistor.

thermal expansion coefficient is 6 (μin/in)/°F. This is also designated as parts per million per degree Fahrenheit and abbreviated to ppm/°F. A graph may be included in the gage packet showing the variation with temperature of both the thermal output, expressed in microstrain, and the gage factor. There may also be a polynomial expression giving the thermal output as a function of temperature. Figure 11.5 shows a typical graph of both thermal output and gage factor variation vs. temperature.

In developing the thermal output vs. temperature curve, the data given are for a foil lot rather than a gage designation. A test gage made from a foil lot is bonded to a test specimen and the procedure given by ASTM is followed (5). The test specimen is placed in a temperature chamber and the gage is connected to a strain indicator, the gage factor is set (usually 2.0) and the instrument is then balanced at the reference temperature of 75°F. The test specimen is unrestrained and allowed to expand or contract freely as the temperature is varied. Since no mechanical or thermal stresses are present at the equilibrium temperatures, the recorded strain at these temperatures is due only to the thermal effects, thus enabling one to plot the thermal output vs. temperature. The thermal output may also be expressed as a polynomial given as

$$\varepsilon_{TO} = A + BT + CT^2 + DT^3 + ET^4 \tag{11.10}$$

where ε_{TO} = thermal output in microstrain
        T = temperature

The coefficients, A, B, C, D, and E may be given for both the Fahrenheit and Celsius temperature scales.

If greater accuracy is required in determining the thermal output when testing an actual structure, the gage, or gages, may be bonded to the structure along with the adjacently placed temperature sensor for each gage. The strain-measuring instrument is balanced at the reference temperature, and the structure in the unrestrained state (no mechanical or thermal stresses

Foil Lot No. A38AD315

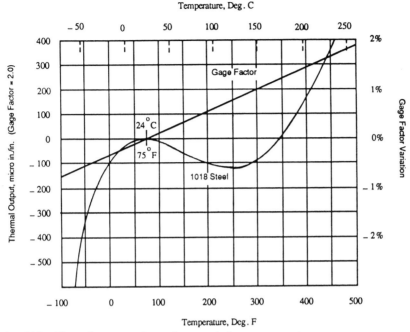

FIG. 11.5. Thermal output and gage factor variation vs. temperature.

present) is subjected to the test temperature, or temperatures. The thermal output (microstrain) at each equilibrium temperature is recorded, which allows subsequent correction in order to obtain the strains due to the loading.

Table 11.1 is a listing of the coefficients for Eq. (11.10) for several foil lot numbers. The temperature coefficient of the gage factor for each foil lot is also included.

Thermal output correction

When the structure carrying the bonded strain gage is loaded and tested at some subsequent temperature, the strain indicator will show an indicated strain, ε'_i, that is made up not only of the thermal output (apparent strain) but also the strain produced by the mechanical and thermal stresses due to the loading. Setting the actual value of the gage factor, G_F, given on the strain gage data sheet, on the strain indicator, a first approximation of the strain at the test temperature is obtained by subtracting the thermal output from the indicated strain. When doing this, care must be taken in using the proper sign of the strain. This gives

$$\varepsilon_i = \varepsilon'_i - \varepsilon_{TO} \tag{11.11}$$

Table 11.1

Foil lot no.	Test material	A	B	C	D	E	
A11BP11	1018 Steel	2.31×10^3	-2.79×10^1	-4.59×10^{-2}	8.60×10^{-5}	-2.9×10^{-8}	°F
		1.37×10^3	-5.51×10^1	-1.23×10^{-1}	4.79×10^{-4}	-3.11×10^{-7}	°C
		Temperature coefficient of gage factor = $(0.7 \pm 0.2)/100°C$					
A12BJ01	1018 Steel	7.03×10^2	-8.31	-1.78×10^{-2}	5.15×10^{-5}	-3.37×10^{-8}	°F
		4.21×10^2	-1.67×10^1	-4.22×10^{-2}	2.75×10^{-4}	-3.53×10^{-7}	°C
	2024-T4 Al	2.98×10^2	-2.94	-1.76×10^{-2}	6.06×10^{-5}	-4.59×10^{-8}	°F
		1.88×10^2	-7.00	-3.89×10^{-2}	3.19×10^{-4}	-4.82×10^{-7}	°C
		Temperature coefficient of gage factor = $(0.8 \pm 0.2)/100°C$					
A12BJ03	1018 Steel	7.47×10^2	-8.80	-1.86×10^{-2}	4.19×10^{-5}	-1.09×10^{-8}	°F
		4.47×10^2	-1.78×10^1	-4.75×10^{-2}	2.36×10^{-4}	-1.14×10^{-7}	°C
	2024-T4 Al	2.40×10^2	-2.26	-1.44×10^{-2}	2.54×10^{-5}	1.26×10^{-8}	°F
		1.54×10^2	-5.58	-3.84×10^{-2}	1.58×10^{-4}	1.32×10^{-7}	°C
		Temperature coefficient of gage factor = $(0.8 \pm 0.2)/100°C$					
A38AD315	1018 Steel	-9.59×10^1	3.02	-2.89×10^{-2}	8.13×10^{-5}	-5.99×10^{-8}	°F
		-2.64×10^1	2.53	-6.96×10^{-2}	4.29×10^{-4}	-6.29×10^{-7}	°C
		Temperature coefficient of gage factor = $(0.8 \pm 0.2)/100°C$					

Source: Courtesy of Measurements Group, Inc.

where ε_i' = strain indicator reading under test conditions
ε_{TO} = thermal output from the data sheet (microstrain)

The value of ε_i given by Eq. (11.11) may be of sufficient accuracy, but ε_i' was obtained with the actual value of G_F set on the strain indicator, while the thermal output, ε_{TO}, was determined with a different value of G_F (generally 2.0). Therefore, if further accuracy is desired, the thermal output should be corrected for the difference in gage factors. This is a simple procedure, as one may recall from Chapter 5. The correction is

$$\varepsilon_2 = \frac{G_{F1}}{G_{F2}} \varepsilon_1 \tag{11.12}$$

where G_{F1} = gage factor set on the strain indicator
G_{F2} = gage factor of the strain gage
ε_1 = indicated strain on the strain indicator
ε_2 = corrected strain

Since ε_{TO} was obtained by setting a gage factor on the strain indicator other than the actual gage factor, a corrected value of the thermal output may be calculated by using Eq. (11.12). By taking $G_{F1} = G_F^*$, $G_{F2} = G_F$, $\varepsilon_1 = \varepsilon_{TO}$, and $\varepsilon_2 = \varepsilon_{TO}'$, we have the corrected thermal output as

$$\varepsilon_{TO}' = \frac{G_F^*}{G_F} \varepsilon_{TO} \tag{11.13}$$

where G_F^* is the gage factor used in determining the thermal output curve. Using Eq. (11.13), Eq. (11.11) can be rewritten as

$$\varepsilon_i = \varepsilon_i' - \frac{G_F^*}{G_F} \varepsilon_{TO} \tag{11.14}$$

If desired, the entire thermal output vs. temperature curve could be corrected beforehand for the actual gage factor by using Eq. (11.13) and plotting a new curve. Note, however, that this correction would be for reference temperature gage factors and would not account for gage factor variation with temperature.

Example 11.1. Two identical gages are attached to a structure at different locations. The gage data and thermal output curve are shown in Fig. 11.5. A gage factor of 2.05 is set on the strain indicator, the instrument is balanced at 75°F, and the structure is loaded and brought to its test temperature of 300°F. The following readings are obtained:

Gage 1 $\varepsilon_1 = 1180$ μin/in
Gage 2 $\varepsilon_2 = -2060$ μin/in

Use Eq. (11.14) to determine the corrected strain for each gage.

Solution. From Fig. (11.5), $\varepsilon_{TO} = -80$ μin/in ($G_F = 2.0$ for this curve). Using Eq. (11.14),

$$\varepsilon_{1i} = \varepsilon'_{1i} - \frac{G_F^*}{G_F}\varepsilon_{TO} = 1180 - \frac{2.0}{2.05}(-80) = 1258 \text{ μin/in}$$

$$\varepsilon_{2i} = \varepsilon'_{2i} - \frac{G_F^*}{G_F}\varepsilon_{TO} = -2060 - \frac{2.0}{2.05}(-80) = -1982 \text{ μin/in}$$

Note that the gage factor correction amounted to only 2 μin/in.

Correcting for thermal output and gage factor variation

In the preceding example, the thermal output was corrected for the actual gage factor at the reference temperature, since that value was set on the strain indicator when making the strain measurements. If the gage factor is also to be corrected as the temperature changes, then the indicated strain as taken from the strain indicator, as well as the thermal output, must be corrected for the change in gage factor. If the thermal output curve was developed using a gage factor of $G_F^* = 2.0$, but a different value of G_F was set on the strain indicator when the strains were measured, then each strain value (indicated and thermal output) would have to be corrected individually before making the final correction.

As long as a few identical gages were used in a test, this process, while inconvenient, could be used. Generally, however, there would likely be a mixture of gages with different gage factors and thermal output curves. Under these conditions it is simpler to use a method correcting both the indicated strain and thermal output in one operation.

A simple method of correcting for the thermal output and the gage factor variation is available. Since the thermal output curves are developed using a particular gage factor of G_F^* (usually 2.0) at the reference temperature, set the same gage factor on the strain indicator when conducting tests. The strain reading can then be corrected for the effect of the thermal output by using Eq. (11.11). The next step is to correct the actual gage factor to its proper value at the test temperature. This can be done by using Eq. (11.7). Combining these two equations gives the actual strain, ε, at the test temperature. Thus,

$$\varepsilon = (\varepsilon'_i - \varepsilon_{TO})\frac{G_F^*}{G_{FR}\left[1 + \frac{\Delta G_F(\%)}{100}\right]} \qquad (11.15)$$

where ε_{TO} = thermal output at the test temperature
ε_i' = strain indicator reading under test conditions
G_F^* = gage factor at which the thermal output was recorded
G_{FR} = actual gage factor at the reference temperature
$\Delta G_F(\%)$ = percent variation in gage factor at test temperature, with the proper sign

Example 11.2. A strain gage having a gage factor of 2.15 is bonded to a steel structure and a gage factor of 2.0 is set on the strain indicator, which is balanced at room temperature. At the test temperature the following data are recorded:

$$\text{Indicated strain} = 2675 \text{ }\mu\text{in/in}$$

$$\text{Thermal output} = -850 \text{ }\mu\text{in/in}$$

$$\Delta G_F(\%) = 0.75 \text{ percent}$$

Solution. Using Eq. (11.15),

$$\varepsilon = (\varepsilon_i' - \varepsilon_{TO}) \frac{G_F^*}{G_{FR}\left[1 + \dfrac{\Delta G_F(\%)}{100}\right]} = [2675 - (-850)] \frac{2.0}{2.15\left[1 + \dfrac{0.75}{100}\right]}$$

$$= 3255 \text{ }\mu\text{in/in}$$

Note that the reference temperature gage factor, G_{FR}, will be the manufacturer's gage factor, G_F, providing the reference temperature for this test is 75°F.

The preceding method is the easiest to use, since the correction can be made in one step. Other ways may be employed.

1. The manufacturer's gage factor can be set on the strain indicator and the instrument balanced. This would require the indicated strain and the thermal output to be corrected separately for gage factor variation.
2. Set any arbitrary gage factor, or the test temperature gage factor, on the instrument and balance it at the reference temperature.
3. Set the test temperature gage factor on the strain indicator. Bring the test structure to the test temperature and balance the indicator before applying the load. The drawback here, of course, is being sure the structure is stress free when balancing the instrument.

11.4. Strain gage-test material mismatch (2)

Self-temperature-compensated strain gages are manufactured for materials that have coefficients of thermal expansion ranging from 0 to 18 parts per million per degree Fahrenheit (ppm/°F). These values cover a range of commonly used engineering materials. Gages used on plastics, however, are

Table 11.2. Thermal expansion coefficients of common materials

Material	Expansion coefficient	
	(per °F)	(per °C)
Aluminum, 2024-T4, 7075-T6	12.9	23.2
Beryllium copper 25	9.3	16.7
Brass, 30-70	11.1	20.0
Bronze, phosphor (10%)	10.2	18.4
Copper	9.3	16.7
Iron, gray cast	6.0	10.8
Magnesium, AZ-31B	14.5	26.1
Molybdenum	2.2	4.0
Monel	7.5	13.5
Steel, 1008, 1018	6.7	12.1
Steel, 4340	6.3	11.3
Steel, 304 stainless	9.6	17.3
Steel, 316 stainless	8.9	16.0
Tin, pure	13.0	23.4
Titanium, pure	4.8	8.6

Source: reference 2.

manufactured with coefficients of 30, 40, and 50 ppm/°F. If a strain gage compensated for steel, for instance, is used on a second material with a different coefficient of thermal expansion, then the thermal output curve furnished with the gage will no longer be directly applicable. The amount of deviation will depend on the difference in the thermal expansion coefficients of the two materials. Table 11.2 is a partial listing of thermal expansion coefficients for some common engineering materials.

If a strain gage is used on a material for which it is not compensated, and if the difference in thermal expansion coefficients is not too large, then over a limited temperature range near the reference temperature the error produced in using the given thermal output curve may be acceptable. As the difference between the thermal expansion coefficients becomes larger, some steps should be taken to determine the thermal output for the gage when it is used on material for which it is not compensated. This is particularly true for plastics, not only because of the wide variety, but also because of the difference between manufacturers for supposedly the same type of plastic. When such gages are to be used on a specific application, it might be advisable to determine the thermal output curve for that particular strain gage–plastic combination. This can be done in accordance with the procedure discussed earlier.

When a strain gage is applied to a material for which it is mismatched, an approximate correction can be made by using Eq. (11.6), which is the expression for thermal output. If the strain gage is applied to the first

TEMPERATURE EFFECTS ON STRAIN GAGES

material, then the thermal output is

$$\varepsilon_{TO1} = \left[\alpha_{m1} - \alpha_g + \frac{\beta}{G_F}\right]\Delta T = \alpha_{m1}\,\Delta T + \left[\frac{\beta}{G_F} - \alpha_g\right]\Delta T \quad \text{(a)}$$

where the subscript 1 refers to the first material tested. If a similar gage is applied to a second material, then the thermal output is

$$\varepsilon_{TO2} = \left[\alpha_{m2} - \alpha_g + \frac{\beta}{G_F}\right]\Delta T = \alpha_{m2}\,\Delta T + \left[\frac{\beta}{G_F} - \alpha_g\right]\Delta T \quad \text{(b)}$$

where the subscript 2 refers to the second material.

In Eqs. (a) and (b), the last term on the right-hand side is the same, since it refers to the strain gage. Thus,

$$\left[\frac{\beta}{G_F} - \alpha_g\right]\Delta T = \varepsilon_{TO1} - \alpha_{m1}\,\Delta T = \varepsilon_{TO2} - \alpha_{m2}\,\Delta T \quad \text{(c)}$$

If the thermal output, ε_{TO1}, is known, then the thermal output, ε_{TO2}, can be approximated by using Eq. (c). This gives

$$\varepsilon_{TO2} = \varepsilon_{TO1} + (\alpha_{m2} - \alpha_{m1})\,\Delta T \quad (11.16)$$

Equation (11.16) gives a first approximation for the thermal output when the strain gage is applied to a second material. This amounts to rotating the given thermal output curve about the reference temperature. If α_{m2} is larger than α_{m1}, the rotation will be counterclockwise; if α_{m2} is less than α_{m1}, the rotation will be clockwise.

Figure 11.6 shows the thermal output and gage factor variation curves for a strain gage manufactured from foil lot number A12BJ03, Table 11.1. Tests gages of this foil lot were bonded to both 1018 steel and 2024-T4 aluminum to produce the curves shown. The thermal output curve for steel shows large changes with temperature, both above and below the reference temperature. The thermal output curve for aluminum, on the other hand, is much flatter. It is evident that gages made of this foil and bonded to aluminum would give good results in the low-temperature region.

Example 11.3. Using Fig. 11.6, assume that only the thermal output data available is for 1018 steel. Make a first approximation, using Eq. (11.16), for the thermal output for 2024-T4 aluminum. Compare it with the actual curve for aluminum in Fig. 11.6.

Solution. From Table 11.2, the thermal expansion coefficient for 1018 steel is 6.7 ppm/°F and 12.9 ppm/°F for 2024-T4 aluminum. Using Eq. (11.16),

$$\varepsilon_{TO2} = \varepsilon_{TO1} + (\alpha_{m2} - \alpha_{m1})\,\Delta T = \varepsilon_{TO1} + (12.9 - 6.7)\,\Delta T \ \mu\text{in/in}$$

Foil Lot No. A12BJ03

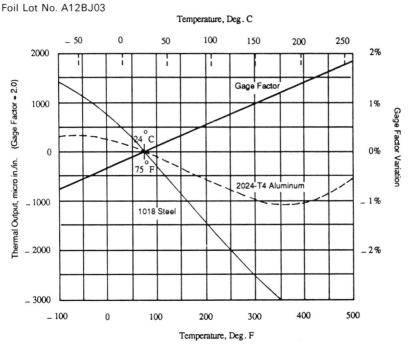

FIG. 11.6. Thermal output and gage factor variation vs. temperature.

where ε_{TO1} is the thermal output for 1018 steel and ε_{TO2} is the calculated value of the thermal output for 2024-T4 aluminum. Determine ε_{TO1} at various temperatures from the curve for 1018 steel, or compute it by using the polynomial coefficients given in Table 11.1. Carrying out the calculations, the results over a temperature range from $-100°F$ to $500°F$ are tabulated.

T, °F	ε_{TO1}, μin/in	ε_{TO2}, μin/in
−100	1398	313
−50	1135	360
0	747	282
50	266	111
100	−278	−123
150	−856	−391
200	−1439	−664
250	−2003	−918
300	−2524	−1129
350	−2979	−1274
400	−3364	−1349
450	−3608	−1283
500	−3747	−1112

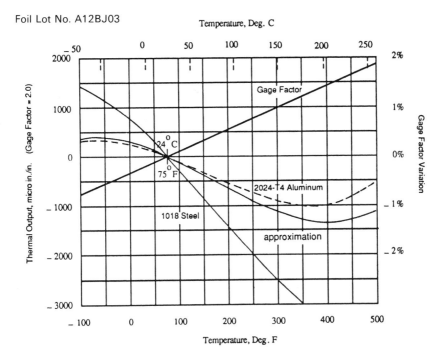

FIG. 11.7. Thermal output and gage factor variation vs. temperature.

The results are plotted in Fig. 11.7. For the values of the thermal expansion coefficients used, a first approximation of the thermal output, ε_{TO2}, shows that the approximated values are positive and slightly higher than the test values at temperatures below 75°F, while the approximated values are negative and below the test values for temperatures above 75°F. The error in the approximate values ranges between 15 and 20 percent up to 300°F, then increases considerably above that temperature. In lieu of other information, however, this correction for the thermal output would not be unreasonable, particularly if the strains imposed by mechanical and thermal stresses were large. It does illustrate, though, the need for an actual test if more exact values of the strain are required.

11.5. Compensating gage

It was pointed out in Chapter 5 that two identical gages placed in adjacent arms of a half-bridge circuit and bonded to the same material would give temperature compensation if both gages were subjected to the same temperature. This also applies to a full-bridge circuit if all four gages, bonded to the same material, were always at the same temperature during the test. Furthermore, in either of these circuits the lead wires must be routed together and be at the same temperature.

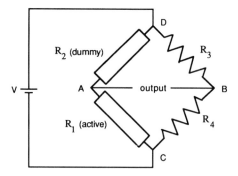

FIG. 11.8. Temperature-compensated circuit using a dummy gage.

A common arrangement for temperature compensation is the use of a dummy gage, identical to the active gage, in a half-bridge circuit. The dummy gage must be bonded to a stress-free piece of material identical to the material on which the active gage is bonded and placed as close to the active gage as possible so that it experiences the same temperature. The lead wires of both the active gage and dummy gage should be routed together. This circuit is shown in Fig. 11.8.

If the bridge is initially balanced, then, from Eq. (5.1),

$$R_1 R_3 = R_2 R_4 \tag{11.17}$$

where R_1 = active stain gage
R_2 = dummy strain gage

If there is a change in temperature only, $\Delta R_1 = \Delta R_2$, since the active and dummy gages are identical, are mounted on the same type of material, and are maintained at identical temperatures. Therefore,

$$(R_1 + \Delta R_1)R_3 = (R_2 + \Delta R_2)R_4 \tag{11.18}$$

This shows that the bridge remains balanced, irrespective of the temperature change, since the influence of temperature has been eliminated, and any unbalance of the bridge will be due solely to mechanical strain on the active gage.

An alternative method would employ the dummy gage and active gage in quarter-bridge circuits. The dummy gage would be placed adjacent to the active gage and records kept of both gage outputs. The thermal output recorded from the dummy gage would be subtracted from the active gage strain. This would, of course, double the required strain gage channels.

If the full bridge is considered, and it is assumed for simplicity that all resistances are identical strain gages of resistance R_g and mounted on the same material, then the circuit output given by Eq. (5.6) is, taking $n = 0$,

$$\Delta E_0 = \frac{V}{4R_g} [\Delta R_1 - \Delta R_2 + \Delta R_3 - \Delta R_4] \quad (11.19)$$

If each of the resistance changes is composed of load-induced change plus temperature-induced change, and all gages have undergone the same temperature change, then Eq. (11.19) can be written as

$$\Delta E_0 = \frac{V}{4R_g} [\Delta R_{1L} + \Delta R_{1T} - \Delta R_{2L} - \Delta R_{2T} + \Delta R_{3L} + \Delta R_{3T} - \Delta R_{4L} - \Delta R_{4T}] \quad (11.20)$$

where   subscript L = load-induced resistance change
subscript T = temperature-induced resistance change

Since $\Delta R_{1T} = \Delta R_{2T} = \Delta R_{3T} = \Delta R_{4T}$, Eq. (11.20) becomes

$$\Delta E_0 = \frac{V}{4R_g} [\Delta R_{1L} - \Delta R_{2L} + \Delta R_{3L} - \Delta R_{4L}] \quad (11.21)$$

This shows that we have temperature compensation for an initially balanced bridge as long as adjacent arms are made up of strain gages of the same type, bonded to the same material, and kept at the same temperature.

Although the basic idea for temperature compensation is simple enough, nevertheless, like many other aspects of strain gage work, attention to detail is essential if optimum results are to be achieved. One must always remember that the observations from what, in other respects, is a practically perfect test can be made quite valueless by faulty temperature compensation. Some points to be kept in mind are the following:

1. The magnitude of the error included in the indicated observation depends upon
   (a) Changes in temperature between active and dummy gages.
   (b) The gages and material upon which they are mounted.
   (c) The operating temperature level.
2. The piece of material upon which the dummy is mounted may be unintentionally subjected to mechanical strain.
3. The thermal connection between the block carrying the dummy gage and the material upon which the active gage is mounted may not be very good so that a temperature differential is set up.
4. There will be a difference between gages of the same lot, particularly at the higher temperatures.

There are conditions under which temperature compensation can be attained by having the dummy gage play an active, rather than a passive,

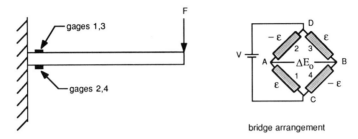

FIG. 11.9. Cantilever beams with strain gages aligned parallel to the longitudinal axis and temperature-compensated.

role in the measurement of stress-induced strains, and at the same time increase the output signal. One such arrangement is shown in Fig. 11.9, which consists of a thin cantilever beam. Since top and bottom surface strains at a given section are equal in magnitude but of opposite sign when the load F is applied, either a half-bridge circuit consisting of gages 1 and 2 (or 3 and 4) in adjacent arms, or a full-bridge circuit, as shown, can be used. This requires, of course, that no thermal gradient exists in the beam and that all gages are at the same temperature. The output signal is either two or four times that of a single gage, depending on the circuit, and will give strains due only to the bending caused by load F.

Figure 11.10 shows a tension member with four gages. Under the requirement that all gages are at the same temperature, the full-bridge circuit shown will be temperature-compensated, and the output signal will be $2(1 + v)$ times the average longitudinal strain. This circuit will read only the

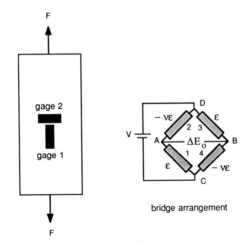

FIG. 11.10. Tension member with strain gages.

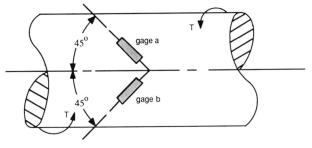

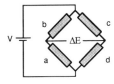

FIG. 11.11. Strain gages arranged for measuring torque.

effect of the axial load F and will cancel bending strains. On the other hand, if only gages 1 and 2 were used in adjacent arms of a half-bridge circuit, the circuit would be temperature-compensated but the relationship between the strains, ε_1 and ε_2, would not be known unless the load, F, was a pure axial load.

A third example is illustrated in Fig. 11.11. If gages a, b, c, and d are arranged in a full-bridge circuit with gages a and b in adjacent arms and gages c and d in arms opposite a and b, respectively, then the bridge (all gages must be at the same temperature), will be temperature-compensated and the output will be a function of the torque only.

As a final example, an instrument was used in which a full-bridge circuit had to be wired to external binding posts that were arranged in a fairly compact rectangle on the instrument's top surface and adjacent to the balancing control. Two active gages were arranged in opposite arms, and since testing took place at room temperature, two precision resistors were placed in the other two arms at the binding posts. When balancing the instrument, drifting was observed and balance could not be obtained. This continued for some time until the operator realized that his arm, when adjusting the instrument, was directly above one of the precision resistors, causing its temperature to change. Replacing the two precision resistors with two dummy gages bonded to a piece of the test material and moving them away from the instrument solved the problem.

358   THE BONDED ELECTRICAL RESISTANCE STRAIN GAGE

Problems

11.1. A strain gage, whose thermal output curve is shown in Fig. 11.5, is bonded to a machine element made of mild steel. The strain indicator is balanced at 75°F using a gage factor of 2.0, then the machine element is heated to the test temperature of 200°F and a load is applied. If $G_F = 2.05$ from the gage package data and the strain indicator shows, after loading, an indicated strain of $\varepsilon'_i = -2465$ µin/in, determine the actual strain, correcting for both the thermal output and the gage factor variation.

11.2. The machine element in Problem 11.1 has its temperature raised to 250°F and a new load is applied. After a reading is taken the temperature is then raised to 400°F and the loading is again changed. For the following data, correct the strains as in Problem 11.1 and determine the difference in strain between the two temperatures:

$T = 250°F \quad \varepsilon'_i = -1875$ µin/in
$T = 400°F \quad \varepsilon'_i = -3628$ µin/in

11.3. A strain gage of the same type as used in Problem 11.1 is bonded to a test specimen and the indicator is balanced at room temperature. A gage factor of 3.0 was inadvertently set on the strain indicator rather than 2.0. When the test specimen was brought to its test temperature of 300°F and loaded, the indicated strain was $\varepsilon'_i = 1936$ µin/in. Determine the actual strain.

11.4. Plot the thermal output curves for the foils given in Table 11.1.

11.5. A strain gage with foil lot number A12BJ01 is to be used on a magnesium member. Determine the approximate thermal output curve, ε_{TO2}, using the thermal output curve for steel for values of ε_{TO1}.

11.6. A strain gage with foil lot number A11BP11 is to be used on a plastic whose thermal expansion coefficient is 40 ppm/°F. Determine the approximate thermal output curve.

11.7. A rectangular rosette has a nominal gage factor of 2.12 for all sections. The thermal output curve associated with this rosette was obtained on 2024-T4 aluminum. The rosette is bonded to a steel test member and a gage factor of 2.0 is set on the instrument, which is then balanced at 75°C. The test member is loaded and brought to a temperature of 300°F. At this temperature the thermal output (for aluminum) is $\varepsilon_{TO} = -950$ µin/in and the gage factor variation is 1 percent. Gage a is aligned along a chosen coordinate axis and all angles are measured from this axis. The following strains were recorded:

$\varepsilon'_a = 875$ µin/in at $\theta = 0°$
$\varepsilon'_b = -1960$ µin/in at $\theta = 45°$
$\varepsilon'_c = -1575$ µin/in at $\theta = 90°$

(a) Determine the principal strains for the uncorrected readings.
(b) Determine the principal strains for the corrected readings.

11.8. A rectangular rosette has a nominal gage factor of 2.145 for all sections. The thermal output curve associated with this rosette was obtained on 1018 steel. The rosette is applied to a steel test member, a gage factor of 2.0 is set on the strain indicator, and the instrument is balanced at 75°F. The test member is loaded and brought to a temperature of $-50°F$, where the thermal output is -500 µin/in and the gage variation factor is -0.5 percent. Gage a is aligned along a chosen coordinate axis and all angles are measured from this axis. The following strains were recorded:

$\varepsilon'_a = -685$ μin/in at $\theta = 0°$
$\varepsilon'_b = -1825$ μin/in at $\theta = 45°$
$\varepsilon'_c = 1335$ μin/in at $\theta = 90°$

(a) Determine the principal strains for the uncorrected readings.
(b) Determine the principal strains for the corrected readings.

11.9. A delta rosette, bonded to aluminum and having a nominal gage factor of 2.08 for all sections, is loaded to its test temperature of 350°F. At this temperature the thermal output is $\varepsilon_{TO} = -90$ μin/in and the gage variation factor is 1.2 percent. The strain indicator was initially balanced with $G_F = 2.0$ at 75°F. Gage a is aligned along a chosen reference axis and all angles are measured from this axis. The following strains were recorded:

$\varepsilon'_a = -535$ μin/in at $\theta = 0°$
$\varepsilon'_b = -845$ μin/in at $\theta = 120°$
$\varepsilon'_c = 180$ μin/in at $\theta = 240°$

(a) Determine the principal strains for the uncorrected readings.
(b) Determine the principal strains for the corrected readings.

REFERENCES

1. Murray, William M. and Peter K. Stein, *Strain Gage Techniques*, Lectures and laboratory exercises presented at MIT, Cambridge, MA: July 8–19, 1963, pp. 95–96.
2. "Temperature-Induced Apparent Strain and Gage Factor Variation in Strain Gages," TN-504, Measurements Group, Inc., P.O. 27777, Raleigh, NC 27611, 1983.
3. "Catalog 500: Part B—Strain Gage Technical Data," Measurements Group, Inc., P.O. Box 27777, Raleigh, NC 27611, 1988.
4. "SR-4 Strain Gage Handbook," BLH Electronics, Inc., 75 Shawmut Road, Canton, MA 02021, 1980.
5. *1986 Annual Book at ASTM Standards*, 1916 Race St., Philadelphia, PA 19103, "Performance Characteristics of Bonded Resistance Strain Gages," Vol. 03.01, Designation: E251-86, pp. 413–428. Copyright ASTM. Reprinted with permission.

12

TRANSDUCERS

12.1. Introduction

When one or more strain gages are used to measure some quantity whose magnitude can be determined by the indication of strain on some load-bearing member, the whole unit is frequently described as a transducer. The load-bearing member may have one, two, three, or more strain gages mounted on it, depending on the quantity to be measured, the precision desired, and the influence of extraneous effects, some of which can be eliminated or reduced to negligible proportions.

In general, the load-bearing elements for transducers may be divided into a few categories which depend upon what is to be measured, as well as being dependent on space requirements. These include direct stress (tension or compression) for the measurement of large forces, members in bending for determining medium or small forces, the indication of torsion, the measurement of fluid pressure, etc. There are many variations of apparatus to accomplish these ends and considerable overlapping of the different procedures, and some of the devices that have been developed for special conditions are exceptionally ingenious. A review of some of the more usual types of transducer is presented in this chapter.

Let us now examine a simple case involving four strain gages (one for each arm of the Wheatstone bridge) as indicated in Fig. 12.1. Here $R_1 = R_2 = R_3 = R_4 = R_g$ and, if idealized bridge conditions are assumed and the bridge is initially balanced, $E = 0$ and $R_1 R_3 = R_2 R_4$. For this case, then, the bridge ratio is

$$a = \frac{R_2}{R_1} = \frac{R_3}{R_4} = 1$$

From Eq. (5.39), the bridge output, ΔE_0, can be written as,

$$\Delta E_0 = \frac{V}{4R_g} [\Delta R_{g1} - \Delta R_{g2} + \Delta R_{g3} - \Delta R_{g4}](1 - n) \qquad (12.1)$$

Since the unit changes in resistance will be small compared to unity, the

TRANSDUCERS

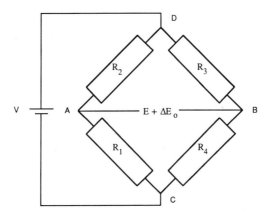

FIG. 12.1. Wheatstone bridge.

nonlinearity term is determined by using Eq. (5.42). Thus,

$$n = \cfrac{1}{1 + \left[\cfrac{2R_g}{\Delta R_{g1} + \Delta R_{g2} + \Delta R_{g3} + \Delta R_{g4}}\right]} \qquad (12.2)$$

Equation (12.1), the general expression for the output from the initially balanced Wheatstone bridge, tells one that the bridge output is directly proportional to the applied voltage, V, and for small unit changes in resistance is nearly proportional to the following:

1. The algebraic difference between the unit changes of resistance in adjacent arms of the bridge.
2. The algebraic sum of the unit resistance changes in opposite arms of the bridge.

In addition

3. If two or more gages happen to be connected in series in one arm of the bridge, the average value of the corresponding strains will be reflected in the bridge output.

This means that if the gages are appropriately located, a bridge output will be produced representing the addition, subtraction, or the average of strains at certain particular locations.

The full bridge

Since Eq. (12.1) for ΔE_0, the output of an initially balanced Wheatstone bridge, contains an equal number of terms with positive and negative signs, this suggests that if one were designing a transducer for full-bridge operation, strain gages with positive and negative gage factors might be considered in order to achieve the maximum output, or indication, per unit load. As it may not be possible, or desirable, to use gages with gage factors of opposite sign, it is fortunate that the same result can be achieved by using like gages and mounting them alternately in regions of tension and compression of the load-carrying element of the transducer. This is common practice, which works best when the strains in tension and compression are of equal magnitude.

Equation (12.1) also tells us that if all gages are alike and the gages in adjacent arms of the bridge are subjected to strains of opposite sign, the indication ΔE_0 will be larger than that from a half bridge. In the same manner, if the strains on the gages are of the same sign, the bridge output will be less than that from a half bridge whose gages in opposite arms are subjected to the two largest strains. In the worst case, there may be no bridge output all ($\Delta E_0 = 0$).

The half bridge

There are certain situations in which it will be more convenient to use a half bridge instead of a full bridge. In this case two active gages are employed instead of four. The two gages represented by R_3 and R_4 in Fig. 12.1 can be replaced by any two equal fixed resistors (for initial bridge balance), or they may be left out. In that event one must be sure that the applied voltage does not send a current through the gages in excess of the normal carrying capacity, which is usually about 30 milliamperes.

With two fixed resistors for R_3 and R_4, $\Delta R_3 = \Delta R_4 = 0$. In this case Eq. (12.1) reduces to

$$\Delta E_0 = \frac{V}{4R_g}[\Delta R_{g1} - \Delta R_{g2}](1 - n) \qquad (12.3)$$

Equation (12.3) is the same as Eq. (4.26) for the potentiometric circuit when $a = 1$. The nonlinearity factor is given by

$$n = \frac{1}{1 + \dfrac{2R_g}{\Delta R_{g1} + \Delta R_{g2}}} \qquad (12.4)$$

The half bridge is particularly useful for bending members with a symmetrical cross section in which the tensile and compressive strains on opposite surfaces are of equal magnitude. For this case $n = 0$ and the half-bridge output becomes

$$\Delta E_0 = \frac{V}{4R_g}[\Delta R_{g1} - \Delta R_{g2}] \qquad (12.5)$$

The quarter bridge

When one wants to measure strain at a single point, or in rare cases, to produce a transducer with a single active gage, the single active gage and three fixed resistors can be used in the Wheatstone bridge. In order to accomplish bridge balance at zero load, one of the three resistors must be equal to that of the gage. The other two, then, can have any resistance values, but they must be equal to each other if a four-arm bridge is being used.

A convenient way to provide the three fixed resistors, although not the only one, is to mount three strain gages, identical to the active one, on a piece of material similar to that upon which the active gage has been mounted. This arrangement gives an equal-arm Wheatstone bridge suitable for both static and dynamic measurements. If the material carrying the three inactive, or dummy, gages has the same thermal characteristics as the material carrying the active gage, the system will be temperature-compensated.

If dynamic measurements only are to be made and temperature compensation is of no concern, it would be preferable to use the potentiometric circuit and change the ballast ratio from 1 to about 10. This increases the circuit efficiency from 50 percent to about 90 percent.

12.2. Axial-force transducers

Tension–compression load cell

This type of transducer, generally called a load cell, is one of the earliest to be used. By proper end connections either tensile or compressive loads, or both, may be measured. The central section where the strain gages are bonded is made long enough so that the strains at the gage location are not affected by the end conditions. This section is designed so that maximum possible strains are reached, yet the member remains within the elastic region and well below the yield point of the material in order to reduce hysteresis. The cross section of the load cell at the strain gage location can have different geometries, with cylindrical, square, or tubular cross sections being common.

Figure 12.2 shows a cylindrical load cell for both tensile and compressive loads. Four gages are shown; gages 1 and 3 are 180° apart and aligned in the longitudinal direction, while gages 2 and 4 are 180° apart and aligned in the transverse direction. The gages are arranged into a full bridge as illustrated, with $\varepsilon_1 = \varepsilon_3 = \varepsilon$ and $\varepsilon_2 = \varepsilon_4 = -\nu\varepsilon$. This bridge arrangement cancels bending strains and is temperature-compensated as long as no temperature gradients exist in the member.

Instead of using Eqs. (12.1) and (12.2) in determining the bridge output, ΔE_0, and the nonlinearity factor, $(1 - n)$, we will return to Eqs. (5.39) and (5.40) for each case. The bridge output, ΔE_0, for this bridge arrangement is

$$\Delta E_0 = V\left(\frac{G_F}{4}\right)[2(1 + \nu)\varepsilon](1 - n) \qquad (12.6)$$

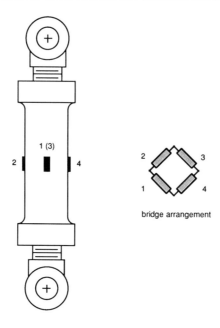

longitudinal gages: 1 and 3
transverse gages: 2 and 4

FIG. 12.2. Load cell for tensile and compressive loads.

The nonlinearity factor, $(1 - n)$, is

$$1 - n = \frac{2}{2 + G_F(1 - v)\varepsilon} \tag{12.7}$$

These results correspond to the results obtained in Example 5.2, where this bridge arrangement was examined.

Equation (12.6) shows that the bridge gives an output of $2(1 + v)$ times as great as that of a single longitudinal gage, considering the nonlinearity factor to be unity. Substituting the nonlinearity factor, $(1 - n)$, given by Eq. (12.7) into Eq. (12.6), the resulting bridge output is

$$\Delta E_0 = \frac{V G_F(1 + v)\varepsilon}{2 + G_F(1 - v)\varepsilon} \tag{12.8}$$

The indicated strain, ε_i, in terms of the actual strain, ε, is

$$\varepsilon_i = \frac{2\varepsilon}{2 + G_F(1 - v)\varepsilon} \tag{12.9}$$

Solving Eq. (12.9) for ε gives

$$\varepsilon = \frac{2\varepsilon_i}{2 - G_F(1 - v)\varepsilon_i} \qquad (12.10)$$

The ratio, $\varepsilon/\varepsilon_i$, is

$$\frac{\varepsilon}{\varepsilon_i} = 1 + \frac{G_F(1 - v)\varepsilon_i}{2 - G_F(1 - v)\varepsilon_i} \qquad (12.11)$$

When using these equations, note that the strains must be entered as $\varepsilon \times 10^{-6}$ in/in.

These equations show that the bridge output is nonlinear, since all gages do not see the same strain magnitudes. The transverse strains, because of the Poisson effect, are about 30 percent of the longitudinal strains. Equation (12.10) shows that the actual strain will be larger than the indicated strain for a tensile force, while the converse is true for a compressive force. The bridge nonlinearity at a strain level of 1000 μin/in, for either tension or compression strains, is about 0.07 percent.

Another nonlinearity factor is present in the geometry of the load cell in that the area changes under load. This can be approximated for a round cross section by considering the change in diameter due to the Poisson effect. The diameter, d, at any load within the elastic region, is

$$d = d_0(1 - v\varepsilon) \qquad (12.12)$$

where d_0 is the diameter at no load. The bridge and geometry nonlinearities, however, are offsetting, with the bridge nonlinearity being the higher of the two.

It may be desirable, at times, to have only gages 1 and 3 active. In that case, precision resistors could be used in arms 2 and 4 to complete the full bridge; however, a more convenient way would be to bond two strain gages to material similar to the transducer and use them as dummy gages. This would give temperature compensation provided all gages remained at the same temperature. The bridge output, ΔE_0, would be double that of a single gage and would be nonlinear.

Ring-type load cell

The proving ring has been in use for years as a standard for the calibration of tensile-testing machines. The diametral deflection of the ring is a measure of the applied load, where the deflection is measured by means of a precision micrometer. The thickness of the cross section, which is the difference between the inner and outer radii, is small compared to the mean radius.

Rather than measure diametral deflection, strain gages may be bonded

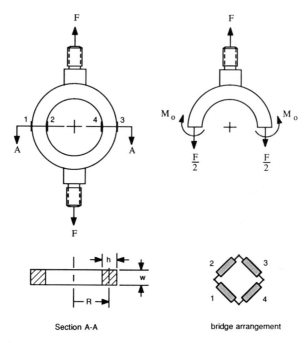

FIG. 12.3. Ring-type load cell.

to the ring, as shown in Fig. 12.3, arranged into a full bridge, and the bridge output used as a measure of the applied load. An axial force and bending moment act at the section containing the strain gages, as illustrated in the free-body diagram of the upper half of the load cell. Since each gage is subjected to the same axial strain due to the axial force, $F/2$, these strains are canceled and the bridge responds only to the strains induced by the bending moment, M_0. For the tensile load, F, gages 1 and 3 will be in compression due to bending, and gages 2 and 4 will be in tension. The converse will be true for a compressive load. Furthermore, this bridge arrangement gives full temperature compensation.

The strain at the gages may be estimated from the bending stresses. The moment, M_0, is

$$M_0 = \frac{FR}{2}\left(1 - \frac{2}{\pi}\right) \tag{12.13}$$

Since the cross section is rectangular, the strain due to M_0 is

$$\varepsilon = \frac{6M_0}{wh^2 E} = \frac{3FR}{wh^2 E}\left(1 - \frac{2}{\pi}\right) \tag{12.14}$$

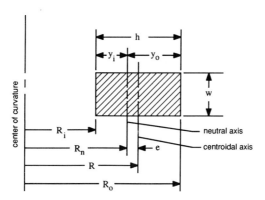

FIG. 12.4. Curved beam cross section.

This reduces to

$$\varepsilon = \frac{1.09FR}{wh^2 E} \qquad (12.15)$$

where $h \ll R$.

If the difference between the inner and outer radii increases, the load cell will no longer be considered a thin ring. In this case, a better estimate of the bending strains due to M_0 may be obtained from curved beam theory (1). In the curved beam, the centroidal axis and neutral axis do not coincide, with the neutral axis shifted inward towards the center of curvature. The geometry, shown in Fig. 12.4, has the following notation:

R_i = radius of inner fiber

R_n = radius of neutral axis

R = radius of centroidal axis

R_0 = radius of outer fiber

e = distance between the centroidal and neutral axes

h = section thickness, $R_0 - R_i$

w = section width

y_i = distance from the neutral axis to the inner fiber

y_0 = distance from the neutral axis to the outer fiber

The bending moment, M_0, now becomes

$$M_0 = \frac{FR}{2}\left(1 - \frac{2}{\pi} + \frac{2e}{\pi R}\right) \qquad (12.16)$$

Since $e = R - R_n$, the radius, R_n, must be computed. It is

$$R_n = \frac{h}{\ln\left(\dfrac{R_0}{R_i}\right)} \qquad (12.17)$$

The stresses at the inner and outer fibers are

$$\sigma_i = \frac{M_0 y_i}{AeR_i} \qquad (12.18)$$

$$\sigma_0 = \frac{M_0 y_0}{AeR_0} \qquad (12.19)$$

where A is the cross-sectional area. For the sense of M_0 shown in Fig. 12.3, σ_i will be a tensile stress and σ_0 will be a compressive stress. The corresponding strains are

$$\varepsilon_i = \frac{\sigma_i}{E} = \frac{M_0 y_i}{AeR_i E} \qquad (12.20)$$

$$\varepsilon_0 = \frac{\sigma_0}{E} = \frac{M_0 y_0}{AeR_0 E} \qquad (12.21)$$

For a tensile force, F, acting on the load cell, we see that $\varepsilon_1 = \varepsilon_3 = -\varepsilon_0$ and $\varepsilon_2 = \varepsilon_4 = \varepsilon_i$ and so the bridge output is nonlinear. For the development of these equations, the reader is referred to Reference 1.

The expressions for the strains at the gage locations are estimates, since the bosses where the load is applied have a stiffening effect. They can be used for design, but calibration is essential.

12.3. Simple cantilever beam

A device often used as a transducer is the cantilever beam. Among its applications, it may be used to measure force, to serve as a comparator, or to determine deflections in areas not readily accessible to other instruments.

Single active gage

The most basic application uses a single active gage and three fixed resistors in the Wheatstone bridge. The beam can be made from a piece of uniform bar stock of rectangular cross section, with the strain gage mounted near the fixed end on the longitudinal center line of the upper surface. The force, F, can then be measured after a suitable calibration of the beam has been performed. This device is subject to the following limitations:

1. The output will be low because only one arm of the Wheatstone bridge is active.
2. The line of action of the applied force, F, must always remain parallel to itself (including calibration) and at the same distance, L, from the center of the gage.
3. Unless a self-temperature-compensated strain gage is used to match the thermal properties of the beam, the apparatus can only be used precisely at the temperature of calibration, otherwise serious errors may occur.
4. No compensation is provided for forces (if any), other than F, which may produce lateral bending, torsion, or direct axial thrust.

The bridge output is given as

$$\Delta E_0 = \frac{G_F V \varepsilon}{4}(1-n) \tag{12.22}$$

The nonlinearity factor is

$$1 - n = \frac{1}{1 + \frac{1}{2}G_F \varepsilon} \tag{12.23}$$

The indicated strain, ε_i, in terms of the actual strain, ε, is

$$\varepsilon_i = \frac{2\varepsilon}{2 + G_F \varepsilon} \tag{12.24}$$

Solving Eq. (12.24) for ε produces

$$\varepsilon = \frac{2\varepsilon_i}{2 - G_F \varepsilon_i} \tag{12.25}$$

The ratio of the actual strain to the indicated strain is

$$\frac{\varepsilon}{\varepsilon_i} = 1 + \frac{G_F \varepsilon_i}{2 - G_F \varepsilon_i} \tag{12.26}$$

Longitudinal and transverse gages on the same side

There may be some cases for which two gages must be mounted on the same side of the beam. Here one can take advantage of the Poisson effect, which produces a lateral strain of opposite sign from the axial strain. This arrangement uses one gage mounted in a longitudinal direction and a second gage bonded in the transverse direction. This arrangement can be used as a half bridge when the gages are connected in adjacent arms. With the two gages at the same temperature, the bridge output will be automatically temperature-compensated. However, all other characteristics of the single gage application also apply to this. Depending upon the value of Poisson's ratio for the bar material, the output for this bridge will be about 30 percent greater than that of a single gage.

For this bridge, the longitudinal strain is $\varepsilon_a = \varepsilon$ and the transverse strain is $\varepsilon_n = -v\varepsilon$. With the gages placed in bridge arms 1 and 2, the unit resistance changes are

$$\frac{\Delta R_1}{R_1} = F_a(\varepsilon_a + K\varepsilon_n) = F_a(1 - vK)\varepsilon \qquad (a)$$

$$\frac{\Delta R_2}{R_2} = F_a(\varepsilon_n + K\varepsilon_a) = F_a(K - v)\varepsilon \qquad (b)$$

From Eq. (7.21),

$$F_a = \frac{G_F}{1 - v_0 K} \qquad (7.21)$$

Using the values of $\Delta R_1/R_1$, $\Delta R_2/R_2$, and F_a from Eqs. (a), (b), and (7.21), the bridge output, ΔE_0, is

$$\Delta E_0 = \frac{V}{4}\left(\frac{\Delta R_1}{R_1} - \frac{\Delta R_2}{R_2}\right)(1 - n)$$

$$= V\left(\frac{G_F}{4}\right)\left(\frac{1}{1 - v_0 K}\right)(1 - vK + v - K)(\varepsilon)(1 - n) \qquad (12.27)$$

The nonlinearity factor, $(1 - n)$, can be written as

$$1 - n = \frac{1}{1 + \frac{1}{2}\left(\frac{\Delta R_1}{R_1} + \frac{\Delta R_2}{R_2}\right)}$$

$$= \frac{1}{1 + \frac{G_F}{2}\left(\frac{1}{1 - v_0 K}\right)(1 - vK + K - v)\varepsilon} \qquad (12.28)$$

If K is ignored ($K = 0$), then ΔE_0 and $(1 - n)$ revert to

$$\Delta E_0 = V\left(\frac{G_F}{4}\right)(1 + v)(\varepsilon)(1 - n) \tag{12.29}$$

$$1 - n = \frac{1}{1 + \dfrac{G_F}{2}(1 - v)\varepsilon} \tag{12.30}$$

Two longitudinal gages on opposite surfaces

Provided that the two sides of the beam are free from any obstruction, a considerable advantage in output can be obtained by mounting the gages back to back on opposite surfaces. Because they are subjected to strains of equal magnitude but of opposite sign, they can be placed in adjacent arms of a half bridge. Since $\varepsilon_1 = \varepsilon$ and $\varepsilon_2 = -\varepsilon$, the bridge output, ΔE_0, will be linear and is

$$\Delta E_0 = \frac{VG_F\varepsilon}{2} \tag{12.31}$$

This arrangement is compensated for temperature changes provided both gages are maintained at like temperatures. It is also compensated for direct axial thrust, which will produce the same resistance changes in both gages, although axial thrust will produce bridge nonlinearity. For metallic gages, the variation of this nature in $(1 - n)$ will usually be small enough to be neglected.

Full bridge, two gages back to back on opposite surfaces

Four gages mounted back to back in pairs on opposite surfaces of the beam and arranged as a full bridge will give the largest bridge output. Since $\varepsilon_1 = \varepsilon_3 = \varepsilon$ and $\varepsilon_2 = \varepsilon_4 = -\varepsilon$, the bridge output, ΔE_0, will be linear and is

$$\Delta E_0 = VG_F\varepsilon \tag{12.32}$$

Provided all gages are maintained at the same temperature, this arrangement gives temperature compensation. As in the two-arm bridge, the strains caused by axial thrust will be nullified, although these strains will produce bridge nonlinearity.

There are several comments in order concerning the cantilever beam.

1. If the line of action of the force remains parallel to itself, the moment at the gage section decreases because of the shortening of the moment arm due to the curvature of the beam.

2. The strain along the length of the strain gage is not constant. This can be alleviated by designing a constant-stress beam of uniform thickness and a triangularly shaped width, or uniform width and a parabolically shaped thickness. The load in each case is applied at the narrowest point of the beam. For a tapered width beam, see Fig. 7.5.

12.4. Bending beam load cells

A variety of load cells can be constructed by using different configurations of beams. Whether or not all make satisfactory load-measuring devices must be determined by a combination of analysis and testing. Several different types, among the many available, will be discussed here.

Fixed-end beam

One may consider a beam with fixed ends and center loading, shown in Fig. 12.5, for use as a load cell. One placement of the strain gages and the bridge arrangement are shown, with gages 1 and 3 being in compression and gages 2 and 4 being in tension for the loading illustrated. Because the beam is symmetrical, the reactions at each built-in end are identical. Furthermore, the supports are very stiff compared to the beam. Expressions for the moments and reactive forces at the fixed ends may be developed or found in a text on mechanics of materials.

Since the ends of the beam are constrained from moving laterally, this influence will not be accounted for in the expressions for the end reactions. Because of this constraint, a horizontal force is produced that affects bending

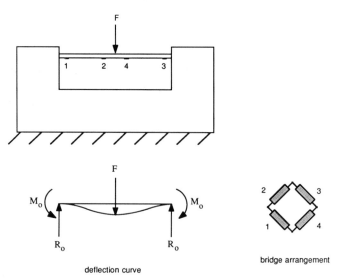

FIG. 12.5. Beam with fixed ends.

moments in the beam and therefore the deflection curve. While the strains produced by this force are canceled by the bridge arrangement, nevertheless, they will produce nonlinearity in the bridge output.

Two fixed-end beams

A beam-type load cell that overcomes the lack of lateral movement of the fixed ends is shown in Fig. 12.6. During loading, either in tension or compression, the ends are free to move laterally and thus eliminate horizontal forces on the beams. The central section where the load is applied and the two end supports are very stiff compared to the thinned beam sections, and so practically all of the deflection is produced in the thin sections. This load cell, however, has twice the deflection of the single beam shown in Fig. 12.5.

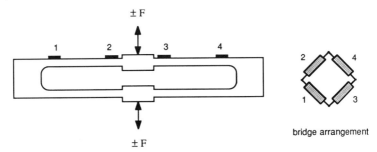

FIG. 12.6. Dual-beam load cell.

One arrangement of four strain gages for a full bridge is shown. Gages 1 and 4 are subjected to strains of like sign, and gages 2 and 3 are subjected to strains of like sign. If the load is compressive, for instance, gages 1 and 4 will have tensile strains and gages 2 and 3 will have compressive strains.

When designing this load cell, one wants to estimate the strain level at the gage locations. In order to accomplish this, a free-body diagram of the upper beam is shown in Fig. 12.7. Sections $A-B$ and $C-D$ have the same moments of inertia, while section $B-C$ has a much larger moment of inertia in order to reduce the deflection in this section. Although section $B-C$ will deflect slightly (dependent on the value of its moment of inertia compared to section $A-B$), most of the deflection will occur in sections $A-B$ and $C-D$. Since the beam is symmetrical, the reactions at both ends are equal; however, the beam is statically indeterminate to the first degree, since M_0 is unknown. Knowing the slope of the deflection curve is horizontal at point A and at the center under the load, M_0 in terms of the beam dimensions and the load, F, can be computed. The moment-area method, for instance, may easily be used.

If the gages can be located so the strains at gages 1 and 4 are equal in

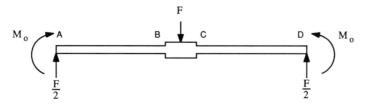

FIG. 12.7. Free-body diagram of the upper beam.

magnitude but opposite in sign to gages 2 and 3, the bridge output, ΔE_0, will be linear. Because of the lateral movement of the ends, though, there may be some nonlinearity effect because of the slight change in geometry. Also, if all gages are subjected to the same temperature, the bridge will be temperature-compensated.

S-shaped dual beam

The S-shaped, dual beam load cell uses two beams attached to sections whose stiffness is much larger than that of the beams. It is used for direct tensile or compressive loads, as shown in Fig. 12.8. For best results, the load cell should be machined from a solid block of material. Eccentric loading errors are minimized and the gages are easily protected.

Figure 12.9 shows the loading (for a tensile force) acting on one of the beams, along with the deflection curve, for estimating the strains in the beam.

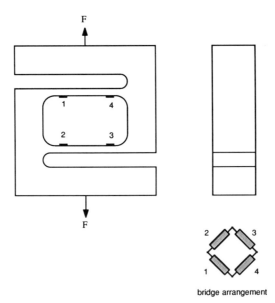

FIG. 12.8. S-shaped, dual-beam load cell.

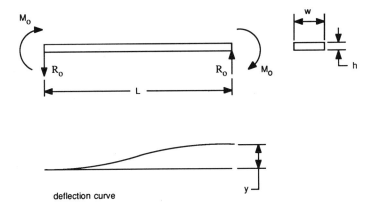

FIG. 12.9. Fixed-end beam with end displacement.

From symmetry, the reactions at each end of the beam are identical. During loading, the right-hand side of the beam, relative to the left, moves up through a distance, y, thereby producing the reactions shown. The shear force, R_0, and the movement, M_0, can be determined in terms of the deflection, y, and the beam dimensions. Once again, the moment-area method lends itself to the determination of the reactions. The values of R_0 and M_0 in terms of the beam deflection and beam dimensions are

$$R_0 = \frac{12EIy}{L^3} \qquad (12.33)$$

$$M_0 = \frac{6EIy}{L^2} \qquad (12.34)$$

For the gage placements shown, gages 1 and 3 are in tension and gages 2 and 4 are in compression for a tensile load, while the converse applies for a compressive load. Furthermore, as long as all gages are subjected to the same temperature, the bridge will be temperature-compensated. The nonlinearity of the bridge will depend on the values of the strains at each gage.

There are a number of other beam-bending load cells in use or that could be constructed for laboratory use. An examination of a manufacturer's catalog will show beam-bending load cells are used for applications involving loads from less than 1-lb at the low end to about 1000 lb at the upper end. For loads in excess of 1000 lb, other designs are generally utilized. For an excellent discussion of strain-gage-based transducers, see Reference 2.

12.5. Shear beam load cell

The shear beam load cell, usually designed for high loads, is in the form of a cantilever beam with a cross section large enough that the beam deflection

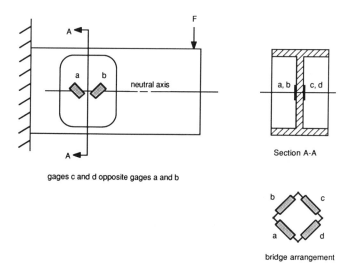

FIG. 12.10. Shear beam load cell.

is kept to a minimum. Since the bending stresses on the outer surface would be quite low under these conditions, a recess is machined on each side so the cross section formed resembles an I-beam. Here, most of the bending moment is resisted by the flanges, while the vertical shear is carried by the web. The shearing stress is maximum at the neutral axis, and so the dimensions of the I-beam section can be chosen so that the strains will produce a desired bridge output. Such a load cell is shown in Fig. 12.10.

Because there is pure shear at the neutral axis, the principal stresses, and therefore the principal strains, are at $\pm 45°$ from the neutral axis. Two pairs of strain gages, bonded back to back on opposite surfaces of the recess, can be centered across the neutral axis at $\pm 45°$. Although the gages are subjected to a slight amount of bending strain because they extend on either side of the neutral axis, this effect tends to be self-canceling. A better arrangement, for instance, would use a two-element 90° gage, generally used for torque measurements, on each surface. Choosing a torque gage with electrically independent elements whose grids are $\pm 45°$ to the gage longitudinal axis allows the gage to be bonded so that its longitudinal axis coincides with the neutral axis of the beam. In this manner, the elements of the gage will experience bending strains of the same magnitude but of opposite sign. When the gages are arranged into a full bridge, the bending strains will cancel. Furthermore, this arrangement will also cancel any bending due to side loading.

As long as the load is to the right side of the recess, as shown in Fig. 12.10, the bridge output is relatively insensitive to the point of load application. Although it is desirable to keep the beam as short as possible,

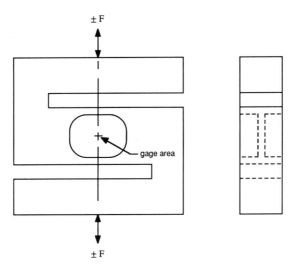

FIG. 12.11. Folded shear beam.

the load must be far enough from the recess that its localized effects will not influence the bridge output.

The shearing stress at the neutral axis, and thereby the shearing strain, must be determined in order to estimate the bridge output, ΔE_0, for a given load. The equation for shear stress in the web, which can be found in texts on mechanics of materials, is

$$\tau = \frac{VQ}{It} \qquad (12.35)$$

where V = vertical shear force on the section
I = moment of inertia about the neutral axis
t = web thickness
Q = first moment of the area above the neutral axis

The principal stresses, at 45° on either side of the neutral axis, are equal in magnitude but opposite in sign, giving $\sigma_1 = -\sigma_2 = \tau$. The principal strains are

$$\varepsilon_1 = \frac{\tau(1 + \nu)}{E} = -\varepsilon_2 \qquad (12.36)$$

The shear beam load cell may also be constructed so that its profile is S-shaped, as shown in Fig. 12.11. This configuration is also referred to as a folded shear beam by some manufacturers. The line of action of the applied force goes through the center of the strain gage bridge, thus eliminating bending at that section.

12.6. The torque meter

Although many different types of torque meters have been devised, probably the most common consists of a shaft of circular cross section with four like strain gages mounted at 45° to the axis of the shaft. Care must be taken in assuring the gages are mounted at precisely 45°, and that companion gages subjected to tension (or compression) are bonded exactly opposite each other. A typical torque meter is shown in Fig. 12.12.

In constructing a torque meter, one should be aware of its characteristics, which are stated as follows:

1. The unit is automatically compensated for changes in temperature. This is due to the fact that a uniform temperature change will produce equal resistance changes in all four arms of the bridge, thereby producing no change in the condition of balance.
2. Theoretically, the instrument will not respond to the effects of axial thrust, if such should exist. This is because axial thrust will produce equal resistance changes in all four arms of the Wheatstone bridge; therefore, there will be no change in the condition of balance.
3. There will be no response to bending, if such should occur, because the resistance change in the two front gages will be equal in magnitude but opposite in sign to the resistance changes in the two gages at the back.
4. The output of the bridge will be linear with respect to the torque, T, because the nonlinearity factor is $(1 - n) = 1$; that is, $n = 0$.

Due to the location of the gages, torque produces resistance changes in

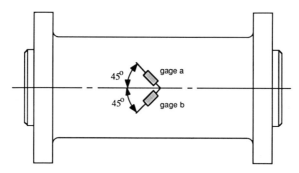

gages c and d diametrically opposite gages a and b, respectively

bridge arrangement

FIG. 12.12. Torque meter.

each bridge arm. Since the gages are alike, the bridge ratio is $a = 1$, and the resistance changes are

$$\Delta R_{g1} = -\Delta R_{g2} = \Delta R_{g3} = -\Delta R_{g4}$$

Using these resistance changes, the bridge output, ΔE_0, given by Eq. (12.1) is

$$\Delta E_0 = \frac{V}{4R_g}[\Delta R_{g1} - (-\Delta R_{g2}) + \Delta R_{g3} - (-\Delta R_{g4})](1-n)$$

This reduces to

$$\Delta E_0 = V\left(\frac{\Delta R_g}{R_g}\right)(1-n) \qquad (12.37)$$

Also, substituting these resistance changes into Eq. (12.2) shows that the nonlinearity term, n, is zero, giving a linear bridge output.

The case can be examined where there is not only torque but an axial load acting on the meter. Under these conditions, the resistance change in each gage is

$$\Delta R_1 = \Delta R_{1T} + \Delta R_{1A} = \Delta R_T + \Delta R_A \qquad \text{(a)}$$

$$\Delta R_2 = \Delta R_{2T} + \Delta R_{2A} = -\Delta R_T + \Delta R_A \qquad \text{(b)}$$

$$\Delta R_3 = \Delta R_{3T} + \Delta R_{3A} = \Delta R_T + \Delta R_A \qquad \text{(c)}$$

$$\Delta R_4 = \Delta R_{4T} + \Delta R_{4A} = -\Delta R_T + \Delta R_A \qquad \text{(d)}$$

The subscripts T and A refer to torque and axial thrust, respectively. The bridge output now becomes

$$\Delta E_0 = \frac{V}{4R_1}[\Delta R_T + \Delta R_A - (-\Delta R_T + \Delta R_A)$$
$$+ \Delta R_T + \Delta R_A - (-\Delta R_T + \Delta R_A)](1-n)$$

Simplifying, this reduces to

$$\Delta E_0 = V\left(\frac{\Delta R_T}{R_1}\right)(1-n) \qquad (12.38)$$

Equation (12.38) shows that the bridge output does not change because of the axial load, providing the nonlinearlity factor, $(1-n)$, is unity. The nonlinearity factor, however, must be examined to see if it affects the bridge output. Rather than use the approximate expression for the nonlinear term given by Eq. (12.2), Eq. (5.40) will be used. If the resistance changes given by Eqs. (a), (b), (c), and (d) are substituted into Eq. (5.40), it will be found the nonlinearity factor reduces to

$$1 - n = \frac{1}{1 + (\Delta R_A/R_1)} \tag{12.39}$$

Equation (12.39) shows that the nonlinearity term will have an effect, although minor, when an axial force is combined with torsion. This means that the nonlinearity term will have a different value for each different combination of axial thrust and torque.

If the torque meter is used in a stationary application, the lead wires from the strain gage bridge may be readily connected to a suitable indicator. For limited angular motion at a low rate of rotation, the lead wires may be of such a length that windup is permitted. If, however, the torque meter rotates, then some arrangement must be made to bring the signal to the instrumentation, either through slip rings, radiotelemetry, or some other method.

12.7. The strain gage torque wrench

Mechanical torque wrenches have been in use for many years, with the most common having a pointer attached to the head end and extending over a scale, calibrated to read torque, attached at the handle end. The handle is pin-connected to the wrench body so that the force is transmitted to the body through the pin, thereby keeping the force at a fixed point. Rather than use a pointer and scale, strain gages could be bonded to the wrench body near the head end, the system calibrated, and the torque read on a suitable strain indicator. The force, however, would still have to be applied at a fixed point. Can one, then, arrange strain gages so that the indicator reading is a measure of the torque and independent of the point of force application?

Meier (3) investigated this problem and arrived at a bridge arrangement so that the bridge output was linearly related to the torque at the wrench-head center line, yet was independent of the point of force application. Figure 12.13 shows the wrench, the strain gage placement, and the bending moment diagram.

The bending moment is maximum at section 3 where the torque is being applied. Since it is impractical to measure bending at this section, the bending moment, M_3, can be related to the bending moments, M_1 and M_2, at sections 1 and 2, respectively. Any forces and moments applied to the wrench must be to the left of section 1, with none applied between sections 1 and 2. The

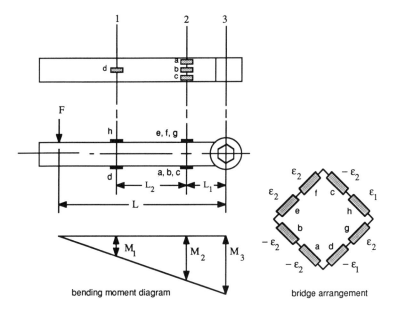

FIG. 12.13. Gage arrangement on torque wrench for direct torque measurement. (From ref. 3.)

moments of the three sections are

$$M_3 = FL \tag{a}$$

$$M_2 = F(L - L_1) = M_3 - FL_1 \tag{b}$$

$$M_1 = F(L - L_1 - L_2) = M_3 - F(L_1 + L_2) \tag{c}$$

If Eqs. (b) and (c) are solved for the force, F, then

$$F = \frac{M_3 - M_2}{L_1} = \frac{M_3 - M_1}{L_1 + L_2} \tag{d}$$

From Eq. (d), M_3 is

$$M_3 = M_2 + \frac{L_1}{L_2}(M_2 - M_1) \tag{12.40}$$

Taking $L_2 = 2L_1$,

$$M_3 = \tfrac{1}{2}(3M_2 - M_1) \tag{12.41}$$

Since the bending moments, M_1 and M_2, can be expressed in terms of strain,

$$M_1 = ZE\varepsilon_1 \qquad (e)$$

$$M_2 = ZE\varepsilon_2 \qquad (f)$$

where Z is the section modulus for bending and E is the modulus of elasticity. Substituting the values of M_1 and M_2, given by Eqs. (e) and (f), respectively, into Eq. (12.41) produces

$$M_3 = \frac{EZ}{2}(3\varepsilon_2 - \varepsilon_1) \qquad (12.42)$$

The eight strain gages bonded to the wrench can be arranged into a full bridge to produce the operation indicated in parentheses in Eq. (12.42). For the bending moment diagram shown, gages a, b, and c will experience a compressive strain of $-\varepsilon_2$, while gage d will have a compressive strain of $-\varepsilon_1$. Gages e, f, and g will have tensile strains of ε_2, while gage h will have a tensile strain of ε_1. The bridge output, ΔE_0, for the given strains is

$$\Delta E_0 = -\frac{VG_F}{2}(3\varepsilon_2 - \varepsilon_1) \qquad (12.43)$$

Comparing Eqs. (12.42) and (12.43), it can be seen that the bridge output is proportional to the torque, M_3.

The torque wrench can easily be calibrated using known weights. Meier found the calibration curve of indicator reading against the torque, M_3, to be very consistent and straight over a wide range of level arms and applied weights. While the unit has been described as a torque wrench, it can be applied to other situations requiring a torque arm. One application, for instance, would be the determination of reaction torque for a cradle-mounted piece of equipment, using the device described as the arm for the measurement of torque about the cradle axis.

12.8. Pressure measurement

The measurement of pressure is often required during the course of a project. There are many devices available using pressure force to act on an elastic mechanical element, thereby causing it to deflect. Among these elements are the Bourdon tube with different configurations, diaphragms, bellows, straight tubes, and flattened tubes. These elements are used in conjunction with some sort of measuring system, so their deflection is an indication of pressure. If the pressure-measuring device is to be constructed rather than purchased, there are several options, depending on project requirements.

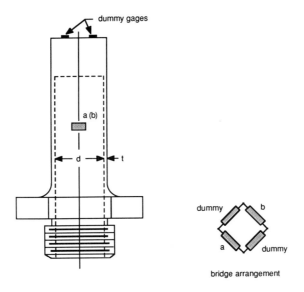

FIG. 12.14. Thin-walled pressure transducer.

Thin-walled cylindrical tube

For static or slowly varying pressures, a simple and effective method is to construct a thin-walled cylindrical tube, with two gages mounted in the circumferential (hoop) direction. A full bridge can be used by placing these two active gages in opposite bridge arms, then completing the bridge by bonding two dummy gages on an unstrained piece of similar material placed adjacent to the cylinder, or by extending the solid upper end of the cylinder and bonding the dummy gages to this unstrained portion. If all gages are maintained at the same temperature, the bridge will be temperature-compensated. A typical transducer of this type is shown in Fig. 12.14.

The circumferential strain, ε_H, and the longitudinal strain, ε_L, are

$$\varepsilon_H = \frac{pd}{4tE}(2 - v) \qquad (12.44)$$

$$\varepsilon_L = \frac{pd}{4tE}(1 - 2v) \qquad (12.45)$$

where p = internal pressure
 d = inner diameter
 t = wall thickness
 E = modulus of elasticity
 v = Poisson ratio

With only the circumferential strain gages active and in opposite bridge arms, the bridge output, ΔE_0, is

$$\Delta E_0 = \frac{VG_F}{4}(2\varepsilon_H) = \frac{VG_F pd}{8tE}(2-v)(1-n) \qquad (12.46)$$

The nonlinearity factor, $(1-n)$, is

$$1 - n = \frac{2}{2 + G_F \varepsilon_H} \qquad (12.47)$$

Although the circumferential stress is twice the longitudinal stress, the same is not true for the strains. Using Eqs. (12.44) and (12.45), the ratio of strains is

$$\frac{\varepsilon_H}{\varepsilon_L} = \frac{2-v}{1-2v} \qquad (12.48)$$

For steel with $v = 0.3$, $\varepsilon_H = 4.25\varepsilon_L$. If all gages were bonded to the cylinder, two circumferential and two longitudinal, and arranged into a fully active bridge, the bridge output would be reduced by approximately 24 percent.

This type of pressure transducer is best used at relatively high pressure for a compact design. As Eq. (12.44) indicates, the diameter, d, must be increased and/or the wall thickness, t, decreased in order to obtain reasonable strain readings for lower pressures. Once the transducer dimensions have been chosen, however, it can be constructed and calibrated by using a deadweight tester, for instance. The frequency response can be improved by reducing the internal volume through the insertion of a solid plug, thus reducing the flow caused by pressure variation.

Diaphragm pressure transducer

A second type of pressure transducer uses a diaphragm. The diaphragm may be made from a thin sheet of flat material clamped between two elements of the transducer body, or it can be machined as an integral part of the transducer body. The information outlined here may be used to arrive at a preliminary design, but the final output of the instrument will have to be obtained by calibration. In determining the characteristics of the diaphragm, the following restrictions apply:

1. The diaphragm is rigidly clamped at its outer edge.
2. The diaphragm is flat and of uniform thickness.
3. The deflection of the center will not exceed one-half of the diaphragm thickness.
4. The natural frequency of the diaphragm must be high enough to respond adequately to fluctuating pressure.

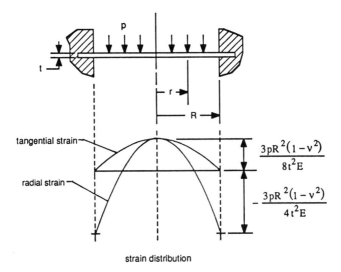

FIG. 12.15. Clamped circular plate with strain distribution.

In determining the characteristics of the diaphragm, the analysis for a uniformly loaded thin circular plate clamped at the edge can be used (4). The plate and its loading are shown in Fig. 12.15. The pressure acts on the upper surface and the strain gages are bonded to the under surface.

The tangential bending moment, M_t, and the radial bending moment, M_r, at any radius are

$$M_t = \frac{p}{16}[R^2(1+v) - r^2(1+3v)] \qquad (12.49)$$

$$M_r = \frac{p}{16}[R^2(1+v) - r^2(3+v)] \qquad (12.50)$$

The corresponding stresses are

$$\sigma_t = \frac{3p}{8t^2}[R^2(1+v) - r^2(1+3v)] \qquad (12.51)$$

$$\sigma_r = \frac{3p}{8t^2}[R^2(1+v) - r^2(3+v)] \qquad (12.52)$$

The strains follow as

$$\varepsilon_t = \frac{3p(1-v^2)}{8t^2E}(R^2 - r^2) \qquad (12.53)$$

$$\varepsilon_r = \frac{3p(1-v^2)}{8t^2E}(R^2 - 3r^2) \qquad (12.54)$$

The strains given by Eqs. (12.53) and (12.54) are also plotted in Fig. 12.15. At $r = 0$ the tangential and radial strains are identical and expressed as

$$\varepsilon_t = \varepsilon_r = \frac{3pR^2(1 - v^2)}{8t^2 E} \qquad (12.55)$$

At $r = R$ the tangential strain is zero and the radial strain becomes

$$\varepsilon_r = -\frac{3pR^2(1 - v^2)}{4t^2 E} \qquad (12.56)$$

Equations (12.55) and (12.56) show where the gages should be placed. A pair of stacked orthogonal gages could be used at the center, while two radial gages could be placed as close to the boundary as possible, then arranged into a full bridge. Although the bridge would be temperature compensated, an examination of Eq. (5.40), using these strains, shows that the nonlinearity factor is not zero.

Special gages, Fig. 12.16, have been designed for use with diaphragms (5). This gage takes advantage of the strain distribution shown in Fig. 12.15. Since the tangential strain decreases more slowly with increasing radius than does the radial strain, the central element is designed to measure tangential strain. The outer elements are then arranged in a radial direction to take advantage of the radial strain at the boundary, where it is maximum. If the strain is averaged over the region covered by each element, and using $G_F = 2.0$, the bridge output is approximately

$$\frac{\Delta E_0}{V} = \frac{0.82pR^2(1 - v^2) \times 10^3}{t^2 E} \quad \text{mv/V} \qquad (12.57)$$

The deflection at any radius is

$$y = \frac{3p(1 - v^2)}{16Et^3}(R^2 - r^2)^2 \qquad (12.58)$$

The maximum deflection, at the center of the plate, is

$$y_{max} = \frac{3pR^4(1 - v^2)}{16Et^3} \qquad (12.59)$$

In order to have the transducer respond satisfactorily to pressure pulses, the natural frequency of the diaphragm must be at least three to five times higher than the forcing frequency (5). The undamped natural frequency of

Fig. 12.16. Diaphragm strain gage for a pressure transducer. (Courtesy of Measurements Group, Inc.)

the diaphragm is

$$f_n = \frac{0.469t}{R^2}\sqrt{\frac{gE}{\gamma(1-v^2)}} \quad \text{cycles/sec} \tag{12.60}$$

where g = gravitational constant, 386.4 in/sec^2
γ = specific weight of diaphragm material, lb/in^3

Comments

The transducers described in this chapter have the intent of do-it-yourself, where such an instrument will be used with existing strain-measuring instrumentation. They are, therefore, not designed to stand alone. With the do-it-yourself transducer, desirable adjustments can be made at the instrument (such as gage factor adjustment) to bring it within the desired limits. For more precise compensation procedures, the reader is referred to the paper by Dorsey (6) or to Reference 2.

Problems

In all problems use steel with $\nu = 0.3$ and $E = 30 \times 10^6$ psi.

12.1. The load cell shown in Fig. 12.2 is used to measure loads between $\pm 75\,000$ lb. The load cell has a diameter of 1.50 in, $G_F = 2.15$, and $R_g = 120$ ohms. With the load at both extremes, determine the following:

(a) The bridge nonlinearity.
(b) The geometric nonlinearity.

12.2. In Problem 12.1 the bridge is rearranged so that gages 1 and 3 are active gages, with $R_2 = R_4 = R_g$ being dummy gages bonded to a similar piece of unstrained material. For the same loading conditions, determine the following:

(a) The nonlinearity factor.
(b) The bridge output, ΔE_0, if the supply voltage is 10 volts.

12.3. A compressive force, F, acts on a ring-type load cell. By considering the strains based on a curved beam, $\varepsilon_1 = \varepsilon_3 = \varepsilon_0$ and $\varepsilon_2 = \varepsilon_4 = -\varepsilon_i$, write the expression for the nonlinearity factor, $1 - n$.

12.4. A ring-type load cell is subjected to a tensile load of $F = 10\,000$ lb. Its dimensions are $R_0 = 3.0$ in, $R_i = 1.5$ in, and $w = 0.75$ in. Determine the following:

(a) The percentage difference in M_0 between Eqs. (12.13) and (12.16).
(b) The strains at each gage using thin-ring equations.
(c) The strains at each gage based on curved-beam equations.

12.5. The cantilever beam in Fig. 12.17 is to have a constant strain of 1200 μin/in along its tapered length when the load is applied at the vertex of the equilateral triangle formed by its two sides. Determine w.

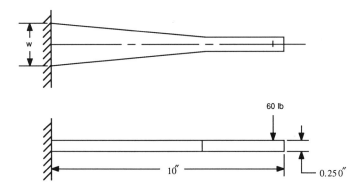

FIG. 12.17.

12.6. Design a load cell having the geometry shown in Fig. 12.6. The width is limited to 0.625 in, the ratio I_{BC}/I_{AB} is 25, the total deflection is not to exceed 0.015 in, and the maximum load is 500 lb. For the final gage location, compute the bridge nonlinearity at the maximum load.

12.7. Design a load cell having the configuration shown in Fig. 12.8. The width is limited to 1 in, and the maximum deflection must not exceed 0.012 in at the rated load of 700 lb. Determine the gage locations and their strains. Calculate the bridge nonlinearity at the maximum load.

12.8. Design a shear beam load cell having the configuration shown in Fig. 12.10. The maximum load of 50 000 lb is to produce $\Delta E_0/V$ of approximately 2 mV/V.

12.9. Using Eq. (5.40), derive Eq. (12.39).

12.10. A torque meter has a diameter of 1.25 in and uses four 350-ohm gages, with $G_F = 2.10$, to form a full bridge. If $\Delta E_0/V = 2$ mV/V at maximum torque, determine the value of the torque.

12.11. If the transverse sensitivity of the gages in Problem 12.10 is 0.9 percent, what will be the percentage change in torque if the transverse sensitivity is considered?

12.12. Design a torque wrench, shown in Fig. 12.13, to the following specifications:

(a) The maximum torque is 200 ft-lb.
(b) The overall length of the wrench must not exceed 18 in.
(c) The maximum strain at section 2 shall be 1000 μin/in at full torque.

12.13. A thin-walled cylindrical pressure transducer has an internal diameter of 1.25 in and a wall thickness of 0.05 in. Two circumferential gages with $R_g = 350$ ohms and $G_F = 2.10$ are bonded to the cylinder. If the hoop strain, ε_H, is limited to 1000 μin/in, determine the maximum internal pressure and the corresponding value of $\Delta E_0/V$.

12.14. Make a preliminary design of a diaphragm pressure transducer to measure a peak pressure of 75 psi at a frequency of 100 cycles/sec. The desired bridge output, $\Delta E_0/V$, is 1.5 mV/V at the peak pressure. Use $G_F = 2.0$.

REFERENCES

1. Cook, Robert D. and Warren C. Young, *Advanced Mechanics of Materials*, New York, Macmillan, 1985, Chap. 10.
2. *Strain Gage Based Transducers, Their Design and Construction*. Measurements Group, Inc., P.O. Box 27777, Raleigh, NC 27611, 1988.
3. Meier, J. H., "Some Phases of the Technique of Recording Performance Data on Large Machines," *SESA Proceedings*, Vol. X, No. 1, 1952, pp. 35–52.
4. Timoshenko, S., *Strength of Materials*, 3d edition, Part II, New York, Van Nostrand Reinhold, 1958, Chap. 4.
5. "Design Considerations for Diaphragm Pressure Transducers," TN-510, Measurements Group, Inc., P.O. Box 27777, Raleigh, NC 27611, 1982.
6. Dorsey, James, "Homegrown Strain-gage Transducers," *Experimental Mechanics*, Vol. 17, No. 7, July 1977, pp. 255–260.

13

STRAIN GAGE SELECTION AND APPLICATION

13.1. General considerations

On first observation, the strain gage appears to be a rather simple instrument that can be applied with minimum effort. This may be so, perhaps, if the gage is to be bonded to a fairly large plane area with ample working room and using a quick-setting cement. The novice soon learns, however, that even the supposedly simple task of satisfactorily soldering lead wires to the gage's solder tabs is not easy. When he moves on to bonding gages in a confined space and then attaching lead wires, his appreciation of the talent of a skilled technician rises rapidly.

When selecting a strain gage, or gages, for a project, the conditions under which the gage will operate must be considered. When all of the variables that go into gage construction are examined (backing material, foil, gage length, configuration, etc.), thousands of types are available. A study of manufacturer's catalogs shows that gages are divided into related groups, or series, of the same basic construction. Since gages belonging to a series have similar characteristics and capabilities, the task of choosing a gage is therefore reduced.

The first step in choosing a strain gage installation is to list as many conditions as possible affecting the system. Some of these are the following:

1. Is the strain to be measured in elastic or plastic region? If the strain is in the plastic region, for instance, then a post-yield gage will be chosen.
2. If the stress field is uniaxial, a single-element gage aligned along the principal stress direction will suffice. If the field is biaxial and the principal strain directions are known, a two-element rectangular rosette can be used. If the principal strain directions are unknown, a three-element rosette will be required.
3. What is the duration of the test? Will it be measured in minutes, hours, or years? The concern here is the shifting of the zero reference point.
4. How difficult will the installation of gages be?
5. Will the tests be static or dynamic? If they are purely dynamic, then consideration has to be given to a foil that exhibits good fatigue properties.
6. The temperature range over which the gage will operate and the choice of its self-temperature-compensation number must be considered.

7. Are strain gradients perpendicular to the test surface or in the plane of the test surface?
8. The choice of an adhesive is important and cannot be over emphasized.

Depending on the special requirements of a given test, other conditions can be added to this list. The cost of the strain gages, however, may have low priority, since the gage cost is generally small when compared to the total cost of a test.

After the conditions that affect the system are listed, a manufacturer's catalog can be consulted in order to choose a specific gage. Here will be found a designation code giving the features of the gage. They are as follows:

1. Gage series and type of strain-sensing alloy
2. Backing or carrier material of the strain-sensing alloy
3. Self-temperature compensation number
4. The active gage length
5. Grid and tab geometry
6. Gage resistance
7. Options, if desired

13.2. Strain gage alloys (1, 2)

Constantan

One of the most common strain gage foils is a copper–nickel alloy generally known as constantan. It finds wide use in static strain measurements as well as being employed in transducers. It also shows good fatigue life when applied to alternating strains, providing the strain levels are kept below ± 1500 µin/in. It has a low and controllable temperature coefficient of resistance as well as good strain sensitivity, which gives a nominal gage factor of 2.0. Furthermore, the gage factor is relatively insensitive to strain level and temperature.

Constantan can be processed for self-temperature compensation so that it matches the thermal expansion characteristics of a number of common engineering materials with thermal expansion coefficients ranging from zero to 50 ppm/°F. We have seen how mismatching of the coefficients of thermal expansion of the gage and test material rotates the thermal output curve around the reference temperature in order to obtain the most favorable results in a desired temperature range. An examination of strain gage catalogs shows that satisfactory gage resistance is obtainable even for very small gages made of this material.

If very large strains, on the order of 5 percent, are to be measured, then an annealed constantan foil is used. If gage lengths of $\frac{1}{8}$ inch or larger are used, strains in excess of 20 percent can be measured. Annealed constantan, however, is not recommended for cyclic strains, since permanent resistance change occurs as a function of number of strain cycles.

Constantan has several disadvantages. If the test temperature is above 150°F (66°C), it shows reference point drift, which is undesirable for tests conducted over a long period of time. The thermal output is also very high at temperatures below −50°F (−45°C) and above 400°F (205°C).

Isoelastic

Isoelastic, most generally used for dynamic strain measurement, is a nickel–chromium–iron alloy with molybdenum added. Its high gage factor of approximately 3.2 improves the signal-to-noise ratio in dynamic testing. This, coupled with superior fatigue life, makes it particularly useful for dynamic strain measurement.

The thermal output of isoelastic is about 80 μin/in/°F and it cannot be self-temperature compensated, thereby making it generally unsuitable for measuring static strains. This feature makes it undesirable for any long-range measurements if a stable reference point must be maintained. Furthermore, its response becomes nonlinear at strains on the order of 0.5 percent; hence, it is confined to strain measurement in the elastic region. In special cases, where a high-output response is desired, it may be used with a full-bridge circuit, thereby obtaining circuit temperature compensation.

Karma

Karma, a nickel–chromium alloy with small percentages of iron and aluminum, is another desirable material, since gages made of this material show minimal reference point drift with time and temperature. Because of this stability, it is a fine choice for long-time static measurements at or near room temperature. It is recommended for static strain measurements from −452°F (−270°C) to 500°F (260°C), but encapsulated gages can be used to 750°F (400°C) for short time periods. The material also exhibits good fatigue life with minimum reference point drift even after being cycled a large number of times. Because of its high resistivity, smaller gages for a given resistance can be manufactured.

Karma can be self-temperature compensated over a broad temperature range, but it is more limited than constantan in the number of thermal expansion coefficients for which it may be compensated. An advantage, however, is a flatter thermal output curve. Another feature is a gage factor that goes negative with increasing temperature, thus compensating for the temperature-induced change in the modulus of elasticity of the test material.

Karma has several disadvantages. It is difficult to solder, and for this reason gages with copper-clad tabs are available. Gages of this material are also more difficult to manufacture, making them more expensive than gages using constantan.

Platinum–tungsten

A platinum–tungsten alloy has been developed for high-temperature use. It has unusual stability and fatigue life at temperatures above 750°F (400°C), does not undergo any metallurgical changes to about 1650°F (900°C), and so its resistance remains essentially unchanged with time. It has a high-temperature coefficient of resistance that is not adjustable, although repeatable; thus, it cannot be self-temperature compensated. If temperature compensation is desired, it should be done through circuit compensation.

This material is used for dynamic strain measurements to 1500°F (815°C) and for static strain measurements to 1200°F (650°C). It has a higher strain sensitivity than copper–nickel or nickel–chromium alloys, but it is nonlinear. The strain range is generally limited to approximately ±0.3 percent.

13.3. Grid backing materials (1–4)

The strain-sensing element (either foil or wire) of a strain gage is mounted on a backing (carrier) material. The backing material serves several purposes.

1. It protects the strain-sensing grid from damage during handling and installation.
2. It provides a bonding surface to the test piece.
3. It transmits strain from the test piece to the strain-sensing alloy. Its stiffness must be low enough so it can follow the strains in the test piece without affecting it. On the other hand, it must be stiff when compared to the strain-sensing alloy so that the conductor material follows the strains without irregular distortion.
4. It provides electrical insulation between the strain-sensing element and the test piece.

Paper carriers

One of the first backing materials, and one still used, is a nitrocellulose paper. Strain gages using this readily available backing material easily conform to the surface of a test specimen. The gages are usually bonded to a test piece with a nitrocellulose cement that impregnates the paper's pores and cures by evaporation. Gages bonded in this manner can operate between −100°F and 180°F (−73 to 82°C), although they can be used for short periods of time beyond the upper temperature limit. At room temperature, this combination of paper and adhesive, when properly applied, can be subjected to strains in excess of 10 percent before breaking down.

Polyimide resins

Polyimide resins can be provided in both cast film and glass-reinforced laminated construction. It is a general-purpose material used for both static and dynamic strains. The cast film types are tough, flexible, and can be elongated up to 20 percent. Because of their flexibility, they can be contoured to fit small radii. This material can be used at temperatures ranging from cryogenic to 400°F (205°C). For higher temperatures, however, the resin can be reinforced with glass and the gage encapsulated for use to 700°F (370°C), although the temperature can be increased to 750°F (400°C) for short-duration tests.

Epoxy resins

Epoxy resins reinforced with glass fibers were developed in order to improve temperature capabilities. This material has an operating temperature range, for both static and dynamic strain measurements, from cryogenic to about 550°F (290°C), with an upper limit of 750°F (400°C) for short-duration tests. This backing also has improved dimensional stability for use in precision transducers. The glass reinforcement, however, reduces the maximum strain to about 1 percent but results in an extremely thin carrier. Since it is more brittle than polyimide, it requires more care in handling in order to prevent damage.

Metallic carriers

Metallic carriers have been discussed in Chapter 1 under weldable gages. Weldable wire gages are covered in Section 1.5, while weldable foil gages are discussed in Section 1.6.

13.4. Gage length, geometry, and resistance (1, 2)

Gage length

When referring to gage length, it is the active or strain-measuring portion that is referred to, not the overall or matrix length. A major purpose of using strain gages is to determine strains at critical points on a structure. Because these points are often where stress concentrations exist, thereby resulting in strain gradients which may be quite steep, consideration must be given to the strain gradient along the gage length. Steep strain gradients may also occur in dynamic measurements, such as occur when the propagation of stress waves in a material is being studied. Since a strain gage averages the strain along its active length, choosing a gage length considerably longer than the peak strain region results in a strain reading on the low side; therefore, a gage length consistent with the peak strain region should be chosen. For nonhomogeneous materials, however, a gage length long enough to span the representative structure of the material should be used in order to average the strain over voids, etc.

When possible, gages with lengths from $\frac{1}{8}$ in to $\frac{1}{4}$ in are preferable, since they are easier to apply, offer the largest number of geometries and options, and are less expensive. For gages of identical resistance and applied voltage, the larger gage will dissipate heat more easily because of the lower heat generated per unit area. This is particularly important when the gage is bonded to a material with poor heat-transfer qualities.

Gage geometry

When choosing a strain gage for a particular test, several elements enter into the decision. Among these are the shape of the strain-sensing grid, the number of grids and their orientation relative to each other, solder tab arrangement, and space available for mounting. If the principal stress is known to be uniaxial and its direction is also known, then a single grid gage may be used. This condition generally does not exist and single gages should be used only when one is absolutely sure one has a uniaxial stress state.

For the biaxial stress state a three-element rosette is used if the principal stress directions are unknown. The grids of the rosette may have any orientation relative to each other, but rosettes have been standardized on the delta and the rectangular configuration. This makes data reduction simpler, particularly for the rectangular rosette. When bonding a rosette to a test specimen, any orientation can be used, but usually one rosette leg is aligned along some chosen axis of the specimen.

If mounting space is confined, stacked rosettes are preferred, particularly when there is a high strain gradient in the plane of the mounting surface. In this case they give a closer approximation of the strain at the point, but heat dissipation may be a problem. They are stiffer than the plane rosette and conform less easily to curved surfaces. On the other hand, plane rosettes are preferred when the strain gradient is normal to the surface, since all grids are as close to the specimen surface as possible.

When the principal stress directions are known, then a two-element rosette may be used. The principal stress directions may be apparent from the geometry of the test specimen, such as a thin-walled tube with internal pressure, for instance. Generally, the principal directions are determined through the use of a brittle lacquer coat or a photoelastic coating. It is obvious that if the principal axes are known, considerable savings in time and labor can be attained in wiring a number of two-element gages rather than three-element gages.

Special-purpose gages, such as gages for residual stress measurement, crack detection, or diaphragm gages for pressure transducers, are available.

When a high strain gradient transverse to the gage axis exists, a gage with a narrow grid width should be chosen in order to give a better strain average. The reduced gage area, though, will reduce the ability to dissipate heat.

Gage resistance

An examination of manufacturer's catalogs shows that strain gages may be obtained with resistances up to 1000 ohms. The two most common resistance values, however, are 120 and 350 ohms. As we saw in Chapter 5, lead-line resistance desensitizes the circuit. If lead-line resistance or other parasitic resistances are present, then choosing a higher-resistance gage will reduce the circuit desensitization. This is illustrated in Example 5.4. For the same applied voltage, a higher-resistance gage reduces the heat generated. A 120-ohm gage, for instance, generates nearly three times the wattage of a 350-ohm gage. Conversely, if the wattage remains the same, a higher voltage may be used on the higher resistance gage in order to increase the output.

Self-temperature compensation

Choosing a gage for the proper self-temperature compensation number is a matter of examining a strain gage catalog in order to determine the available thermal expansion coefficients. When the desired number is chosen, it is a matter of adding it in the proper place in a manufacturer's strain gage designation code. Self-temperature compensation and its use have been covered in Chapter 11.

Options

Both standard options and special options are available. Among these are attached lead wires, gage encapsulation, solder dots, and etched integral terminals, to name a few. For a complete description of options, both standard and special, consult a manufacturer's catalog.

13.5. Adhesives (1, 2, 4)

Successful use of strain gages is very much dependent upon satisfactorily bonding the gage to the test specimen. The chosen adhesive must have sufficient shear strength in order to transmit strains in the test specimen to the strain-sensing grid, yet it must be compatible with both the gage backing material and the test material so that neither is damaged. Further, the adhesive should have long-term stability so that it does not decompose or show appreciable creep over the test's lifetime.

The manufacturer's instructions in the use of the adhesive must be followed carefully, particularly if the adhesive calls for mixing a resin and hardener. The adhesive must be capable of forming a thin glue line free of voids, with minimum curing time being a desirable feature. It also helps to electrically isolate the grid from the test material. When checking the resisance between the grid and test specimen, the recommended resistance is 10 000 megohms minimum, but preferably higher. An adhesive should be capable of high elongation as well as have the ability to operate over a wide temperature range.

A large number of adhesives are available, each with detailed techniques for its application. Manufacturers will supply the user with instructions. One thing that is crucial to satisfactory strain gage performance, regardless of the adhesive used, is cleanliness. The test surface must be free of grease, rust, or other contaminants so that the bare base material is exposed. During the preparation process the hands must be kept clean and care taken not to touch the surface. After the surface is prepared the gages should be applied without undue delay.

Although there are numerous adhesives available, only the more commonly used ones will be discussed. Detailed information on specific adhesives and their methods of application can be obtained from information bulletins supplied by manufacturers.

Nitrocellulose

Nitrocellulose adhesives (such as Duco) were once widely used when paper-backed gages were prevalent. This type of adhesive sets by solvent evaporation; thus, its use today is limited to paper-backed gages or gages with a porous backing. A minimum pressure has to be applied during the curing process, which is usually in excess of 24 hours, depending on humidity and temperarure. Application of heat will accelerate the curing process, however. The curing process may be monitored by periodically checking the gage resistance to ground, since the resistance increases as the adhesive sets.

Gages bonded with nitrocellulose adhesives may be used up to 180°F (82°C). They are hygroscopic (i.e., they absorb moisture from the air) and must be protected with a moisture-resistant coating once the adhesive is fully cured to ensure electrical and dimensional stability. Because adhesives of this type are vulnerable to ketonic solutions, they are easily removed without surface damage by using a ketonic solution.

Cyanoacrylate

Cyanoacrylate adhesives are widely used as general-purpose cements that are fast curing and simple to use, since no mixing is required. While the life of an unopened container of cyanoacrylate is approximately 9 months when stored at room temperature, the life can be extended by refrigeration at 40°F (4°C). When removed from the refrigerator, the adhesive should be allowed to come to room temperature before opening in order to prevent condensation and possible damage to the material. Once the container has been opened it should be stored in a cool, dark area rather than returning it to the refrigerator.

In preparing a gage for bonding, the gage backing material is treated with a catalyst, sparingly used, and allowed to dry for approximately 1 minute. A thin coat of adhesive is put on, the gage is placed on the test specimen, and thumb pressure is applied to the gage. Polymerization takes place in the adhesive film in approximately 1 minute. The bonded gage is ready for use by the time lead wires are attached.

The glue line is sensitive to moisture and must be protected by a coating. A properly protected gage, however, can be used in wet atmospheres for short-duration tests. These adhesives are excellent for short-term tests but are seldom used for tests extending over long time periods, since the bond is subject to embrittlement with age. Gages bonded with these cements can be used to measure strains of the order of 15 percent and can operate over a temperature range of $-25°$ to $150°F$ (-32 to $65°C$).

Epoxies

Epoxy adhesives, in use for many years, come in a wide selection of two types; namely, one where polymerization takes place at room temperature and another that requires the application of external heat for correct polymerization. With epoxies, there is no solvent evaporation involved, very little shrinkage, and a good permanent bond is formed with a wide variety of materials. Epoxies also exhibit excellent moisture and chemical resistance, and can be used over a temperature range from cryogenic to $600°F$ ($315°C$).

One type of epoxy, using an amine catalyst, cures at room temperature through the exothermic reaction produced when the adhesive components, hardener and resin, are mixed together. Another type of epoxy, activated by an acid anhydride catalyst, requires external heat for polymerization to occur properly. A temperature of at least $250°F$ ($120°C$) must be maintained for several hours. Both types require a clamping pressure during the curing process. Furthermore, if either type is to be used at a temperature higher than the curing temperature, then a post-cure temperature above the expected maximum test temperature should be maintained for several hours. For the room-temperature-curing epoxy, the post-cure temperature should be 70 degF to 85 degF (40 to 47 degC) above the maximum test temperature. For the hot-cure epoxy, the post-cure temperature should be 85 degF to 115 degF (47 to 64 degC) above the maximum test temperature.

Other adhesives

Other available adhesives are generally used for more specialized applications. Among these are phenolic, polyimide, and ceramic adhesives. Phenolic adhesives are little used because they require complicated, long curing cycles and high clamping pressure. Polyimide adhesives are difficult to work with and the solvents in them are not easily removed. Remaining solvents degrade the adhesive properties. Ceramic adhesives are applied to free-filament gages and thermocouples for temperatures that exceed the limits of organic materials. Again, for special applications, consult the manufacturers and their application departments.

13.6. Bonding a strain gage to a specimen

Bonding a strain gage appears to be a simple process, but close attention must be paid to each step. This involves surface preparation of the test specimen, cementing the gage to that surface, soldering lead wires, and finally applying a protective coat to the installation. Cleanliness cannot be overemphasized; the hands should be washed frequently during the process or cleaned with neutralizer, no material should be reused, and the work area must be kept clean.

Surface preparation

1. Using a degreasing agent, such as trichloroethylene or carbon tetrachloride, clean the test surface, being sure to have adequate ventilation.
2. Sand the degreased surface in order to remove all scale, dirt, or dust particles.
3. Clean the surface with a sponge or tissue saturated with the cleaning solvent.
4. Using a metal conditioner, wet lap the area with silicon-carbide paper.
5. Using a clean tissue, wipe the area dry with one stroke. Do not reuse the tissue. Repeat several times.
6. Using a ballpoint pen or 4-H pencil, locate and mark reference lines for gage alignment. *Do not use a scribe: make certain you do not scratch the surface.*
7. Using a cotton-tip swab, dip it into metal conditioner and scrub the surface. Wipe dry with one stroke using a clean tissue. Using a clean swab and tissue, repeat several times until the cotton-tip shows no foreign material.
8. Dip a cotton swab into neutralizer and scrub the surface. Wipe clean with one stroke using a clean tissue. Repeat several times to ensure the surface is neutralized.
9. Install the gage as soon as possible.

Bonding the gage

Since bonding techniques will differ depending on the adhesive, the method for a cyanoacrylate adhesive will be described, since it is a widely used cement.

1. Remove the gage from its packet and place it, bonding side down, on a clean surface. Position a separate terminal strip relative to the gage tabs.
2. Using a piece of cellophane tape, place it over the gage and terminal strip. Pull the tape from the surface at a shallow angle, being certain the gage and terminal strip are firmly attached.
3. Place the gage on the test specimen, aligning the reference tabs on the gage with the marked reference system on the test surface. The tape and gage are now in the desired position.
4. Lift one end of the tape from the test surface until the gage and terminal strip are just clear. The remainder of the tape is still attached to the specimen.

5. Pull the free end of the tape back until the bonding surfaces of the gage and terminal strip are exposed. Brush catalyst sparingly onto the bonding surfaces. Allow to dry for 1 minute.
6. Apply one or two drops of adhesive at the boundary line of the tape and test surface. Pull the free end of the tape taut and towards the test specimen, making a shallow angle. At the same time, using a clean tissue, wipe over the tape from the boundary line towards the free end so that the cement spreads under the gage and terminal strip, bonding them to the surface. Apply thumb pressure to the gage and terminal strip for approximately 1 minute.
7. After several minutes, grasp one end of the tape and slowly and carefully pull back on itself until it is removed. The gage is now ready for soldering.

Completing the installation

Now that the gage is successfully bonded to the test specimen, there remains the task of attaching the lead wires and then applying a protective coat to the entire installation. The procedure is outlined in the following:

1. If the gage has an open grid, cover the grid area with a piece of masking tape, leaving the solder tabs exposed.
2. A 30–40-watt soldering iron with a smooth, tinned tip is required.
3. Use a fine rosin-core solder whose melting temperature is compatible with the test environment.
4. With the soldering iron at the proper temperature, lay the solder across the gage tab and apply the iron firmly for a second. Lift the solder and iron at the same time, leaving behind a shiny mound of solder on the tab.
5. Lightly tin the terminal strips.
6. Separate the individual leads of the composite lead wire and remove about $\frac{1}{2}$ inch to $\frac{3}{4}$ inch insulation from each. On each individual lead, separate one strand, twist the remaining strands together and tin for a short distance at the insulation. Snip off the remaining end, leaving about $\frac{1}{8}$ inch of the tinned bundle. The single strand will be used as a jumper wire from the terminal strip to the gage tab.
7. Solder the tinned lead wires to the terminal strip.
8. Using the single strand of each lead, solder to the gage soldering tab, arranging it so there is some slack between the terminal strip and gage tab. (Fine insulated wire may be used in place of the single strand.)
9. Clean all solder joints with rosin solvent, remove the masking tape, and clean the gage with rosin solvent.
10. Secure the lead wires so they cannot accidently be pulled loose.
11. Check the resistance between the gage and the specimen. It should be at least 10 000 megohms.
12. Apply a protective coating to the gage, terminal strips, and a short distance onto the lead wire insulation.

REFERENCES

1. "Catalog 500: Part B—Strain Gage Technical Data," Measurements Group, Inc., P.O. Box 27777, Raleigh, NC 27611, 1988.
2. "SR-4 Strain Gage Handbook," BLH Electronics, Inc., 75 Shawmut Road, Canton, MA 02021, 1980.
3. "Weldable and Embedable Integral Lead Strain Gages," Applications and Installation Manual, Eaton Corp., Ailtech Strain Gage Products, 1728 Maplelawn Rd., Troy, MI 48084, 1985.
4. Vaughn, John, *Application of B & K Equipment to Strain Measurements*, Bruel & Kjaer, Naerum, Denmark, 1975, Chaps. 3 and 4.

ANSWERS TO SELECTED PROBLEMS

2.2. $\sigma_1 = 20\,500$ psi; $\sigma_2 = 4500$ psi; $\theta = 45°$.

2.4. $\sigma_1 = \sigma_2 = 9500$ psi; Mohr's circle is a point.

2.6. $\sigma_1 = -\sigma_2 = 7500$ psi; $\theta = 45°$.

2.7. $\sigma_1 = -7000$ psi; $\sigma_2 = -23\,000$ psi; $\theta = 135°$.

2.9. (a) Point A: $\sigma_1 = -\sigma_2 = 6258$ psi; Point B: $\sigma_1 = 26\,510$ psi; $\sigma_2 = -1477$ psi.
(b) Point A: $\tau_{max} = 6258$ psi; Point B: $\tau_{max} = 13\,944$ psi.

2.11. Left bearing reactions: $F_y = 306$ lb; $F_z = 217$ lb.
Right bearing reactions: $F_y = 310$ lb; $F_z = -569$ lb; $\tau_{max} = 8614$ psi.

2.12. $\sigma_{max} = 33\,096$ psi; $\tau_{max} = 16\,799$ psi.

2.14. $\sigma_1 = 11\,701$ psi; $\sigma_2 = -4701$ psi; $\theta = 18.8°$.

2.16. $\varepsilon_1 = 1818$ μin/in; $\varepsilon_2 = 197$ μin/in; $\theta = 161.8°$.

2.18. $\varepsilon_1 = 753$ μin/in; $\varepsilon_2 = -183$ μin/in; $\theta = 102.3°$.

2.20. $\varepsilon_1 = -\varepsilon_2 = 250$ μin/in; $\theta = -45°$.

2.22. $\varepsilon_1 = 400$ μin/in; $\varepsilon_2 = -800$ μin/in; $\theta = 90°$.

2.24. $\varepsilon_1 = 1258$ μin/in: $\varepsilon_2 = -188$ μin/in; $\theta = 139.2°$.

2.25. (a) $\varepsilon_1 = 2654$ μin/in; $\varepsilon_2 = -1154$ μin/in; $\gamma_{max} = 1904$ μradians.
(b) Rectangular rosette: $\varepsilon_a = 2500$ μin/in; $\varepsilon_b = 1500$ μin/in; $\varepsilon_c = -1000$ μin/in.
Delta rosette: $\varepsilon_a = 2500$ μin/in; $\varepsilon_b = 525$ μin/in; $\varepsilon_c = -775$ μin/in.

2.26. $\varepsilon_1 = 1244$ μin/in; $\varepsilon_2 = -844$ μin/in; $\theta = 101.7°$.

2.28. $\varepsilon_1 = 1027$ μin/in; $\varepsilon_2 = -2027$ μin/in; $\theta = 65.5°$ ccw from gage b.

2.30. $\sigma_1 = 32\,664$ psi; $\sigma_2 = 15\,521$ psi.

2.32. $\sigma_1 = 13\,810$ psi; $\sigma_2 = -56\,667$ psi.

2.34. $\sigma_3 = 10\,385$ psi.

2.36. (b) $\varepsilon_1 = 1791$ μin/in; $\varepsilon_2 = -693$ μin/in; $\varepsilon_3 = -471$ μin/in.
(d) $\sigma_1 = 52\,203$ psi; $\sigma_2 = -5145$ psi; $\theta = 17.4°$.

2.38. $\varepsilon_a = 0$ μin/in; $\varepsilon_b = 68$ μin/in; $\varepsilon_c = 951$ μin/in.

3.2. $I = 0.020\,15$ amps; $n = 0.0074$.

3.4. $I = 0.028$ amps; $n = -0.12$.

3.5. $R_p = 5880$ ohms.

4.5. $\eta = 90$ percent; $\varepsilon = 0.102\,041$ in/in; No.

4.8. (a) $\varepsilon_a = -844$ μin/in; $\varepsilon_b = 1950$ μin/in; $\varepsilon_c = 3844$ μin/in; $\varepsilon_d = 1050$ μin/in.
(b) $\Delta E = 0.031\,25$ volts.
(c) $n = 0.000\,625$.

4.9. (a) $\varepsilon_g = 1855$ μin/in; $\varepsilon_b = -757$ μin/in.
(b) $\Delta R_g = 0.4630$ ohms; $\Delta R_b = -0.1889$ ohms.
(c) $\Delta E = 0.033\,96$ volts.

ANSWERS TO SELECTED PROBLEMS

5.2. $\Delta E_0/V = (G_F\varepsilon)/(4 + 2G_F\varepsilon)$; $\varepsilon/\varepsilon_i = 2/(2 - G_F\varepsilon_i)$.

5.4. $\Delta E_0/V = (G_F\varepsilon)/2$; $\varepsilon/\varepsilon_i = 1$.

5.6. $\Delta E_0/V = G_F(1 + v)\varepsilon/2$; $\varepsilon/\varepsilon_i = 1$.

5.8. $\sigma_1 = 32\,490$ psi; $\sigma_2 = -19\,560$ psi; $F = 340\,923$ lb.

5.11. $P_i = 15\,000$ psi.

5.12. $F_x = 2209$ lb $(+x)$; $F_y = 115$ lb $(+y)$; $F_z = 104.7$ lb $(-z)$.
$\varepsilon_1 = -180$ μin/in; $\varepsilon_2 = 240$ μin/in; $\varepsilon_3 = 300$ μin/in; $\varepsilon_4 = -120$ μin/in.

5.14. (a) $F_1 = 53\,014$ lb; $F_2 = 62\,027$ lb.
(b) $M_1 = 15\,904$ in-lb (cw); $M_2 = 14\,910$ in-lb (ccw).

5.16. Increase $= 11.75$ percent.

5.18. (a) $\Delta E_{m0} = 3.83$ mV; (b) $\Delta E_{m0} = 52.22$ mV.

5.21. (b) $\Delta E_{m0} = 14.85$ mV; $\Delta E_{m0} = 14.28$ mV; $\Delta E_{m0} = 11.72$ mV.

5.23. $W = 2.94$ lb.

5.25. $\varepsilon_1 = 2000$ μin/in; $\varepsilon_2 = 1200$ μin/in; $\varepsilon_3 = -500$ μin/in; $\varepsilon_4 = 937$ μin/in.

6.2. $R_p = 468.76$ ohms.

6.3. $R_p = 1114.1$ ohms; $R_s = 14.48$ ohms.

6.5. $R_s = 40$ ohms.

6.7. $R_p = 327.8$ ohms; $R_s = 69.3$ ohms.

6.9. $R_s = 77.4$ ohms.

6.11. $R_p = 690$ ohms; $R_s = 30.96$ ohms.

7.1. (a) $\varepsilon_H = 680$ μin/in; $\varepsilon_L = 160$ μin/in.
(b) $\varepsilon_H' = 693$ μin/in; $\varepsilon_L' = 186$ μin/in.
(c) $\eta_H = 1.91$ percent; $\eta_L = 16.25$ percent.

7.3. $\varepsilon_a' = 1015$ μin/in.

7.5. $G_F = 0.939$.

8.2. $\varepsilon_H/\varepsilon_L = 4.25$.

8.4. $\sigma_1 = 63\,718$ psi; $\sigma_2 = -4576$ psi; $\tau_{max} = 34\,147$ psi; $\theta = 129.3°$.

8.6. $\sigma_1 = 10\,890$ psi; $\sigma_2 = -19\,290$ psi; $\tau_{max} = 15\,090$ psi; $\theta = 98.3°$.

8.8. $\sigma_1 = -11\,611$ psi; $\sigma_2 = -38\,703$ psi; $\tau_{max} = 19\,352$ psi; $\theta = 120°$.

8.10. $\sigma_1 = 4474$ psi; $\sigma_2 = -18\,188$ psi; $\tau_{max} = 11\,331$ psi; $\theta = 101.8°$.

8.12. $\sigma_1 = -14\,120$ psi; $\sigma_2 = -29\,166$ psi; $\tau_{max} = 14\,583$ psi; $\theta = 65.6°$.

8.14. $\varepsilon_a = -295$ μin/in; $\varepsilon_b = 752$ μin/in; $\varepsilon_c = -75$ μin/in; $\varepsilon_d = 550$ μin/in.

9.2. $\varepsilon_a' = 146$ μin/in; $\varepsilon_b' = 533$ μin/in.

9.4. $\varepsilon_a = 951$ μin/in; $\varepsilon_b = 141$ μin/in; $\varepsilon_c = 431$ μin/in.

9.6. $\varepsilon_a = 117$ μin/in; $\varepsilon_b = -870$ μin/in; $\varepsilon_c = 858$ μin/in.

9.8. $\varepsilon_a = \varepsilon_b = \varepsilon_c = 782$ μin/in.

9.10. $\varepsilon_a = 807$ μin/in; $\varepsilon_b = 401$ μin/in; $\varepsilon_c = -219$ μin/in.

9.12. $\varepsilon_a = \varepsilon_c = 807$ μin/in; $\varepsilon_b = -8$ μin/in.

9.14. (b) $\varepsilon_a = 171$ μin/in; $\varepsilon_b = 1571$ μin/in; $\varepsilon_c = 784$ μin/in.
(c) $\varepsilon_1 = 1612$ μin/in; $\varepsilon_2 = -656$ μin/in.
(d) $\sigma_1 = 46\,655$ psi; $\sigma_2 = -5684$ psi; $\theta = 52.8°$.
(e) $\tau_{max} = 26\,170$ psi.

10.2. (a) $\phi = 26.6°$; (b) $\phi = 28.7°$; (c) $\phi = 30.2°$.

10.4. $\sigma_H = \dfrac{E\varepsilon_\phi}{1 - v} = \dfrac{pd}{8t(1 - v)}[3(1 - v) + (1 + v)\cos 2\phi]$.

10.6. $\gamma_{xy} = 276$ μradians.
11.1. $\varepsilon = -2300$ μin/in.
11.3. $\varepsilon = 2885$ μin/in.
11.7. (a) $\varepsilon_1 = 1673$ μin/in; $\varepsilon_2 = -2373$ μin/in.
(b) $\varepsilon_1 = 3754$ μin/in; $\varepsilon_2 = -26$ μin/in.
11.9. (a) $\varepsilon_1 = 207$ μin/in; $\varepsilon_2 = -1007$ μin/in.
(b) $\varepsilon_1 = 283$ μin/in; $\varepsilon_2 = -871$ μin/in.
12.1. (a) Tension: nonlinearity factor = 0.9989. Compression: nonlinearity factor = 1.0011.
(b) Tension: nonlinearity factor = 1.008. Compression: nonlinearity factor = 0.9992.
12.3. $1 - n = \dfrac{1}{1 + \dfrac{G_F(\varepsilon_0 - \varepsilon_i)}{2}}.$
12.5. $w = 1.6$ in.
12.10. $T = 2106$ in-lb.
12.11. Percent difference = 0.86.
12.13. $\Delta E_0/V = 1.05$ mV/V.

INDEX

Adhesives, 396–8
   cyanoacrylate, 397–9
   epoxies, 398
   nitrocellulose, 397
Axial strain sensitivity, 236

Backing material, 393–4
Baker, M. A., 9
Ballast circuit, *see* Potentiometric circuit
Ballast resistor, 100–1
Biaxial stress, 45
Biermasz, A. J., 16
Bonded wire strain gage, 24–7
Bridge input resistance, 151–2, 173–5
Bridge output resistance, 151–2, 176–7
Bridge ratio, 155, 159–61
Brittle lacquer coatings, 3, 36–8

Calibration
   potentiometric circuit, 141–4
   strain gage use under other conditions, 246–8
   Wheatstone bridge, 193–5
Circuits, elementary
   constant current, 94–6
      advantages, 96–7
   constant voltage, 91–4
   nonlinearity, 93–4
Circuits, potentiometric
   advantages and limitations, 105–6
   applications, 104–5
   ballast resistor, 100–1
   calibration, 141–4
   characteristics, 102–3
   circuit analysis, 106–9
   circuit efficiency, 111–12
   circuit equations, 101–2
   components, 100
   dynamic strains, 118–19
   gages in series, 112–14
   linearity considerations
      fixed ballast resistance, 123–4
      variable ballast resistance, 121–3
   measurements, static vs. dynamic, 114–15
   nonlinearity, 102, 108–9
   signal measurement, 147–9
   static strain, 118
   temperature effects, 129–41
      ballast and gage leads, 133–5
   voltage limitation, 110
Coatings
   brittle lacquer, 3, 36–8
   photoelastic, 38
Compensating strain gage, 27–8, 353–7, 363
Constantan, 14, 339–41, 391–2
Crack measuring gage, 34–5
Cyanoacrylate cement, 397–8

Data analysis, 253–6
Delta rosette
   analysis, 267–9
   Mohr's circle, 269–73
   principal stress directions, 269
   T-delta, 278–81
   transverse sensitivity, 301–6
Desensitization of circuits
   full bridge, 227–31
   half bridge, 218–25
   kinds, 207
   meter resistance, 175–9
   power supply resistance, 173–5
   reasons for varying, 205–6
   single gage, 207–17
      combination, series and parallel, 211–16

INDEX

Desensitization of circuits (*contd.*)
  single gage (*contd.*)
    resistance in parallel, 209–11
    resistance in series, 207–9
    temperature effects, 216
Dorsey, J., 8, 387
Dummy gage, *see* Compensating strain gage

Embedment gage, 36
Epoxy cement, 398
Equiangular rosette, *see* Delta rosette

Four-element rosette
  rectangular, 275–8
  T-delta, 278–81
Friction gage, 35–6

Gage factor
  determination, 242–3
  manufacturers, 26, 236
    relation with axial and normal strains, 240–2
  variation with temperature, 340–3
Gages
  crack measuring, 34–5
  embedment, 36
  friction, 35–6
  semiconductor, 32–3
  temperature, 33–4

Hines, F. F., 318
Hydrostatic strain component, 69
Hydrostatic stress component, 56

Indicated vs. actual strain, 165–8, 206
Invariants
  strain, 81
  stress, 81
Isoelastic, 240, 392

Jones, E., 10

Karma, 392
Kelvin, Lord, 5
Kern, R. E., 320, 322

Lateral effect, 234–51
  basic equations, 236–42
  transverse sensitivity factor, K, 238–40
Lead-line resistance, 180–91
  full bridge, 180–1
  half bridge—four wire, 181–4
  half bridge—three wire, 184–6
  quarter bridge—three wire, 187–8
  quarter bridge—two wire, 188–90
Load cell
  axial force, 363–5
  bending beam, 372–5
  ring type, 365–8
  shear beam, 375–7

Maslen, K. R., 10
Material, backing, 393–4
  epoxy resins, 394
  metallic, 394
  paper carrier, 393
  polyimide resins, 393–4
Material, strain gage
  constantan, 14, 339–41, 391–2
  isoelastic, 392
  karma, 392
  platinum tungsten, 393
  properties desired, 10
McClintock, F. A., 281
Measurements, fundamental laws, 97–8
Meier, J. H., 10, 278, 380, 382
Meter resistance, 175–9
Mohr's circle
  delta rosette, 269–73
  rectangular rosette, 261–5
  strain, 68–70
  stress, 54–7
Multiple circuits, 195–7

Nitrocellulose cement, 397
Nonlinearity of circuits
  elementary, 93–4
  potentiometric, 102, 108–9
  Wheatstone bridge, 150–1, 163–4
Normal strain sensitivity, 236

Perry, C. C., 330–1
Photoelastic coating, 38
Plane shearing stress, determination, 327–30
Plane strain, 62–5
Plane stress, *see* Biaxial stress
Platinum tungsten, 393

INDEX

Poisson's ratio, 17, 73
Potentiometric circuit
  advantages and limitations, 105–6
  applications, 104–5
  ballast resistor, 100–1
  calibration, 141–4
  characteristics, 102–3
  circuit analysis, 106–9
  circuit efficiency, 111–12
  circuit equations, 101–2
  components, 100
  dynamic strains, 118–19
  gages in series, 112–14
  linearity considerations
    fixed ballast resistance, 123–4
    variable ballast resistance, 121–3
  measurements, static vs. dynamic, 114–18
  nonlinearity, 102, 108–9
  signal measurement, 147–9
  static strains, 118
  temperature effects, 129–41
    ballast and gage leads, 133–5
  voltage limitation, 110
Pressure transducer
  diaphragm, 384–7
  thin-walled cylinder, 383–4
Principal strains, 64–5
Principal stresses, 48–53, 260–5, 269–73

Rectangular rosette
  analysis, 258–61
  four element, 275–8
  Mohr's circle, 261–5
  principal stress directions, 260–1
  transverse sensitivity, 296–300
Resistance, basic equations for unit change, 236–8
Resistor, ballast, 100–1, 119–26
Rosettes
  delta
    analysis, 267–9
    Mohr's circle, 269–73
    principal stress directions, 269
    transverse sensitivity, 301–6
  geometry, 256–8
  graphical solutions, 281–7
  rectangular
    analysis, 258–61
    four element, 275–8
    Mohr's circle, 261–5
    principal stress directions, 261
    transverse sensitivity, 296–300
  stress equations, summary, 280–1
  T-delta, 278–81
  transverse sensitivity
    delta, 301–6
    rectangular, three-element, 296–300
    two different orthogonal gages, 294–6
    two identical orthogonal gages, 291–4

Sanchez, J. C., 126
Semiconductor gages, 32–3
Semiconductor materials, 8–9
Sensitivity variation
  full bridge, 227–31
  half bridge, 218–25
  reasons, 205–6
  single gage, 207–16
Shear gage, 330–4
Shear strain, Mohr's circle sign convention, 68–9
Shear stress
  biaxial stress state, 51–2
  determination of plane, 327–30
  Mohr's circle sign convention, 54–6
Shoub, H., 14, 22
Stein, P. K., 157
Strain
  apparent, *see* Thermal output
  basic concepts, 61–2
  correcting for thermal output and gage factor variation, 348–9
  elastic, in metals, 7–8
  indicated vs. actual, 165–8, 206
  invariants, 81
  Mohr's circle, 68–70
  nonlinearity, 102, 108–9, 163–4
  plastic, in metals, 8
  principal, 64–5
  shear, sign, 63–4
  small vs. large, 20–4
  temperature-induced, 337–40
  thermal output correction, 344–7
  transformation equations, 63–5
Strain gage
  alloys, 391–3
  basic principle, 5
  bonding, 398–400
  characteristics, 4–5
  compensating, 27–8, 353–7, 363
  foil, 29–31
  gage length, 394–5
  general considerations, 390–1
  geometry, 395
  lateral effect, 234–51
  orthogonally crossed pair, 248–51

Strain gage (*contd.*)
  properties desired, 10
  resistance, 395–6
  self-temperature compensated, 131–2, 343–5
  self-temperature compensation, 396
  temperature effects, 337
  test material mismatch, 349–51
  use under conditions differing from calibration, 246–8
  weldable
    foil, 31
    wire, 27–9
  wire, 24–9
Strain sensitivity
  analysis, 14–24
    general case, 14–17
    small vs. large strain, 17–24
    uniform straight wire, 17–20
  axial and normal, 236–8
    gage factor relation, 240–2
  definition, 5–8, 236
  material properties, desired, 10
  numerical values, 11–13
  reasons for varying, 205–6
Strain transformation equations, 63–5
Stress
  basic concepts, 43–4
  biaxial, 45
  circuit, indication of normal stress, 320
  fields, 253–6
  invariants, 81
  Mohr's circle, 54–7
  principal, 48–53, 260–5, 269–73
    using a single gage, 326–7
  sign convention, 45
  transformation equations, 45–53
Stress gage
  normal, 310–12
  single round wire, L configuration, 312–14
  two orthogonal gages, 314–16
  V-type, 321–5
Stress–strain gage, 316–20
Stress–strain relations, 72–7
Stress transformation equations, 45–53

Temperature gages, 33–4
Temperature-induced strain, 337–40
Thermal expansion coefficients, 350
Thermal output, 338, 344
  correction, 344–7
Thévenin's theorem, 173

Torque meter, 378–80
Torque wrench, 380–2
Transducers
  axial force, 363–5
  bending beam, 372–5
  cantilever beam, 368–72
  full bridge, 361–2
  half bridge, 362
  pressure measurement
    diaphragm, 384–7
    thin-walled cylinder, 383–4
  quarter bridge, 363
  ring-type, 365–8
  shear beam, 375–7
  torque meter, 378–80
  torque wrench, 380–2
Transformation equations
  strain, 63–5
  stress, 45–53
  summary, strain, 65
  summary, stress, 52–3
Transverse sensitivity
  definition, 238–40
  delta rosette, 301–6
  determination, 244–6
  rectangular rosette, 296–300
  two different orthogonal gages, 294–6
  two identical orthogonal gages, 291–4
  typical values, 239–40

Unbonded wire strain gage, 24

Weibull, W., 12, 14, 20, 22
Weldable strain gage
  foil, 31
  wire, 27–9
Weymouth, L. J., 141
Wheatstone bridge
  bridge input resistance, 151–2, 173–5
  bridge output resistance, 151–2, 176–7
  bridge ratio, 155, 159–61
  calibration, 193–5
  derivation of elementary bridge equations, 157–65, 169–72
  elementary bridge equations, 149–52
  general bridge equations, 172–9
  lead-line resistance, 180–91
  meter current, 152–3, 179
  meter resistance, 175–9
  nonlinearity, 150–1, 163–4
  null balance reference bridge, 154
  null balance system, 153

reference system, 153–4
resistance in series with bridge, 172–5
summary of properties, 155–7
unbalance reference bridge, 154

unbalance system, 153
Williams, S. B., 320, 335
Wnuk, S. P., Jr., 141
Wright, W. V., 126